Introduction to Combinatorial Mathematics

Introduction to Combinatorial Mathematics

C. L. Liu

Department of Electrical Engineering
Massachusetts Institute of Technology

McGraw-Hill Book Company

New York
St. Louis
San Francisco
Toronto
London
Sydney

Introduction to Combinatorial Mathematics

Library of Congress Catalog Card Number 68-22763

ISBN 07-038124-0

4567890MAMM754321

Preface

Combinatorial theory is a fascinating branch of mathematics with numerous applications in engineering, the physical sciences, the social sciences, economics, and operations research. This is an introductory book on the subject. It is an outgrowth of a set of class notes I used in teaching a course in applied combinatorial mathematics offered by the Electrical Engineering Department at the Massachusetts Institute of Technology.

The topics covered may be divided into four groups which are representative of the various aspects of applied combinatorial mathematics: enumerative analysis (Chapters 1 to 5), theory of graphs (Chapters 6 to 9), optimization techniques (Chapters 10 to 13), and design of experiments (Chapter 14). The level of presentation is appropriate for advanced undergraduate and first-year graduate students. No prior knowledge of combinatorial mathematics or modern algebra is assumed. Sections marked with ★ may be omitted without loss of continuity. Necessary algebraic concepts are introduced as needed, and those sections marked with • may be skipped by the student who has had a course in modern algebra.

In view of the fact that this is an introductory text, I did not attempt to compile an extensive bibliography. Rather, it is hoped that the student will make more effective use of a small number of selected references accompanying each chapter.

An integral part of the book is the collection of problems at the end of each chapter. Some of these problems require the student to apply the abstract concepts developed in the text, whereas others require him to extend these concepts in a natural way. A book containing the solutions to all problems has been prepared.

I wish to express my sincere thanks to Professor Gian-Carlo Rota, who taught me combinatorial mathematics in a most inspiring way. I want to thank Professor David A. Huffman, who helped to initiate and organize the course, and has also made many useful technical suggestions. I am indebted to Professor Robert M. Fano, Director of Project MAC at the Massachusetts Institute of Technology, for his encouragement and support throughout the writing of this book. Under his leadership, the atmosphere and environment at Project MAC have been most favorable and enjoyable. To Mr. Murray Edelberg I owe a sincere debt of gratitude. He made numerous valuable suggestions for the book, helped with the proofreading work, contributed to the problem sets, and prepared a most thorough solution manual. Thanks are also due to Dr. Donald R. Haring for reviewing the entire manuscript carefully and to Professor Shimon Even and Messrs. Peter J. Denning and John A. Williams for their suggestions. Special thanks should go to Miss Kathleen Dimond for the excellent job in typing the manuscript several times over. Last, but not least, I wish to thank my wife Jane, who contributed to the problem sets, helped with the proofreading work, provided many useful criticisms and suggestions related to the text, and also has been a constant source of encouragement and understanding.

C. L. Liu

Contents

Chapter 1
Permutations and Combinations

1-1 INTRODUCTION

When groups of dots and dashes are used to represent alphanumeric symbols in telegraph communication, a communication engineer may wish to know the total number of distinct representations consisting of a fixed number of dots and dashes. To study the physical properties of materials, a physicist may wish to compute the number of ways molecules can be arranged in molecular sites or to compute the number of ways electrons are distributed among different energy levels. A transportation engineer may wish to determine the number of different acceptable train schedules. A computer scientist may wish to have some idea about the number of possible moves his chess-playing program should examine in responding to each of the opponent's moves. Enumerating problems such as these are discussed in the basic theory of combinations and permutations, the topic studied in this chapter.

The words *selection* and *arrangement* will be used in the ordinary sense. Thus, there should be no ambiguity in the meanings of statements such as "*to select* two representatives from five candidates," "there are 10 possible outcomes when two representatives *are selected* from five candidates," "the books *are arranged* on the shelf," and "there are 120 ways *to arrange* five different books on the shelf." The word *combination* has the same meaning as the word *selection*, and the word *permutation* has the same meaning as the word *arrangement*. Formally, an *r-combination* of n objects is defined as an unordered selection of r of these objects, and an *r-permutation* of n objects is defined as an ordered arrangement of r of these objects. For example, to form a committee of 20 senators is an unordered selection of 20 senators from the 100 senators and is therefore a 20-combination of the 100 senators. On the other hand, the outcome of a horse race can be viewed as an ordered arrangement of the t horses in the race and is therefore a t-permutation of the t horses. Notice that we are simply defining the terms r-combination and r-permutation here and have not mentioned anything about the properties of these n objects because there is no need to do so as far as the definitions are concerned.

We are interested here in enumerating the number of combinations or permutations of a given set of objects. The notation $C(n,r)$ denotes the number of *r-combinations* of *n distinct* objects, and the notation $P(n,r)$ denotes the number of *r-permutations* of *n distinct* objects. In the following sections, both $C(n,r)$ and $P(n,r)$ shall be evaluated. However, it is quite clear that $C(n,n) = 1$ (there is just one way to select n objects out of n objects), $C(n,1) = n$ (there are n ways to select one object out of n objects), $C(3,2) = 3$ (for three objects A, B, and C, the selections of two objects are AB, AC, and BC), and $P(3,2) = 6$ (for three objects A, B, and C, the arrangements of two objects are AB, BA, AC, CA, BC, and CB).

1-2 THE RULES OF SUM AND PRODUCT

Among the five Roman letters a, b, c, d, and e and the three Greek letters α, β, and γ, it is clear that there are $5 \times 3 = 15$ ways to select two letters, one from each alphabet. On the other hand, since there are five ways to select a Roman letter and three ways to select a Greek letter, there are $5 + 3 = 8$ ways to select one letter that is either a Roman *or* a Greek letter. These notions are stated formally as the following basic rules:

Rule of product If one event can occur in m ways and another event can occur in n ways, there are $m \times n$ ways in which these two events can occur.

Rule of sum If one event can occur in m ways and another event can occur in n ways, there are $m + n$ ways in which one of these two events can occur.

Clearly, the occurrence of an event can mean either the selection or the arrangement of a certain number of objects. Consider the following illustrative examples.

Example 1-1 To choose two books of different languages among five books in Latin, seven books in Greek, and ten books in French, there are $5 \times 7 + 5 \times 10 + 7 \times 10 = 155$ ways since there are 5×7 ways to choose a book in Latin *and* a book in Greek, 5×10 ways to choose a book in Latin *and* a book in French, and 7×10 ways to choose a book in Greek *and* a book in French. However, there are $22 \times 21 = 462$ ways if we just want to choose two books from the twenty-two books. ■

Example 1-2 By the rule of product

$$P(n,r) = P(r,r) \times C(n,r)$$

because one can make an ordered arrangement of r of n distinct objects by first selecting r objects from the n objects and then arranging these r objects in order.

By the rule of sum

$$C(n,r) = C(n-1, r-1) + C(n-1, r)$$

This can be seen from the following argument. Suppose that one of the n distinct objects is marked as a special object. The number of ways to select r objects from these n objects is equal to the sum of the number of ways to select r objects so that the special object is always included [there are $C(n-1, r-1)$ such ways] and the number of ways to select r objects so that the special object is always excluded [there are $C(n-1, r)$ such ways]. ■

1-3 PERMUTATIONS

Let us now derive an expression for $P(n,r)$, the number of ways of arranging r of n distinct objects. Observe that arranging r of n objects into some order is the same as putting r of the n objects into r distinct (marked) positions. There are n ways to fill the first position (to choose one out of the n objects), $n - 1$ ways to fill the second position (to choose one out of the $n - 1$ remaining objects), . . . , and $n - r + 1$ ways to fill the last position (to choose one out of the $n - r + 1$ remaining objects).

Thus, according to the rule of product, we have

$$P(n,r) = n(n-1) \cdots (n-r+1)$$

Using the notation

$$n! = n(n-1)(n-2) \cdots 3 \times 2 \times 1$$

for $n \geq 1$ ($n!$ is read n factorial),†

$$P(n,r) = \frac{n(n-1)(n-2) \cdots (n-r+1)(n-r) \cdots 3 \times 2 \times 1}{(n-r) \cdots 3 \times 2 \times 1}$$

$$= \frac{n!}{(n-r)!}$$

Example 1-3 As we mentioned earlier, the number of ways of arranging two of three distinct objects is

$$P(3,2) = \frac{3!}{(3-2)!} = \frac{3!}{1!} = 6 \quad \blacksquare$$

Example 1-4 There is an alternative way to derive the formula

$$P(n,r) = \frac{n!}{(n-r)!}$$

We shall show that $P(n,n) = n!$ by induction. Clearly, as the basis of induction, $P(1,1) = 1 = 1!$. As the induction hypothesis, suppose that $P(n-1, n-1) = (n-1)!$. To arrange n distinct objects in order, we single out a special object and arrange the remaining $n-1$ objects first. For each ordered arrangement of these $n-1$ objects, there are n positions for the special object (the $n-2$ positions between the arranged objects and the two end positions). Therefore, according to the rule of product, we have

$$P(n,n) = n \times P(n-1, n-1) = n \times (n-1)! = n!$$

Suppose that we divide n marked positions into two groups, the first r positions and the remaining $n-r$ positions. According to the rule of product, the number of ways of putting n objects in these n positions is equal to the product of the number of ways of putting r of the n objects in the first r positions and the number of ways of putting the remaining $n-r$ objects in the remaining

† The definition of $n!$ given here is valid only for positive integers. The definition of $n!$ when n is a fraction or a negative number is discussed under the topic "gamma functions" in advanced calculus. (See, for example, Buck [1].) It should also be pointed out that the value of $0!$ is 1.

$n - r$ positions. Therefore, we have

$$P(n,n) = P(n,r) \times P(n - r, n - r)$$

that is,

$$P(n,r) = \frac{P(n,n)}{P(n - r, n - r)} = \frac{n!}{(n - r)!} \quad \blacksquare$$

Example 1-5 In how many ways can n people stand to form a ring? Observe that there is a difference between a linear arrangement and a circular arrangement of objects. In the case of circular arrangement, the n people are not assigned to absolute positions, but are only arranged relative to one another. There are two ways of looking at this problem.

Method 1 If the n people are arranged linearly and then the two ends of the line are closed to form a circular arrangement, we have a total of $P(n,n)$ such arrangements. However, since only the relative positions of the n people are important, two of the circular arrangements obtained in this manner are actually the same if one arrangement can be changed into a second arrangement by rotating the first arrangement by one position, or two positions, . . . , or n positions. Consequently, the number of circular arrangements is equal to

$$\frac{P(n,n)}{n} = (n - 1)!$$

Method 2 If we pick a particular person and let him occupy a fixed position, the remaining $n - 1$ people will be arranged using this fixed position as reference in a ring. Again, there are $(n - 1)!$ ways of arranging these $n - 1$ people. $\blacksquare$

Let there be n objects that are not all distinct. Specifically, let there be q_1 objects of the first kind, q_2 objects of the second kind, . . . , and q_t objects of the tth kind. Then the number of n-permutations of these n objects is given by the formula

$$\frac{n!}{q_1!q_2! \cdots q_t!} \tag{1-1}$$

To derive this formula, imagine that the n objects are marked so that objects of the same kind become distinguishable from one another. There are, of course, $n!$ ways in which these n "distinct" objects can be permuted. However, two permutations will be the same when the marks are erased if they differ only in the arrangement of marked objects that are of the same kind. Therefore, each permutation of the unmarked objects will correspond to $q_1!q_2! \cdots q_t!$ permutations of the marked objects. The formula in (1-1) follows.

Example 1-6 Five dashes and eight dots can be arranged in

$$\frac{13!}{5!8!} = 1,287$$

different ways. Also, to use only seven of the thirteen dashes and dots, there are

$$\frac{7!}{5!2!} + \frac{7!}{4!3!} + \frac{7!}{3!4!} + \frac{7!}{2!5!} + \frac{7!}{1!6!} + \frac{7!}{7!} = 120$$

distinct representations. ■

Example 1-7 To show that $(k!)!$ is divisible by $(k)!^{(k-1)!}$ for any integer k, we consider a collection of $k!$ objects among which there are k of the first kind, k of the second kind, . . . , and k of the $(k-1)!$th kind. The total number of ways of permuting these objects is given by

$$\frac{(k!)!}{k!k! \cdots k!} = \frac{(k!)!}{(k!)^{(k-1)!}}$$

Since the total number of permutations must be an integral value, $(k!)^{(k-1)!}$ must divide $(k!)!$. ■

The number of ways to arrange r objects when they are selected out of n distinct objects with unlimited repetitions is

$$n^r$$

This result follows directly from an application of the rule of product, since there are n ways to choose an object to fill the first position, n ways to choose an object to fill the second position, . . . , and n ways to choose an object to fill the rth position.

Example 1-8 Among the 10 billion numbers between 1 and 10,000,000,000, how many of them contain the digit 1? How many of them do not? Among the 10 billion numbers between 0 and 9,999,999,999, there are 9^{10} numbers that do not contain the digit 1. Therefore, among the 10 billion numbers between 1 and 10,000,000,000, there are $9^{10} - 1$ numbers that do not contain the digit 1 and $10^{10} - (9^{10} - 1)$ numbers that do. ■

Example 1-9 A binary sequence is a sequence of 0's and 1's. What is the number of n-digit binary sequences that contain an even number of 0's (zero is considered as an even number)? The problem is immediately solved if we observe that because of symmetry half of the 2^n n-digit binary sequences contain an even number of 0's, and the other half of the sequences contain an odd number of 0's. Another way to look at the problem is to consider the 2^{n-1} $(n-1)$-digit binary sequences. If an $(n-1)$-digit binary sequence contains an even number of 0's, we can append to it a 1 as the nth digit to yield an n-digit binary sequence that contains an even number of 0's. If an $(n-1)$-digit binary sequence contains an odd number of 0's, we can append to it a 0 as the nth digit to yield an n-digit binary sequence that contains an even number of 0's. Therefore, there are 2^{n-1} n-digit binary sequences which contain an even number of 0's.

As an extension, let us consider the n-digit quaternary sequences (sequences that have 0's, 1's, 2's, and 3's as the digits). Again, because of symmetry, there are $4^n/2$ sequences in each of which the total number of 0's and 1's is even.

To find the number of quaternary sequences that contain an even number of 0's, we divide the 4^n sequences into two groups: the 2^n sequences that contain only 2's and 3's and the $4^n - 2^n$ sequences that contain one or more 0's or 1's. The sequences in the first group are, of course, sequences that have an even number of 0's. The sequences in the second group can be subdivided into categories according to the patterns of 2's and 3's in the sequences. (For instance, sequences of the pattern $23xx2x3xxx$ will be in one category where the x's are 0's and 1's.) Since half of the sequences in each category have an even number of 0's, the total number of sequences that have an even number of 0's in the second group is $(4^n - 2^n)/2$. Therefore, among the 4^n n-digit quaternary sequences, there are $2^n + (4^n - 2^n)/2$ sequences that have an even number of 0's.

It will be left as an exercise for the reader to show that the number of n-digit quaternary sequences that have an even number of 0's and an even number of 1's is $4^n/4 + 2^n/2$. ∎

1-4 COMBINATIONS

According to the result in Example 1-2, the number of r-combinations of n objects is

$$C(n,r) = \frac{P(n,r)}{r!} = \frac{n!}{r!(n-r)!}$$

It is immediately obvious from this formula that

$$C(n,r) = C(n, n - r)$$

This indeed is what one would expect since selecting r objects out of n objects is equivalent to picking the $n - r$ objects that are not to be selected.

Example 1-10 If no three diagonals of a convex decagon meet at the same point inside the decagon, into how many line segments are the diagonals divided by their intersections? First of all, the number of diagonals is equal to

$$C(10,2) - 10 = 45 - 10 = 35$$

as there are $C(10,2)$ straight lines joining the $C(10,2)$ pairs of vertices but 10 of these 45 lines are the sides of the decagon. Since for every four vertices we can count exactly one intersection between the diagonals as Fig. 1-1 shows (the decagon is convex), there is a total of $C(10,4) = 210$ intersections between the diagonals.

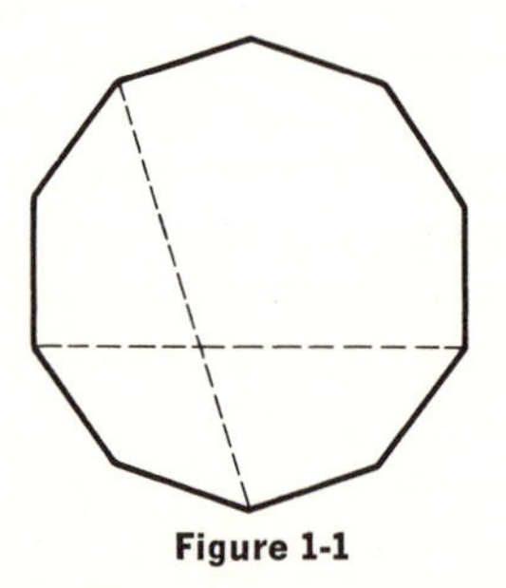

Figure 1-1

Since a diagonal is divided into $k + 1$ straight-line segments when there are k intersecting points lying along it and since each intersecting point lies along two diagonals, the total number of straight-line segments into which the diagonals are divided is

$$35 + 2 \times 210 = 455 \quad \blacksquare$$

Example 1-11 Eleven scientists are working on a secret project. They wish to lock up the documents in a cabinet such that the cabinet can be opened if and only if six or more of the scientists are present. What is the smallest number of locks needed? What is the smallest number of keys to the locks each scientist must carry? To answer the first question, observe that for any group of five scientists, there must be at least one lock they cannot open. Moreover, for any two different groups of five scientists, there must be two different locks they cannot open, because if both groups cannot open the same lock, there is a group of six scientists among these two groups who will not be able to open the cabinet. Thus, at least $C(11,5) = 462$ locks are needed.

As to the number of keys each scientist must carry, let A be one of the scientists. Whenever A is associated with a group of five other scientists, A should have the key to the lock(s) that these five scientists were not able to open. Thus, A carries at least $C(10,5) = 252$ keys. (Although we have only shown that these are lower bounds on the numbers of locks and keys, a scheme can actually be designed using these many locks and with each scientist carrying these many keys.) ■

Example 1-12 In how many ways can three numbers be selected from the numbers 1, 2, . . . , 300 such that their sum is divisible by 3? The 300 numbers 1, 2, . . . , 300 can be divided into three groups: those that are divisible by 3, those that yield the remainder 1 when divided by 3, and those that yield the remainder 2 when divided by 3. Clearly, there are 100 numbers in each of these groups. If three numbers from the first group are selected, or if three numbers from the second group are selected, or if three numbers from the third group are selected, or if three numbers, one from each of the three groups, are selected, their sum will be divisible by 3. Thus, the total number of ways to select three desired numbers is

$$C(100,3) + C(100,3) + C(100,3) + (100)^3 = 1{,}485{,}100 \quad ■$$

We shall now show that when repetitions in the selection of the objects are allowed, the number of ways of selecting r objects from n distinct objects is

$$C(n + r - 1, r) \tag{1-2}$$

Let the n objects be identified by the integers $1, 2, \ldots, n$, and let a specific selection of r objects be identified by a list of the corresponding integers $\{i,j,k, \ldots ,m\}$ arranged in increasing order. For example, the selection in which the first object is selected thrice, the second object is not selected, the third object is selected once, the fourth object is selected once, the fifth object is selected twice, etc., is represented as $\{1,1,1,3,4,5,5, \ldots\}$. To the r integers in such a list we add 0 to the first integer, 1 to the second integer, . . . , and $r - 1$ to the rth integer. Thus, $\{i,j,k, \ldots ,m\}$ becomes $\{i, j + 1, k + 2, \ldots , m + (r - 1)\}$.

For example, the selection {1,1,1,3,4,5,5, . . .} becomes {1,2,3,6,8,10,11, . . .}. Since each selection will then be identified uniquely as a selection of r distinct integers from the integers $1, 2, \ldots, n + (r - 1)$, we have proved the formula in (1-2).

Example 1-13 Out of a large number of pennies, nickels, dimes, and quarters, in how many ways can six coins be selected? The answer is

$$C(4 + 6 - 1, 6) = C(9,6) = 84$$

because this is the same as selecting six coins from a penny, a nickle, a dime, and a quarter with unlimited repetitions. ■

Example 1-14 When three distinct dice are rolled, the number of outcomes is $6 \times 6 \times 6 = 216$. If the three dice are indistinguishable, the number of outcomes is $C(6 + 3 - 1, 3) = 56$. This can be seen by considering the selection of three numbers from the six numbers 1, 2, 3, 4, 5, 6 when repetitions are allowed. ■

When the objects are not all distinct, the number of ways to select one or more objects from them is equal to

$$(q_1 + 1)(q_2 + 1) \cdots (q_t + 1) - 1$$

where there are q_1 objects of the first kind, q_2 objects of the second kind, . . . , and q_t objects of the tth kind. This result follows directly from the rule of product. There are $q_1 + 1$ ways of choosing the object of the first kind, i.e., choosing none of them, one of them, two of them, . . . , or q_1 of them. Similarly, there are $q_2 + 1$ ways of choosing objects of the second kind, . . . , and $q_t + 1$ ways of choosing objects of the tth kind. The term -1 corresponds to the "selection" in which no object at all is chosen and should be discounted.

Example 1-15 How many divisors does the number 1400 have? Since $1400 = 2^3 \times 5^2 \times 7$, the number of its divisors is

$$(3 + 1)(2 + 1)(1 + 1) = 24$$

which is equal to the number of ways to select the prime factors of 1400. (Both 1 and 1400 are considered to be divisors of the number 1400.) ■

Example 1-16 For n given weights, what is the greatest number of different amounts that can be made up by the combinations of these weights? Since a weight can either be selected or not be selected in a combination, there are $2^n - 1$ combinations. If the values of the given weights are properly chosen, we will, at the most, have $2^n - 1$ different combined weights. (As a matter of fact, this number of combinations can always be achieved by choosing the values of the weights as 2^0 units, 2^1 units, 2^2 units, . . . , 2^{n-1} units.)

As an extension of this example, what is the greatest number of different amounts that can be weighed by using a set of n weights and a balance? Each of the weights may be disposed of in one of three ways; that is, it can be placed in the weight pan, in the pan with the substance to be weighed, or it can be unused. Therefore, there are $3^n - 1$ ways of using the n weights. However, in at least half of these $3^n - 1$ ways, the total weight placed in the weight pan is less than or equal to the total weight placed in the other pan. Hence, we can weigh at most $(3^n - 1)/2$ different amounts on a balance when a set of n weights is available. [Indeed, choosing the weights as 3^0 units, 3^1 units, 3^2 units, . . . , 3^{n-1} units enables us to weigh from 1 through $(3^n - 1)/2$ units.] ■

1-5 DISTRIBUTIONS OF DISTINCT OBJECTS

In Secs. 1-3 and 1-4 we have discussed the arrangements of objects (permutations) and the selections of objects (combinations). In this and the following sections, we shall discuss the distribution of objects into distinct or nondistinct positions. As will be seen, the problems of distributions are very closely related to, and in many cases are equivalent to, the problems of permutations and combinations.

In our previous discussion about the permutation of objects, we introduced the notion of placing distinct objects into distinct cells. Two cases must be considered. First, for $n \geq r$, there are $P(n,r)$ ways to place r distinct objects into n distinct cells, where each cell can hold only one object. As was shown before, the first object can be placed in one of the n cells, the second object can be placed in one of the $n - 1$ remaining cells, etc. On the other hand, for $r \geq n$, there are $P(r,n)$ ways to place n of r distinct objects into n distinct cells, where each cell can hold only one object. The argument is similar to the one above; that is, there are r ways to select an object to be placed in the first cell, $r - 1$ ways to select an object to be placed in the second cell, etc.

The distribution of r distinct objects in n distinct cells where each cell can hold any number of objects is equivalent to the arrangement of r of the n cells when repetitions are allowed. In terms of the distribution of distinct objects in distinct cells, since the first object can be placed in one of the n cells, the second object can again be placed in one of the

n cells, etc., there are n^r ways of distributing the objects. The reader should convince himself that n^r is the number of ways no matter whether n is larger or smaller than r.

Notice that in the above case, when more than one object is placed in the same cell, the objects are not ordered inside the cell. When the order of objects in a cell is also considered, the number of ways of distribution is

$$\frac{(n+r-1)!}{(n-1)!} = (n+r-1)(n+r-2) \cdots (n+1)n$$

To prove this result, we imagine such a distribution as an ordered arrangement of the r (distinct) objects and the $n - 1$ (nondistinct) intercell partitions. Using the previously derived formula for the permutation of $r + n - 1$ objects where $n - 1$ of them are of the same kind, we obtain the result $(n + r - 1)!/(n - 1)!$.

There is an alternative way to derive this formula. There are n ways to distribute the first object. After the first object is placed in a cell, it can be considered as an added partition that divides the cell into two cells. Therefore, there are $n + 1$ ways to distribute the second object. Similarly, there are $n + 2$ ways to distribute the third object, . . . , and $n + r - 1$ ways to distribute the rth object.

Example 1-17 The number of ways of arranging seven flags on five masts when all the flags must be displayed but not all the masts have to be used is $5 \times 6 \times 7 \times 8 \times 9 \times 10 \times 11$. The argument is as follows. If there is a single flag on a mast, we assume that it is raised to the top of the mast; however, if there is more than one flag on a mast, the order of the flags on the mast is important. Similarly, seven cars can go through five toll booths in $5 \times 6 \times 7 \times 8 \times 9 \times 10 \times 11$ ways. ■

Observe that the distribution of n objects (q_1 of them are of one kind, q_2 of them are of another kind, . . . , and q_t of them are of the tth kind) into n distinct cells (each of which can hold only one object) is equivalent to the permutation of these objects. Thus, according to the formula in (1-1) the number of ways of distribution is

$$\frac{n!}{q_1!q_2! \cdots q_t!}$$

It is instructive to derive this result from an alternative point of view. Among the n distinct cells, we have $C(n,q_1)$ ways to pick q_1 cells for the objects of the first kind, $C(n - q_1, q_2)$ ways to pick q_2 cells for the objects of the second kind, etc. The number of ways of distribution is, therefore,

$$C(n,q_1)C(n-q_1,q_2)C(n-q_1-q_2,q_3)\cdots C(n-q_1-q_2\cdots-q_{t-1},q_t)P(n-q_1-q_2\cdots-q_t,\ n-q_1-q_2\cdots-q_t)$$
$$=\frac{n!}{q_1!(n-q_1)!}\frac{(n-q_1)!}{q_2!(n-q_1-q_2)!}\frac{(n-q_1-q_2)!}{q_3!(n-q_1-q_2-q_3)!}\cdots\frac{(n-q_1-q_2\cdots-q_{t-1})!}{q_t!(n-q_1-q_2\cdots-q_t)!}(n-q_1-q_2\cdots-q_t)!$$

The factor $P(n-q_1-q_2\cdots-q_t,\ n-q_1-q_2\cdots-q_t)$ is the number of ways of permuting those objects that are one of a kind.

It follows that the number of ways of distributing r objects ($r \le n$), with q_1 of them of one kind, q_2 of them of another kind, etc., into n distinct cells is†

$$C(n,q_1)C(n-q_1,q_2)C(n-q_1-q_2,q_3)\cdots C(n-q_1-q_2\cdots-q_{t-1},q_t)P(n-q_1-q_2\cdots-q_t,\ r-q_1-q_2\cdots-q_t)$$
$$=\frac{n!}{q_1!q_2!\cdots q_t!}\frac{1}{(n-r)!}$$

1-6 DISTRIBUTIONS OF NONDISTINCT OBJECTS

In terms of the distribution of objects into cells, there are $C(n,r)$ ways of placing r nondistinct objects into n distinct cells with at most one object in each cell ($n \ge r$); this follows because the distribution can be visualized as the selection of r cells from the n cells for the r *nondistinct* objects.

The number of ways to place r nondistinct objects into n distinct cells where a cell can hold more than one object is $C(n+r-1,\ r)$. This result comes from the observation that distributing the r nondistinct objects is equivalent to selecting r of the n cells for the r objects with repeated selections of cells allowed. A different argument can be used to derive the result. Imagine the distribution of the r objects into n cells as an arrangement of the r objects and the $n-1$ intercell partitions. Since both the objects and the partitions are nondistinct, the number of ways of arrangement is

$$\frac{(n-1+r)!}{(n-1)!r!}=C(n+r-1,\ r)$$

† This result can be derived by using another argument. We can first select r cells from the n cells and then distribute the r objects into these r cells; that is,

$$C(n,r)\cdot\frac{r!}{q_1!q_2!\cdots q_t!}=\frac{n!}{q_1!q_2!\cdots q_t!}\frac{1}{(n-r)!}$$

If none of the n cells can be left empty (that means r must be larger than or equal to n), the number of ways of distribution is

$$C(r-1, n-1)$$

Since we can first distribute one object in each of the n cells and then distribute the remaining $r - n$ objects arbitrarily, the number of ways of distribution is

$$C((r-n)+n-1, r-n) = C(r-1, r-n) = C(r-1, n-1)$$

A direct extension of this result is the calculation of the number of ways of distributing r nondistinct objects into n distinct cells with each cell containing at least q objects. After placing q objects in each of the n cells, we have

$$C((r-nq)+n-1, r-nq) = C(n-nq+r-1, n-1)$$

Example 1-18 Five distinct letters are to be transmitted through a communications channel. A total of 15 blanks are to be inserted between the letters with at least three blanks between every two letters. In how many ways can the letters and blanks be arranged? There are 5! ways of arranging the letters. For each arrangement of the letters, we can consider the insertion of the blanks as placing 15 nondistinct objects into four distinct interletter positions with at least three objects in each interletter position. Therefore, the total number of ways of arranging the letters and blanks is

$$5! \times C(4-12+15-1, 4-1) = 5! \times C(6,3) = 2{,}400 \quad \blacksquare$$

Example 1-19 In how many ways can $2n + 1$ seats in a congress be divided among three parties so that the coalition of any two parties will ensure them of a majority? This, of course, is a problem of distributing $2n + 1$ nondistinct objects into three distinct cells. Without any restriction on the number of seats each party can have, there are

$$C(3+(2n+1)-1, 2n+1) = C(2n+3, 2n+1) = C(2n+3, 2)$$

ways of distributing the seats. However, among these distributions, there are some in which a party gets $n + 1$ or more seats. For a particular party to have $n + 1$ or more seats, there are

$$C(3+n-1, n) = C(n+2, n) = C(n+2, 2)$$

ways of distributing the seats. The ways of distribution are enumerated by giving the particular party $n + 1$ seats first and then dividing the remaining n seats among the three parties arbitrarily. Therefore, the total number of ways to divide the seats so that no party alone will have a majority is

$$C(2n+3, 2) - 3 \times C(n+2, 2) = \tfrac{1}{2}(2n+3)(2n+2) - \tfrac{3}{2}(n+2)(n+1)$$
$$= \frac{n}{2}(n+1)$$

When there are $2n$ seats, the total number of ways of dividing the seats becomes

$$C(2n+2, 2) - 3 \times C(n+2, 2) + 3 = \tfrac{1}{2}(n-1)(n-2)$$

The term $C(2n+2, 2)$ is the total number of ways of distributing the $2n$ seats. Similarly, $C(n+2, 2)$ is the number of ways of distributing the $2n$ seats such that a particular party gets n or more seats. The term $+3$ is due to the fact that each of the three distributions $(n,n,0)$, $(n,0,n)$, $(0,n,n)$ is accounted for twice in the term $3 \times C(n+2, 2)$. ■

In Secs. 1-5 and 1-6 we have discussed the distribution of (distinct and nondistinct) objects into distinct cells. The distribution of objects into nondistinct cells will be discussed in the next chapter after the development of the concept of generating functions.

★1-7 STIRLING'S FORMULA

In enumerating the permutations and combinations of objects, one frequently needs to calculate the value of $n!$. Even for a moderately large n the evaluation of $n!$ becomes tedious. In this section, we derive an approximation formula for the value of $n!$ which is called *Stirling's formula.* This formula is useful in the computation of the numerical value of $n!$ as well as in many theoretical considerations involving $n!$. Stirling's formula of approximation is

$$n! \approx \sqrt{2\pi n}\left(\frac{n}{e}\right)^n$$

Although the absolute error of such an approximation increases as n increases, the percentage error decreases monotonically. In other words. although

$$\lim_{n\to\infty}\left[n! - \sqrt{2\pi n}\left(\frac{n}{e}\right)^n\right] \to \infty$$

we have

$$\lim_{n\to\infty} \frac{n!}{\sqrt{2\pi n}\,(n/e)^n} \to 1$$

For instance, Stirling's formula approximates 1! by 0.9221 with an 8 percent error, 2! by 1.919 with a 4 percent error, and 5! by 118.019 with a 2 percent error. The percentage error for the approximation of 100! is only 0.08 percent. However, the absolute error is about 1.7×10^{155}.

To prove Stirling's formula, we let

$$a_n = \log (n!) - \tfrac{1}{2} \log n = \log 2 + \log 3 + \cdots + \log (n-1) + \tfrac{1}{2} \log n$$

Consider the curve $y = \log x$. The area under the curve and between the two lines $x = 1$ and $x = n$ is

$$\int_1^n \log x \, dx$$

This area can be approximated by the sum of the areas of n trapezoids which are bounded by the lines $x = k - 1$ and $x = k$ for $k = 2, 3,$

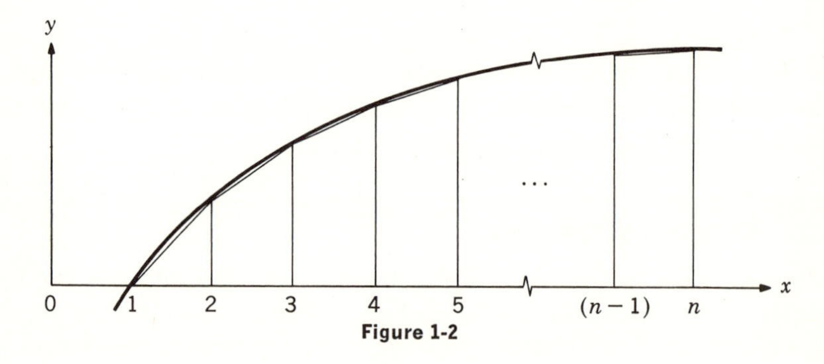

Figure 1-2

$\ldots, n$ as shown in Fig. 1-2. The approximated area is

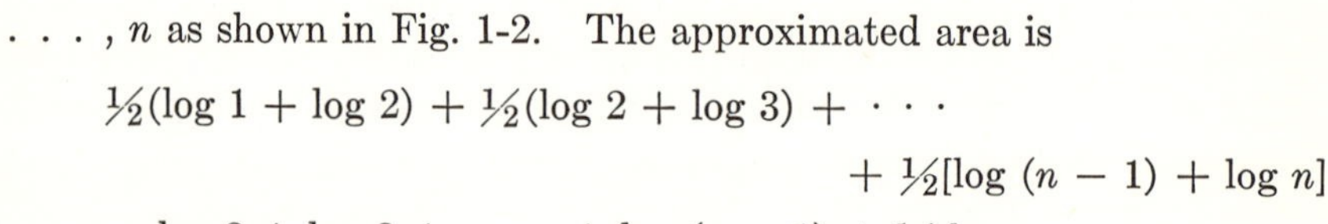

$$\begin{aligned}
&\tfrac{1}{2}(\log 1 + \log 2) + \tfrac{1}{2}(\log 2 + \log 3) + \cdots + \tfrac{1}{2}[\log (n-1) + \log n] \\
&= \log 2 + \log 3 + \cdots + \log (n-1) + \tfrac{1}{2} \log n \\
&= \log (n!) - \tfrac{1}{2} \log n \\
&= a_n
\end{aligned}$$

which is smaller than the exact value of the area, because the curve $y = \log x$ is convex. Therefore, we write

$$a_n < \int_1^n \log x \, dx \tag{1-3}$$

On the other hand, the area under the curve $y = \log x$ and between the two lines $x = 3/2$ and $x = n$ is

$$\int_{3/2}^{n} \log x \, dx$$

which can be approximated by the sum of the areas of the $n - 1$ trapezoids bounded by the tangent at the point $(k, \log k)$ and the lines $x = k - 1/2$ and $x = k + 1/2$ for $k = 2, 3, \ldots, n - 1$, together with the area of the rectangle bounded by the horizontal line at the point

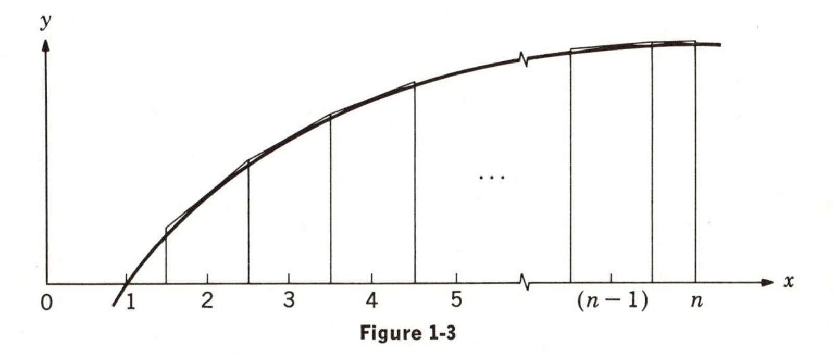

Figure 1-3

$(n, \log n)$ and the two lines $x = n - 1/2$ and $x = n$ as shown in Fig. 1-3. The approximated area is

$$\log 2 + \log 3 + \cdots + \log (n - 1) + 1/2 \log n = a_n$$

Again, because the curve $y = \log x$ is convex, we have

$$\int_{3/2}^{n} \log x \, dx < a_n \tag{1-4}$$

Combining the inequalities in (1-3) and (1-4), we write

$$\int_{3/2}^{n} \log x \, dx < a_n < \int_{1}^{n} \log x \, dx$$

That is,

$$n \log n - n - 3/2 \log 3/2 + 3/2 < a_n < n \log n - n + 1$$

after evaluating the integrals. Recalling that

$$\log (n!) = a_n + 1/2 \log n$$

we have

$$(n + 1/2) \log n - n + 3/2(1 - \log 3/2) < \log (n!)$$
$$< (n + 1/2) \log n - n + 1$$

Therefore, we can write

$$\log (n!) = (n + \tfrac{1}{2}) \log n - n + \delta_n$$

where

$$\tfrac{3}{2}(1 - \log \tfrac{3}{2}) = 0.893 < \delta_n < 1$$

It follows that

$$\begin{aligned}\delta_n &= \log (n!) - (n + \tfrac{1}{2}) \log n + n \\ &= a_n - \int_1^n \log x \, dx + 1 \\ &= 1 - \left(\int_1^n \log x \, dx - a_n\right)\end{aligned}$$

Notice that $\left(\int_1^n \log x \, dx - a_n\right)$ increases monotonically when n increases since $\left(\int_1^n \log x \, dx - a_n\right)$ represents the difference between the area under the curve $y = \log x$ and the sum of the areas of the trapezoids in Fig. 1-2. Therefore, we conclude that δ_n decreases monotonically as n increases. However, since δ_n has a lower bound (0.893), the limit of δ_n as n approaches ∞, denoted by δ, is a constant having a value between 0.893 and 1. Using δ to approximate δ_n for all n's, we have

$$\log (n!) \approx (n + \tfrac{1}{2}) \log n - n + \delta$$

or

$$\begin{aligned} n! &\approx e^{(n+\frac{1}{2}) \log n} e^{-n} e^{\delta} \\ &= n^{(n+\frac{1}{2})} e^{-n} e^{\delta} \\ &= e^{\delta} \sqrt{n} \left(\frac{n}{e}\right)^n \end{aligned}$$

The term e^{δ} is a constant that lies within the range $e^{0.893} = 2.45$ and $e^1 = 2.72$. As it turns out, the value of e^{δ} is equal to $\sqrt{2\pi} = 2.507$.

Except for the last step of determining the constant e^{δ} [which can be looked up in a textbook on advanced calculus (for example, Buck [1])], we have proved Stirling's approximation formula for n factorial.

1-8 SUMMARY AND REFERENCES

A large class of problems in combinatorial mathematics is concerned with computing the number of ways in which some well-defined operation can be performed. In this chapter, the notions of combinations and permutations which are the simplest and yet most fundamental concepts in the study of the theory of enumeration were introduced. On the one hand, we have seen the application of these basic notions to the solution of many problems. On the other hand, we shall study in the subsequent chapters more powerful enumerative techniques which can be compared with the elementary approach in this chapter.

The classical textbook by Whitworth [5] gives a thorough treatment of many topics in combinations and permutations. The companion exercise book, also by Whitworth [6], contains a large collection of problems together with their solutions. Chapter 2 of Feller [2], Chap. 1 of Riordan [3], and Chap. 1 of Ryser [4] also cover the subject material in this chapter.

1. Buck, R. C.: "Advanced Calculus," 2d ed., McGraw-Hill Book Company, New York, 1965.
2. Feller, W.: "An Introduction to Probability Theory and Its Applications," vol. 1, 2d ed., John Wiley & Sons, Inc., New York, 1957.
3. Riordan, J.: "An Introduction to Combinatorial Analysis," John Wiley & Sons, Inc., New York, 1958.
4. Ryser, H. J.: "Combinatorial Mathematics," published by the Mathematical Association of America, distributed by John Wiley & Sons, Inc., New York, 1963.
5. Whitworth, W. A.: "Choice and Chance," reprint of the 5th ed. (1901), Hafner Publishing Company, Inc., New York, 1965.
6. Whitworth, W. A.: "DCC Exercises in Choice and Chance," reprint of the edition of 1897, Hafner Publishing Company, Inc., New York, 1965.

PROBLEMS

1-1. (*a*) Use the relation $C(n,r) = C(n-1, r) + C(n-1, r-1)$ to prove the identity

$$C(n+1, m) = C(n,m) + C(n-1, m-1) + C(n-2, m-2) + \cdots + C(n-m, 0)$$

from $m \leq n$.

(*b*) Prove this identity using combinatorial arguments.

1-2. (*a*) Prove the identity

$$1 \times 1! + 2 \times 2! + 3 \times 3! + \cdots + n \times n! = (n+1)! - 1$$

(*b*) Discuss the combinatorial significance of this identity.

(*c*) Show that any integer m can be expressed uniquely in the following form (factorial representation):

$$m = a_1 \times 1! + a_2 \times 2! + a_3 \times 3! + \cdots + a_i \times i! + \cdots$$

where $0 \leq a_i \leq i$ for $i = 1, 2, \ldots$.

1-3. It is clear that

$$P(n,n) = P(n, n-1)$$

but

$$P(n,n) \neq P(n, n-2)$$

Give a combinatorial explanation of these two relations.

1-4. Use a combinatorial argument to prove the identity

$$C(n,0) + C(n,1) + C(n,2) + \cdots + C(n,n) = 2^n$$

1-5. (*a*) Show that

$$n \times C(n-1, r) = (r+1) \times C(n, r+1)$$

What is the combinatorial significance of this identity?

(*b*) Prove the identity

$$C(n,1) + 2 \times C(n,2) + 3 \times C(n,3) + \cdots + n \times C(n,n) = n \times 2^{n-1}$$

1-6. For a given n, show that $C(n,k)$ is maximum when

$$k = \frac{n-1}{2}, \frac{n+1}{2} \qquad \text{if } n \text{ is odd}$$

$$k = \frac{n}{2} \qquad \text{if } n \text{ is even}$$

1-7. (*a*) Use a combinatorial argument to prove that $(2n)!/2^n$ and $(3n)!/(2^n \times 3^n)$ are integers.

(*b*) Prove that $(n^2)!/(n!)^{n+1}$ is an integer.

1-8. Three integers are selected from the integers 1, 2, . . . , 1,000. In how many ways can these integers be selected such that their sum is divisible by 3?

1-9. (*a*) Among $2n$ objects, n of them are identical. Find the number of ways to select n objects out of these $2n$ objects.

(*b*) Among $3n + 1$ objects, n of them are identical. Find the number of ways to select n objects out of these $3n + 1$ objects.

1-10. From n distinct integers, two groups of integers are to be selected with k_1 integers in the first group and k_2 integers in the second group, where k_1 and k_2 are fixed and $k_1 + k_2 \leq n$. In how many ways can the selection be made such that the smallest integer in the first group is larger than the largest integer in the second group?

1-11. Suppose that no three of the diagonals of a convex n-gon meet at the same point inside of the n-gon. Find the number of different triangles the sides of which are made up of the sides of the n-gon, the diagonals, and segments of the diagonals.

1-12. Consider the set of words of length n generated from the alphabet $\{0,1,2\}$.

(*a*) Show that the number of words in each of which the digit 0 appears an even number of times is $(3^n + 1)/2$.

(*b*) Prove the identity

$$\binom{n}{0} 2^n + \binom{n}{2} 2^{n-2} + \cdots + \binom{n}{q} 2^{n-q} = \frac{3^n + 1}{2}$$

where $q = n$ when n is even, and $q = n - 1$ when n is odd.

1-13. An alphabet of m letters can be transmitted through a communication channel. Find the number of different messages of n letters, if

(*a*) The letters can be used repeatedly in a message.

(*b*) l of the m letters can be used only as the first and the last letters in a message; the other letters can appear anywhere with unrestricted repetitions in a message.

(*c*) l of the m letters can be used only as the first and the last letters in a message; the other letters can appear anywhere, except the two ends, with unrestricted repetitions in a message.

1-14. Five teaching machines are to be used by a group of m students. If the same number of students should be assigned to use the first and the second machines, in how many ways can the assignment be made?

1-15. Among the set of 10^n n-digit integers, two integers are considered to be equivalent if one can be obtained by a permutation of the digits of the other.

(*a*) How many nonequivalent integers are there?

(*b*) If the digits 0 and 9 can appear at most once, how many nonequivalent integers are there, for $n \geq 2$?

1-16. In how many ways can the letters $a, a, a, a, a, b, c, d, e$ be permuted such that no two a's are adjacent?

1-17. (*a*) A Boolean function of n variables is defined by the assignment of a value of either 0 or 1 to each of the 2^n n-digit binary numbers. How many different Boolean functions of n variables are there?

(*b*) A Boolean function can be represented conveniently in tabular form where all the n-digit binary numbers and their values are listed. Such a tabular form is called the truth table of a Boolean function. For example, the following table is the truth table of a Boolean function of three variables:

Three-digit binary number	*Value*
0 0 0	0
0 0 1	1
0 1 0	1
0 1 1	0
1 0 0	1
1 0 1	0
1 1 0	1
1 1 1	1

A self-dual Boolean function is one the truth table of which will remain unchanged after all the 0's and 1's in the table are interchanged. How many self-dual Boolean functions of n variables are there?

(*c*) A symmetric Boolean function is one the truth table of which will remain unchanged for any permutation of the n columns of the binary digits. How many symmetric Boolean functions of n variables are there?

1-18. (*Statistical Mechanics: Bose-Einstein Counting*) A system consists of four identical particles. The total energy of the system is equal to $4E_0$ where E_0 is a positive constant. Each of the particles can have an energy level equal to kE_0 ($k = 0, 1, 2, 3, 4$). A particle of energy kE_0 can occupy one of the $k^2 + 1$ distinct energy states at that energy level. How many different configurations, in terms of energy states occupied by the particles, can the system have?

1-19. (*Statistical Mechanics: Fermi-Dirac Counting*) Consider a system that is the same as the one described in Prob. 1-18 except for the following:

1. At the energy level kE_0 there are $2(k^2 + 1)$ distinct energy states.
2. No two particles can be in the same energy state.

How many different configurations can the system have?

1-20. (*Statistical Mechanics*) A system consists of three identical particles, two of them in one potential well, the other in a separate potential well. The total energy of the system is equal to $3E_0$. Each of the particles can have an energy level equal to kE_0 ($k = 0, 1, 2, 3$). A particle of energy kE_0 can occupy one of the $2(k^2 + 1)$

distinct energy states at that energy level. How many different configurations, in terms of the occupation of energy states in the two potential wells, can the system have?

1-21. A system has N molecules. Each molecule has an A-atom at a fixed location and two identical B-atoms which can occupy any two of four possible positions in the molecule.

(*a*) How many different configurations can the system have?

(*b*) If some molecules can have one or two extra B-atoms by taking them from other molecules, how many different configurations can the system have?

(*c*) Repeat part (*b*) if, instead of two identical B-atoms, each molecule contains a B-atom and a C-atom.

1-22. A system consists of N distinct particles each of which can have an energy level equal to kE_0 for $k = 0, 1, 2, \ldots$. If the total energy of the system is equal to ME_0, where M is a positive integer, in how many ways can the total energy be distributed among the particles?

1-23. (*Information Theory: Sphere Packing*) A set of code words that are of the same length is called a block code. In binary block codes, the *distance* between two code words is defined as the number of digits where the two code words differ. For example, the distance between 0110 and 1011 is 3 and the distance between 0110 and 0111 is 1. When the minimum distance between any two code words in a set of n-digit code words is $2r + 1$, the set forms an *r-error-correcting code of length n*. Let $A(n,r)$ denote the maximum number of code words in an *r-error-correcting code of length n*.

(*a*) Show that

$$A(n,r) \leq \frac{2^n}{\sum_{j=0}^{r} C(n,j)} \qquad \text{(Hamming bound)}$$

(*b*) Show that

$$A(n,r) \geq \frac{2^n}{\sum_{j=0}^{2r} C(n,j)} \qquad \text{(Gilbert bound)}$$

Hint: Consider the following geometric interpretation. In an n-dimensional space in which the cartesian coordinates can assume only the two values 0 and 1, an n-digit binary word can be represented as a point. If a code is an r-error-correcting one, there is at most one code word inside any "sphere" of diameter $2r$.

1-24. (*Information Theory: Group Codes*) A binary group code is a set of binary code words with the property that the modulo 2 sum of any two code words in the set is also a code word in the set. Let $A(r)$ be the total number of code words in an r-error-correcting binary group code of length n. Show that

$$A(r) \leq \frac{2(2r+1)}{4r+2-n}$$

Hint: Show that the following are true:

1. Any column in the list of code words contains either half ones and half zeros or all zeros.
2. The list of code words contains at least $[A(r) - 1](2r + 1)$ ones.

1-25. How many permutations of the integers 1, 2, . . . , n are there such that every integer is followed by (but not necessarily immediately followed by) an integer which differs from it by 1? For example, with $n = 4$, 1432 is an acceptable permutation but 2431 is not.

1-26. Let $a_1, a_2, \ldots, a_{2n}$ denote an ordered sequence of n 1's and n (-1)'s. Let $f(k)$ be the sum of its first k digits. That is,

$$f(k) = \sum_{i=1}^{k} a_i$$

Find the number of such sequences that have the property

$$f(k) \geq 0 \qquad \text{for } k = 1, 2, \ldots, 2n$$

Hint: Define a new function $g(k)$ such that

$$g(k) = \begin{cases} f(k) & 1 \leq k \leq m \\ -(f(k) + 2) & m < k \leq 2n \end{cases}$$

where m is the smallest integer for which $f(m) = -1$, if there is such an integer. Otherwise $m = 2n$.

Chapter 2
Generating Functions

2-1 INTRODUCTION

From three distinct objects a, b, and c, there are three ways to choose one object, namely, to choose either a or b or c. Let us represent these possible choices symbolically as $a + b + c$. Similarly, from these three objects, there are three ways to choose two objects, namely, to choose either a and b, or b and c, or c and a, which can be represented symbolically as $ab + bc + ca$. There is only one way to choose three objects, which can be represented symbolically as abc. Examining the polynomial

$$(1 + ax)(1 + bx)(1 + cx) = 1 + (a + b + c)x + (ab + bc + ca)x^2 + (abc)x^3$$

we discover that all these possible ways of selection are exhibited as the coefficients of the powers of x. In particular, the coefficient of x^i is the representation of the ways of selecting i objects from the three objects. This, of course, is not sheer coincidence. We have an interpretation of the polynomial according to the rule of sum and the rule of product. Symbolically, the factor $1 + ax$ means that for the object a, the two ways of selection are "not to select a" or "to select a." The variable x is a formal variable and is used simply as an indicator. The coefficient of x^0 shows the ways no object is selected, and the coefficient of x^1 shows the ways one object is selected. Similar interpretation can be given to the factors $1 + bx$ and $1 + cx$. Thus, the product $(1 + ax)(1 + bx)(1 + cx)$ indicates that for the objects a, b, and c, the ways of selection are "to select *or* not to select a" *and* "to select *or* not to select b" *and* "to select *or* not to select c." It is clear that the powers of x in the polynomial indicate the number of objects that are selected, and the corresponding coefficients show all the possible ways of selection.

This example motivates the formal definition of the *generating function* of a sequence. Let $(a_0, a_1, a_2, \ldots, a_r, \ldots)$ be the symbolic representation of a sequence of events, or let it simply be a sequence of numbers. The function

$$F(x) = a_0\mu_0(x) + a_1\mu_1(x) + a_2\mu_2(x) + \cdots + a_r\mu_r(x) + \cdots$$

is called the *ordinary generating function* of the sequence $(a_0, a_1, a_2, \ldots, a_r, \ldots)$, where $\mu_0(x)$, $\mu_1(x)$, $\mu_2(x)$, $\ldots$, $\mu_r(x)$, $\ldots$ is a sequence of functions of x that are used as indicators. (Another kind of generating function called the *exponential generating function* will be discussed later in this chapter.) The *indicator functions*, the $\mu(x)$'s, are usually chosen in such a way that no two distinct sequences will yield the same generating function. Clearly, the generating function of a sequence is just an alternative representation of the sequence. For example, using 1, $\cos x$, $\cos 2x$, $\ldots$, $\cos rx$, $\ldots$ as the indicator functions, we see that the ordinary generating function of the sequence $(1, \omega, \omega^2, \ldots, \omega^r, \ldots)$ is

$$F(x) = 1 + \omega \cos x + \omega^2 \cos 2x + \cdots + \omega^r \cos rx + \cdots$$

On the other hand, using 1, $1 + x$, $1 - x$, $1 + x^2$, $1 - x^2$, $\ldots$, $1 + x^r$, $1 - x^r$, $\ldots$ as the indicator functions, the ordinary generating function of the sequence $(3,2,6,0,0)$ is

$$3 + 2(1 + x) + 6(1 - x) = 11 - 4x$$

However, the sequences $(1,3,7,0,0)$ and $(1,2,6,1,1)$ will also yield the same ordinary generating function; that is,

$$1 + 3(1 + x) + 7(1 - x) = 11 - 4x$$

and

$$1 + 2(1 + x) + 6(1 - x) + (1 + x^2) + (1 - x^2) = 11 - 4x$$

Hence, we see that the functions 1, $1 + x$, $1 - x$, $1 + x^2$, $1 - x^2$, $\ldots$ should not be used as indicator functions. The most usual and useful form of $\mu_r(x)$ is x^r. In that case, for the sequence $(a_0, a_1, a_2, \ldots,$

$a_r, \ldots$), we have

$$F(x) = a_0 + a_1x + a_2x^2 + \cdots + a_rx^r + \cdots$$

We shall limit our discussion to indicator functions of this form. From now on, when we talk about the generating functions of a sequence, we shall mean the generating function of the sequence with the powers of x as indicator functions. Notice that the sequence $(a_0,a_1,a_2, \ldots ,a_r, \ldots)$ can be an infinite sequence, and $F(x)$ will then be an infinite series. However, because x is just a formal variable, there is no need to question whether the series converges.†

As will be seen, in addition to being an alternative representation, the generating function is also a useful representation that leads to some very powerful techniques in enumerations and other types of combinatorial problems.

2-2 GENERATING FUNCTIONS FOR COMBINATIONS

We have seen that the polynomial $(1 + ax)(1 + bx)(1 + cx)$ is the ordinary generating function of the different ways to select the objects a, b, and c. Instead of the different ways of selection, we may only be interested in the number of ways of selection. By setting $a = b = c = 1$, we have

$$(1 + x)(1 + x)(1 + x) = (1 + x)^3 = 1 + 3x + 3x^2 + x^3$$

Clearly, we see that there is one way to select no objects from the three objects, $C(3,0)$, three ways to select one object out of three, $C(3,1)$, etc. Usually, a generating function that gives the number of combinations or permutations is called an *enumerator*. In particular, an ordinary generating function that gives the number of combinations or permutations is called an *ordinary enumerator*.

This notion can be extended immediately. To find the number of combinations of n distinct objects, we have the ordinary enumerator

$$\begin{aligned}(1 + x)^n &= 1 + nx + \frac{n(n-1)}{2!}x^2 + \cdots \\ &\qquad + \frac{n(n-1)\cdots(n-r+1)}{r!}x^r + \cdots + x^n \\ &= C(n,0) + C(n,1)x + C(n,2)x^2 + \cdots \\ &\qquad + C(n,r)x^r + \cdots + C(n,n)x^n\end{aligned}$$

† An alternative point of view can also be taken. Except for the case $a_0 = \infty$, $F(x)$ converges at $x = 0$. Therefore, with the understanding that the value of x is set to be 0, we can carry the expression for $F(x)$ along in our computation without concerning ourselves further with the convergence problem.

In the expansion of $(1 + x)^n$, the coefficient of the term x^r is the number of ways the term x^r can be formed by taking r x's and $n - r$ 1's among the n factors $1 + x$. It is for this reason that the $C(n,r)$'s are called the binomial coefficients. In a binomial expansion, $\binom{n}{r}$ is a common alternative notation for $C(n,r)$.

Example 2-1 From

$$\binom{n}{0} + \binom{n}{1} x + \binom{n}{2} x^2 + \cdots + \binom{n}{r} x^r + \cdots + \binom{n}{n} x^n = (1 + x)^n$$

we have the identity

$$\binom{n}{0} + \binom{n}{1} + \binom{n}{2} + \cdots + \binom{n}{r} + \cdots + \binom{n}{n} = 2^n$$

by setting x equal to 1. The combinatorial significance of this identity is that both sides give the number of ways of selecting none, or one, or two, . . . , or n objects out of n distinct objects.

We also have the identity

$$\binom{n}{0} - \binom{n}{1} + \binom{n}{2} - \cdots + (-1)^r \binom{n}{r} + \cdots + (-1)^n \binom{n}{r} = 0$$

by setting x equal to -1. Writing this as

$$\binom{n}{0} + \binom{n}{2} + \binom{n}{4} + \cdots = \binom{n}{1} + \binom{n}{3} + \binom{n}{5} + \cdots$$

we see that the number of ways of selecting an even number of objects is equal to the number of ways of selecting an odd number of objects from n distinct objects. ■

Example 2-2 The identity

$$\binom{n}{0}^2 + \binom{n}{1}^2 + \binom{n}{2}^2 + \cdots + \binom{n}{r}^2 + \cdots + \binom{n}{n}^2 = \binom{2n}{n}$$

can be proved in two ways.

Method 1 We observe that the expression on the left-hand side is the constant term in $(1 + x)^n(1 + x^{-1})^n$. Since

$$(1 + x)^n(1 + x^{-1})^n = (1 + x)^n(1 + x)^n x^{-n} = x^{-n}(1 + x)^{2n}$$

and the constant term in $x^{-n}(1+x)^{2n}$ is $\binom{2n}{n}$, we have proved the identity.

Method 2 We rewrite the identity to be proved as

$$\binom{n}{0}\binom{n}{n}+\binom{n}{1}\binom{n}{n-1}+\binom{n}{2}\binom{n}{n-2}+\cdots$$
$$+\binom{n}{r}\binom{n}{n-r}+\cdots+\binom{n}{n}\binom{n}{0}=\binom{2n}{n}$$

and use a combinatorial argument. To select n objects out of $2n$ objects, we shall first divide them (in any arbitrary manner) into two piles with n objects in each pile. There are $\binom{n}{i}$ ways to select i objects from the first pile and $\binom{n}{n-i}$ ways to select $n-i$ objects from the second pile to make up a selection of n objects. Therefore, the number of ways to make the selection is $\sum_{i=0}^{n}\binom{n}{i}\binom{n}{n-i}$ which is also equal to $\binom{2n}{n}$.

To see an application of this result, let us consider the problem of finding the number of $2n$-digit binary sequences which are such that the number of 0's in the first n digits of a sequence is equal to the number of 0's in the last n digits of the sequence. Since the number of n-digit binary sequences containing r 0's is $\binom{n}{r}$, the number of $2n$-digit binary sequences containing r 0's in the first n digits as well as in the last n digits is $\binom{n}{r}^2$. Therefore, the number of $2n$-digit binary sequences which are such that the number of 0's in the first n digits of a sequence is equal to the number of 0's in the last n digits of the sequence is

$$\binom{n}{0}^2+\binom{n}{1}^2+\binom{n}{2}^2+\cdots+\binom{n}{r}^2+\cdots+\binom{n}{n}^2=\binom{2n}{n}$$

It is instructive for the reader to rephrase the combinatorial argument used in the second method of proof above for this particular problem. ■

Example 2-3 Prove the identity

$$\binom{n}{1} + 2\binom{n}{2} + 3\binom{n}{3} + \cdots + r\binom{n}{r} + \cdots + n\binom{n}{n} = n2^{n-1}$$

Differentiating both sides of the identity

$$\binom{n}{0} + \binom{n}{1}x + \binom{n}{2}x^2 + \cdots + \binom{n}{r}x^r + \cdots + \binom{n}{n}x^n = (1+x)^n$$

we have

$$\binom{n}{1} + 2\binom{n}{2}x + 3\binom{n}{3}x + \cdots + r\binom{n}{r}x^{r-1} + \cdots + n\binom{n}{n}x^{n-1} = n(1+x)^{n-1}$$

The given identity is obtained by setting x equal to 1. ■

Example 2-4 What is the coefficient of the term x^{23} in $(1 + x^5 + x^9)^{100}$? Since $x^5x^9x^9 = x^{23}$ is the only way the term x^{23} can be made up in the expansion of $(1 + x^5 + x^9)^{100}$ and there are $C(100,2)$ ways to choose the two factors x^9 and then $C(98,1)$ ways to choose the factor x^5, the coefficient of x^{23} is

$$C(100,2) \times C(98,1) = \frac{100 \times 99}{2} \times 98 = 485{,}100 \quad ■$$

Example 2-5 Show that the ordinary generating function of the sequence $\binom{0}{0}, \binom{2}{1}, \binom{4}{2}, \binom{6}{3}, \ldots, \binom{2r}{r}, \ldots$ is $(1 - 4x)^{-1/2}$. According to the binomial theorem,† we have

$$(1 - 4x)^{-1/2} = 1 + \sum_{r=1}^{\infty} \frac{(-\tfrac{1}{2})(-\tfrac{1}{2} - 1) \cdots (-\tfrac{1}{2} - r + 1)}{r!}(-4x)^r$$

$$= 1 + \sum_{r=1}^{\infty} \frac{4^r\,(1/2)(3/2)(5/2) \cdots [(2r-1)/2]}{r!}x^r$$

† As a reminder, the binomial theorem is

$$(1 + x)^n = 1 + \sum_{r=1} \frac{n(n-1)(n-2) \cdots (n-r+1)}{r!}x^r$$

where the upper limit of the summation is n if n is a positive integer, and is ∞ otherwise.

$$= 1 + \sum_{r=1}^{\infty} \frac{2^r[1 \times 3 \times 5 \times \cdots \times (2r-1)]}{r!} x^r$$

$$= 1 + \sum_{r=1}^{\infty} \frac{(2^r \times r!)[1 \times 3 \times 5 \times \cdots \times (2r-1)]}{r!r!} x^r$$

$$= 1 + \sum_{r=1}^{\infty} \frac{(2 \times 4 \times 6 \times \cdots \times 2r)[1 \times 3 \times 5 \times \cdots \times (2r-1)]}{r!r!} x^r$$

$$= 1 + \sum_{r=1}^{\infty} \frac{(2r)!}{r!r!} x^r$$

$$= 1 + \sum_{r=1}^{\infty} \binom{2r}{r} x^r$$

As an application of this result we evaluate the sum

$$\sum_{i=0}^{t} \binom{2i}{i} \binom{2t-2i}{t-i}$$

for a given t. Since $\binom{2i}{i}$ is the coefficient of the term x^i in $(1-4x)^{-1/2}$ and $\binom{2t-2i}{t-i}$ is the coefficient of the term x^{t-i} in $(1-4x)^{-1/2}$, $\sum_{i=0}^{t} \binom{2i}{i} \binom{2t-2i}{t-i}$ is the coefficient of the term x^t in $(1-4x)^{-1/2}(1-4x)^{-1/2}$. Since

$$(1-4x)^{-1/2}(1-4x)^{-1/2} = (1-4x)^{-1}$$
$$= 1 + 4x + (4x)^2 + (4x)^3 + \cdots + (4x)^r + \cdots$$

we have

$$\sum_{i=0}^{t} \binom{2i}{i} \binom{2t-2i}{t-i} = 4^t \quad \blacksquare$$

When repetitions are allowed in the selections (or equivalently, when there is more than one object of the same kind), the extension is immediate. For example, the polynomial

$$(1 + ax + a^2x^2)(1 + bx)(1 + cx) = 1 + (a + b + c)x$$
$$+ (ab + bc + ac + a^2)x^2 + (abc + a^2b + a^2c)x^3 + (a^2bc)x^4$$

is the ordinary generating function for the combinations of the objects a, b, and c, where a can be selected twice. The reader should notice the difference between the combinatorial significance of this polynomial and that of the polynomial $(1 + ax)(1 + a^2x^2)(1 + bx)(1 + cx)$, which can be written as $(1 + ax + a^2x^2 + a^3x^3)(1 + bx)(1 + cx)$.

As another example, let us consider the generating function

$$(1 + ax)(1 + a^2x)(1 + bx)(1 + cx) = 1 + (a + b + c + a^2)x + (ab + bc + ac + a^3 + a^2b + a^2c)x^2 + (abc + a^3b + a^2bc + a^3c)x^3 + (a^3bc)x^4$$

We can imagine that there are four boxes, one containing a, one containing two a's, one containing b, and one containing c. The generating function gives the outcomes of the selection of the boxes.

Similarly, the ordinary enumerator for the combinations of the objects a, b, and c, where a can be selected twice, is

$$(1 + x + x^2)(1 + x)^2 = 1 + 3x + 4x^2 + 3x^3 + x^4$$

The significance of the factor $1 + x + x^2$ is that for the object a, there is one way not to select it, one way to select it once, and also one way to select it twice. In the following, we have more illustrative examples.

Example 2-6 Given two each of p kinds of objects and one each of q additional kinds of objects, in how many ways can r objects be selected? The ordinary enumerator for the combinations is

$$(1 + x + x^2)^p(1 + x)^q$$

The coefficient of x^r in the enumerator is

$$\sum_{i=0}^{[r/2]} \binom{p}{i}\binom{p + q - i}{r - 2i}$$

where $[r/2]$ denotes the integral part of $r/2$ (that is, $[r/2] = r/2$ if r is even, and $[r/2] = (r - 1)/2$ if r is odd), because among the p factors of the form $1 + x + x^2$ we can select i x^2's, and among the $p - i$ remaining factors of the form $1 + x + x^2$ and the q factors of the form $1 + x$ we can select $r - 2i$ x's. ■

Example 2-7 The ordinary enumerator for the selection of r objects out of n objects with unlimited repetitions is

$$(1 + x + x^2 + \cdots + x^k + \cdots)^n = \left(\frac{1}{1 - x}\right)^n = (1 - x)^{-n}$$

$$= 1 + \sum_{r=1}^{\infty} \frac{(-n)(-n-1)\cdots(-n-r+1)}{r!}(-x)^r$$

$$= 1 + \sum_{r=1}^{\infty} \frac{(n)(n+1)\cdots(n+r-1)}{r!}x^r$$

$$= \sum_{r=0}^{\infty} \binom{n+r-1}{r} x^r$$

The fact that there are $\binom{n+r-1}{r}$ ways to select r objects from n objects with unlimited repetitions is a result we have proved in Chap. 1. ■

Example 2-8 The ordinary enumerator for the selection of r objects out of n objects ($r \geq n$), with unlimited repetitions but with each object included in each selection, is

$$(x + x^2 + \cdots + x^k + \cdots)^n = x^n \left(\frac{1}{1-x}\right)^n$$

$$= x^n(1-x)^{-n}$$

$$= x^n \sum_{i=0}^{\infty} \binom{n+i-1}{i} x^i$$

$$= \sum_{i=0}^{\infty} \binom{n+i-1}{i} x^{n+i}$$

$$= \sum_{r=n}^{\infty} \binom{r-1}{r-n} x^r \qquad (\text{let } r = n + i)$$

$$= \sum_{r=n}^{\infty} \binom{r-1}{n-1} x^r \quad ■$$

Example 2-9 Show that the number of ways in which r nondistinct objects can be distributed into n distinct cells, with the condition that no cell contains less than q nor more than $q + z - 1$ objects, is the coefficient of x^{r-qn} in the expansion of $[(1 - x^z)/(1 - x)]^n$. Since the ordinary enumerator for the ways a particular cell can be filled is

$$x^q + x^{q+1} + \cdots + x^{q+z-1}$$

the ordinary enumerator for the distributions is

$$(x^q + x^{q+1} + \cdots + x^{q+z-1})^n$$
$$= x^{qn}(1 + x + \cdots + x^{z-1})^n$$
$$= x^{qn}\left(\frac{1 - x^z}{1 - x}\right)^n$$

As an application of this result, we shall find the number of ways in which four persons, each rolling a single die once, can have a total score of 17. That is, for $r = 17$, $n = 4$, $q = 1$, and $z = 6$, the ordinary enumerator is $x^4[(1 - x^6)/(1 - x)]^4$. Since

$$(1 - x^6)^4 = 1 - 4x^6 + 6x^{12} - 4x^{18} + x^{24}$$

$$(1 - x)^{-4} = 1 + \frac{4}{1!}x + \frac{4 \times 5}{2!}x^2 + \frac{4 \times 5 \times 6}{3!}x^3 + \cdots$$

the coefficient of x^{13} in $(1 - x^6)^4(1 - x)^{-4}$ is

$$\frac{4 \times 5 \times 6 \times \cdots \times 16}{13!} - 4 \cdot \frac{4 \times 5 \times 6 \times \cdots \times 10}{7!} + 6 \cdot \frac{4}{1!}$$
$$= \frac{14 \times 15 \times 16}{3!} - 4 \cdot \frac{8 \times 9 \times 10}{3!} + 6 \cdot \frac{4}{1!}$$
$$= 104$$

It is suggested that the reader review Example 1-19 and solve it by finding the enumerator. ■

2-3 ENUMERATORS FOR PERMUTATIONS

It is natural now for us to turn to the generating functions for permutations. However, there is an obvious difficulty when we try to extend our previous results. Since multiplication in the ordinary algebra in the field of real numbers (with which we are so familiar) is commutative (that is, $ab = ba$), we cannot quite handle the case of permutations using ordinary algebra. The situation can be illustrated by an example of the permutations of the two objects a and b. What we want to have as a generating function for the permutations is

$$1 + (a + b)x + (ab + ba)x^2$$

However, this polynomial is equivalent to

$$1 + (a + b)x + (2ab)x^2$$

in which the two distinct permutations ab and ba can no longer be recognized. Instead of introducing a new algebra that is noncommutative for the case of permutations, we shall limit ourselves to the discussion of the enumerators for permutations which can still be handled by the ordinary algebra in the field of real numbers.

A direct extension of the notion of the enumerators for combinations indicates that an enumerator for the permutations of n distinct objects would have the form

$$\begin{aligned} F(x) &= P(n,0)x^0 + P(n,1)x + P(n,2)x^2 + P(n,3)x^3 + \cdots \\ &\qquad + P(n,r)x^r + \cdots + P(n,n)x^n \\ &= 1 + \frac{n!}{(n-1)!}x + \frac{n!}{(n-2)!}x^2 + \frac{n!}{(n-3)!}x^3 + \cdots \\ &\qquad + \frac{n!}{(n-r)!}x^r + \cdots + n!x^n \end{aligned}$$

Unfortunately, there is no simple closed-form expression for $F(x)$, and to carry along the polynomial in our manipulations certainly defeats the purpose of using the generating function representation. However, when we recall the binomial expansion

$$\begin{aligned} (1+x)^n &= 1 + C(n,1)x + C(n,2)x^2 + C(n,3)x^3 + \cdots \\ &\qquad + C(n,r)x^r + \cdots + C(n,n)x^r \\ &= 1 + \frac{P(n,1)}{1!}x + \frac{P(n,2)}{2!}x^2 + \frac{P(n,3)}{3!}x^3 + \cdots \\ &\qquad + \frac{P(n,r)}{r!}x^r + \cdots + \frac{P(n,n)}{n!}x^n \end{aligned}$$

we see the key to defining another kind of generating function, the *exponential generating function*. Let $(a_0,a_1,a_2, \ldots ,a_r, \ldots)$ be the symbolic representations of a sequence of events or simply be a sequence of numbers. The function

$$F(x) = \frac{a_0}{0!}\mu_0(x) + \frac{a_1}{1!}\mu_1(x) + \frac{a_2}{2!}\mu_2(x) + \cdots + \frac{a_r}{r!}\mu_r(x) + \cdots$$

is called the *exponential generating function* of the sequence $(a_0,a_1,a_2, \ldots ,a_r, \ldots)$ with $\mu_0(x)$, $\mu_1(x)$, $\mu_2(x)$, $\ldots$, $\mu_r(x)$, $\ldots$ as the indicator functions. Thus, $(1+x)^n$ is the exponential generating function of the $P(n,r)$'s with the powers of x as the indicator functions. Similarly, an exponential generating function that gives the number of combinations or permutations is called an *exponential enumerator*.

Example 2-10 According to the result in Example 2-5, $(1-4x)^{-1/2}$ is the exponential generating function of the sequence $(P(0,0),P(2,1), P(4,2), \ldots ,P(2r,r), \ldots)$.

The exponential generating function of the sequence $(1, 1 \times 3, 1 \times 3 \times 5, \ldots, 1 \times 3 \times 5 \times \cdots \times (2r+1), \ldots)$ is $(1 - 2x)^{-3/2}$.

The exponential generating function of the sequence $(1,1,1, \ldots,1, \ldots)$ is e^x. ■

Clearly, the exponential enumerator for the permutations of a single object with no repetitions is $1 + x$. We also see in the above that the exponential enumerator for the permutations of n distinct objects with no repetitions is $(1 + x)^n$. (The definition of the exponential enumerator is actually chosen in such a way that the result will come out correctly.)

When repetitions are allowed in the permutations, the extension is immediate. The exponential enumerator for the permutations of *all p of p* identical objects is $x^p/p!$ since there is only one way of doing so. Thus, the exponential enumerator for the permutations of none, one, two, . . . , p of p identical objects is

$$1 + \frac{1}{1!}x + \frac{1}{2!}x^2 + \cdots + \frac{1}{p!}x^p$$

Similarly, the exponential enumerator for the permutation of *all* $p + q$ *of* $p + q$ objects, with p of them of one kind and q of them of another kind, is

$$\frac{x^p}{p!}\frac{x^q}{q!} = \frac{x^{p+q}}{p!q!}$$

which agrees with the known result that the number of permutations is $(p + q)!/p!q!$. It follows that the exponential enumerator for the permutations of none, one, two, . . . , $p + q$ of $p + q$ objects, with p of them of one kind and q of them of another kind, is

$$\left(1 + \frac{1}{1!}x + \frac{1}{2!}x^2 + \cdots + \frac{1}{p!}x^p\right)\left(1 + \frac{1}{1!}x + \frac{1}{2!}x^2 + \cdots + \frac{1}{q!}x^q\right)$$

For instance, the exponential enumerator for the permutations of two objects of one kind and three objects of another kind is

$$\left(1 + \frac{x}{1!} + \frac{x^2}{2!}\right)\left(1 + \frac{x}{1!} + \frac{x^2}{2!} + \frac{x^3}{3!}\right)$$
$$= 1 + \left(\frac{1}{1!} + \frac{1}{1!}\right)x + \left(\frac{1}{1!1!} + \frac{1}{2!} + \frac{1}{2!}\right)x^2 + \left(\frac{1}{1!2!} + \frac{1}{1!2!} + \frac{1}{3!}\right)x^3 + \left(\frac{1}{1!3!} + \frac{1}{2!2!}\right)x^4 + \left(\frac{1}{2!3!}\right)x^5$$

The reader can again check the results by referring to Example 1-6 in Chap. 1.

Example 2-11 The number of r-permutations of n distinct objects with unlimited repetitions is given by the exponential enumerator

$$\left(1 + x + \frac{x^2}{2!} + \frac{x^3}{3!} + \cdots\right)^n = e^{nx} = \sum_{r=0}^{\infty} \frac{n^r}{r!} x^r \quad ■$$

Example 2-12 Find the number of r-digit quaternary sequences in which each of the digits 1, 2, and 3 appears at least once. This problem is the same as that of permuting four distinct objects with the restriction that three of the four objects must be included in the permutations. The exponential enumerator for the permutations of the digit 0 is

$$\left(1 + x + \frac{x^2}{2!} + \frac{x^3}{3!} + \cdots\right) = e^x$$

The exponential enumerator for the permutations of the digit 1 (or 2, or 3) is

$$\left(x + \frac{x^2}{2!} + \frac{x^3}{3!} + \cdots\right) = e^x - 1$$

It follows that the exponential enumerator for the permutations of the four digits is

$$\begin{aligned} e^x(e^x - 1)(e^x - 1)(e^x - 1) &= e^x(e^{3x} - 3e^{2x} + 3e^x - 1) \\ &= e^{4x} - 3e^{3x} + 3e^{2x} - e^x \\ &= \sum_{r=0}^{\infty} \frac{(4^r - 3 \times 3^r + 3 \times 2^r - 1)}{r!} x^r \end{aligned}$$

Therefore, the number of r-digit quaternary sequences in which each of the digits 1, 2, and 3 appears at least once is

$$4^r - 3 \times 3^r + 3 \times 2^r - 1 \quad ■$$

Example 2-13 Find the number of r-digit quaternary sequences that contain an even number of 0's. The exponential enumerator for the permutations of the digit 0 is

$$\left(1 + \frac{x^2}{2!} + \frac{x^4}{4!} + \frac{x^6}{6!} + \cdots\right) = \frac{1}{2}(e^x + e^{-x})$$

The exponential enumerator for the permutations of each of the digits 1, 2, and 3 is

$$\left(1 + \frac{x}{1!} + \frac{x^2}{2!} + \frac{x^3}{3!} + \cdots\right) = e^x$$

It follows that the exponential enumerator for the number of quaternary sequences containing an even number of 0's is

$$\begin{aligned} \tfrac{1}{2}(e^x + e^{-x})e^x e^x e^x &= \tfrac{1}{2}(e^{4x} + e^{2x}) \\ &= 1 + \sum_{r=1}^{\infty} \frac{1}{2} \frac{(4^r + 2^r)}{r!} x^r \end{aligned}$$

Therefore, the number of r-digit quaternary sequences that contain an even number of 0's is $(4^r + 2^r)/2$.

Similarly, to find the number of r-digit quaternary sequences that contain an even number of 0's and an even number of 1's, we have the exponential enumerator

$$\begin{aligned} \tfrac{1}{2}(e^x + e^{-x})\tfrac{1}{2}(e^x + e^{-x})e^x e^x &= \tfrac{1}{4}(e^{2x} + 2 + e^{-2x})e^{2x} \\ &= \tfrac{1}{4}(e^{4x} + 2e^{2x} + 1) \\ &= 1 + \sum_{r=1}^{\infty} \frac{1}{4} \frac{(4^r + 2 \times 2^r)}{r!} x^r \end{aligned}$$

The reader is encouraged to review the method of solution presented in Example 1-9 in Chap. 1. ■

Example 2-14 Find the exponential enumerator for the number of ways to choose r or less objects from r distinct objects and distribute them into n distinct cells, with objects in the same cell ordered. Notice that there are $C(r,m)$ ways to select m objects out of r objects and $n(n + 1) \cdots (n + m - 1)$ ways to arrange them in the n distinct cells. Since the value of m ranges from 0 to r, the total number of ways is

$$\begin{aligned} &C(r,0) + C(r,1) \times n + C(r,2) \times n(n + 1) \\ &\qquad + C(r,3) \times n(n + 1)(n + 2) + \cdots \\ &\qquad + C(r,r) \times n(n + 1) \cdots (n + r - 1) \\ &= r!\left[\frac{1}{r!} \times 1 + \frac{1}{(r - 1)!1!} \times n + \frac{1}{(r - 2)!2!} \times n(n + 1)\right. \\ &\qquad + \frac{1}{(r - 3)!3!} \times n(n + 1)(n + 2) + \cdots \\ &\qquad \left. + \frac{1}{r!} \times n(n + 1) \cdots (n + r - 1)\right] \end{aligned}$$

The expression in the square brackets is the coefficient of the term x^r in the product of the two series

$$e^x = 1 + \frac{x}{1!} + \frac{x^2}{2!} + \cdots + \frac{x^r}{r!} + \cdots$$

and

$$(1-x)^{-n} = 1 + \frac{n}{1!} + \frac{n(n+1)}{2!}x^2 + \cdots + \frac{n(n+1)\cdots(n+r-1)}{r!}x^r + \cdots$$

Therefore, $e^x/(1-x)^n$ is the exponential enumerator for the distributions of r or less objects into n distinct cells, with objects in the same cell ordered. ∎

2-4 DISTRIBUTIONS OF DISTINCT OBJECTS INTO NONDISTINCT CELLS

As examples on the use of exponential generating functions, we shall derive some results on the distribution of distinct objects into nondistinct cells. First we shall derive the number of ways of distributing r distinct objects into n distinct cells so that no cell is empty and the order of objects within a cell is not important. This problem can be viewed as finding the number of the r-permutations of the n distinct cells with each cell included at least once in a permutation. The exponential enumerator for the permutations is

$$\left(x + \frac{x^2}{2!} + \frac{x^3}{3!} + \cdots\right)^n = (e^x - 1)^n$$

$$= \sum_{i=0}^{n} \binom{n}{i} (-1)^i e^{(n-i)x}$$

$$= \sum_{i=0}^{n} \binom{n}{i} (-1)^i \sum_{r=0}^{\infty} \frac{1}{r!} (n-i)^r x^r$$

$$= \sum_{r=0}^{\infty} \frac{x^r}{r!} \sum_{i=0}^{n} (-1)^i \binom{n}{i} (n-i)^r$$

Thus, the number of ways of placing r distinct objects into n distinct cells with no cell left empty is equal to

$$\sum_{i=0}^{n} (-1)^i \binom{n}{i} (n-i)^r = n!S(r,n)$$

where $S(r,n)$ is defined as $(1/n!) \sum_{i=0}^{n} (-1)^i \binom{n}{i} (n-i)^r$ and is called the *Stirling number of the second kind.* Table 2-1 shows some of the Stirling

Table 2-1 Stirling numbers of the second kind, $S(r,n)$

r	n = 1	2	3	4	5	6	7	8	9	10
1	1									
2	1	1								
3	1	3	1							
4	1	7	6	1						
5	1	15	25	10	1					
6	1	31	90	65	15	1				
7	1	63	301	350	140	21	1			
8	1	127	966	1701	1050	266	28	1		
9	1	255	3025	7770	6951	2646	462	36	1	
10	1	511	9330	34105	42525	22827	5880	750	45	1

numbers of the second kind. [It is suggested that the reader convince himself that the result $P(r,n) \times n^{r-n}$, which is obtained by distributing one object in each of the n cells and then distributing the remaining $r - n$ objects in an arbitrary manner, is incorrect.]

It follows that the number of ways of placing r distinct objects into n nondistinct cells with no cell left empty is equal to

$$S(r,n)$$

Previously we proved that there are n^r ways of placing r distinct objects into n distinct cells, when empty cells are allowed. However, the reader should convince himself that when the cells become nondistinct, the number of ways is not equal to $n^r/n!$. As a matter of fact, the number of ways of distributing r distinct objects into n nondistinct cells with empty cells allowed is

$$S(r,1) + S(r,2) + \cdots + S(r,n) \qquad \text{for } r \geq n$$

and is

$$S(r,1) + S(r,2) + \cdots + S(r,r) \qquad \text{for } r \leq n \tag{2-1}$$

These come directly from the argument that the number of ways of distributing r distinct objects into n nondistinct cells with empty cells allowed is equal to the number of ways of distributing these r objects so that one cell is not empty, or two cells are not empty, etc.

For the case of $r \leq n$ (i.e., there are at least as many cells as objects), there is a closed-form expression for the ordinary generating function of the numbers of ways of distributing the objects. Since $S(i,j) = 0$ for $i < j$, the count in the expression in (2-1) does not change if we add to it an infinite number of terms as follows:

$$S(r,1) + S(r,2) + \cdots + S(r,r) + S(r, r+1) + S(r, r+2) + \cdots \tag{2-2}$$

Observe that

$$\frac{e^x - 1}{1!} = S(0,1) + \frac{S(1,1)}{1!}x + \frac{S(2,1)}{2!}x^2 + \cdots + \frac{S(r,1)}{r!}x^r + \cdots$$

$$\frac{(e^x - 1)^2}{2!} = S(0,2) + \frac{S(1,2)}{1!}x + \frac{S(2,2)}{2!}x^2 + \cdots + \frac{S(r,2)}{r!}x^r + \cdots$$

. . . .

$$\frac{(e^x - 1)^k}{k!} = S(0,k) + \frac{S(1,k)}{1!}x + \frac{S(2,k)}{2!}x^2 + \cdots + \frac{S(r,k)}{r!}x^r + \cdots$$

. . . .

$$\frac{(e^x - 1)^r}{r!} = S(0,r) + \frac{S(1,r)}{1!}x + \frac{S(2,r)}{2!}x^2 + \cdots + \frac{S(r,r)}{r!}x^r + \cdots$$

$$\frac{(e^x - 1)^{r+1}}{(r+1)!} = S(0, r+1) + \frac{S(1, r+1)}{1!}x + \frac{S(2, r+1)}{2!}x^2 + \cdots + \frac{S(r, r+1)}{r!}x^r + \cdots$$

. . . .

Therefore, the coefficient of $x^r/r!$, which is the number of ways of distributing r distinct objects into r or more nondistinct cells, in

$$\frac{e^x - 1}{1!} + \frac{(e^x - 1)^2}{2!} + \cdots + \frac{(e^x - 1)^k}{k!} + \cdots + \frac{(e^x - 1)^r}{r!} + \frac{(e^x - 1)^{r+1}}{(r+1)!} + \cdots \tag{2-3}$$

is equal to the expression in (2-2). However, the generating function in (2-3) can be written as

$$e^{e^x - 1} - 1$$

2-5 PARTITIONS OF INTEGERS

As another illustration of the use of generating functions, we shall discuss the distribution of nondistinct objects into nondistinct cells.

A *partition* of an integer is a division of the integer into positive integral parts, in which the order of these parts is not important. For example, 4, $3 + 1$, $2 + 2$, $2 + 1 + 1$, and $1 + 1 + 1 + 1$ are the five different partitions of the integer 4. It is clear that a partition of the integer n is equivalent to a way of distributing n nondistinct objects into n nondistinct cells with empty cells allowed. We shall conduct our discussion in the context of the partitions of integers mainly because it is also an important topic in number theory.

Observe that in the polynomial

$$1 + x + x^2 + x^3 + x^4 + \cdots + x^n$$

the coefficient of x^k is the number of ways of having k 1's in a partition of the integer n. Clearly, there is one way for $0 \leq k \leq n$ and no way for $k > n$ because in a partition of n there can be from no 1's to at most n 1's. It follows that in the infinite sum

$$1 + x + x^2 + x^3 + x^4 + \cdots + x^r + \cdots = \frac{1}{1 - x}$$

the coefficient of x^k is the number of ways of having k 1's in a partition of any integer larger than or equal to k. Similarly, in the polynomial

$$1 + x^2 + x^4 + x^6 + x^8 + \cdots + x^{[n/2]}$$

the coefficient of x^{2k} is the number of ways of having k 2's in a partition of the integer n. Also, in the infinite sum

$$1 + x^2 + x^4 + x^6 + x^8 + \cdots + x^{2r} + \cdots = \frac{1}{1 - x^2}$$

the coefficient of x^{2k} is the number of ways of having k 2's in a partition of any integer larger than or equal to $2k$. Notice that a 2 in a partition will be accounted for by the term x^2, two 2's in a partition will be accounted for by the term x^4, etc. It follows then that

$$\begin{aligned} F(x) &= (1 + x + x^2 + x^3 + \cdots + x^r + \cdots) \\ &\quad (1 + x^2 + x^4 + x^6 + \cdots + x^{2r} + \cdots) \\ &\quad (1 + x^3 + x^6 + x^9 + \cdots + x^{3r} + \cdots) \\ &\quad (1 + x^4 + x^8 + x^{12} + \cdots + x^{4r} + \cdots) \\ &\qquad \cdots (1 + x^n + x^{2n} + x^{3n} + \cdots + x^{nr} + \cdots) \\ &= \frac{1}{(1 - x)(1 - x^2)(1 - x^3)(1 - x^4) \cdots (1 - x^n)} \end{aligned}$$

is the ordinary generating function of the sequence $(p(0),p(1), \ldots ,p(n))$, where $p(i)$ denotes the number of partitions of the integer i. However, notice that $F(x)$ does not enumerate the $p(j)$'s for $j > n$; rather, it enumerates the number of partitions of the integer j that *have no part exceeding* n. For example, from

$$\frac{1}{(1-x)(1-x^2)(1-x^3)} = 1 + x + 2x^2 + 3x^3 + 4x^4 + 5x^5 + 7x^6 + \cdots$$

we observe that there are three ways to partition the integer 3 and there are seven ways to partition the integer 6 such that the parts do not exceed 3. The ordinary generating function of the infinite sequence $(p(0),p(1),p(2), \ldots ,p(n), \ldots)$ is

$$F(x) = \frac{1}{(1-x)(1-x^2)(1-x^3) \cdots}$$

It is immediately clear that in

$$\frac{1}{(1-x)(1-x^3)(1-x^5) \cdots (1-x^{2n+1})}$$

the coefficient of x^k for $k \leq 2n+1$ is the number of partitions of the integer k into odd parts, and the coefficient of x^k for $k > 2n+1$ is the number of partitions of the integer k into odd parts not exceeding $2n+1$. Similarly, in

$$\frac{1}{(1-x)(1-x^3)(1-x^5) \cdots}$$

the coefficient of x^k is the number of partitions of the integer k into odd parts. Also in

$$\frac{1}{(1-x^2)(1-x^4)(1-x^6) \cdots (1-x^{2n})}$$

the coefficient of x^k for $k \leq 2n$ is the number of partitions of the integer k into even parts, and the coefficient of x^k for $k > 2n$ is the number of partitions of the integer k into even parts not exceeding $2n$. Again, in

$$\frac{1}{(1-x^2)(1-x^4)(1-x^6) \cdots}$$

the coefficient of x^k is the number of partitions of the integer k into even parts.

Also, the polynomial

$$(1+x)(1+x^2)(1+x^3) \cdots (1+x^n)$$

enumerates the partitions of integers no larger than n into distinct (unequal) parts and the partitions of integers larger than n into distinct parts not exceeding n, and

$$(1 + x)(1 + x^2)(1 + x^3) \cdots (1 + x^n) \cdots$$

enumerates the partitions of the integers into distinct parts.

In the following examples it will be shown that the usefulness of these results goes beyond the simple enumeration of the number of partitions of integers.

Example 2-15 Since

$$\begin{aligned}(1 + x)(1 + x^2)(1 + x^3)(1 + x^4) \cdots (1 + x^r) \cdots \\ = \frac{1 - x^2}{1 - x} \cdot \frac{1 - x^4}{1 - x^2} \cdot \frac{1 - x^6}{1 - x^3} \cdot \frac{1 - x^8}{1 - x^4} \cdot \cdots \cdot \frac{1 - x^{2r}}{1 - x^r} \cdot \cdots \\ = \frac{1}{(1 - x)(1 - x^3)(1 - x^5) \cdots}\end{aligned}$$

we conclude that the number of partitions of an integer into distinct parts is equal to the number of partitions of the integer into odd parts. For instance, the integer 6 can be partitioned into distinct parts in four different ways, namely,

$$6 \qquad 5 + 1 \qquad 4 + 2 \qquad 3 + 2 + 1$$

There are also exactly four different ways in which 6 can be partitioned into odd parts. They are

$$5 + 1 \qquad 3 + 3 \qquad 3 + 1 + 1 + 1 \qquad 1 + 1 + 1 + 1 + 1 + 1$$ ■

Example 2-16 Since

$$\begin{aligned}(1 - x)(1 + x)(1 + x^2)(1 + x^4)(1 + x^8) \cdots (1 + x^{2^r}) \cdots \\ = (1 - x^2)(1 + x^2)(1 + x^4)(1 + x^8) \cdots (1 + x^{2^r}) \cdots \\ = (1 - x^4)(1 + x^4)(1 + x^8) \cdots (1 + x^{2^r}) \cdots \\ = 1\end{aligned}$$

we have the identity

$$\frac{1}{1 - x} = (1 + x)(1 + x^2)(1 + x^4)(1 + x^8) \cdots (1 + x^{2^r}) \cdots$$

Recalling that

$$\frac{1}{1 - x} = 1 + x + x^2 + x^3 + x^4 + \cdots$$

we conclude that any integer can be expressed as the sum of a selection of the integers 1, 2, 4, 8, . . . , 2^r, . . . (without repetition) in exactly one way. (This is the well-known fact that a decimal number can be represented uniquely as a binary number.)

Directly from this identity we have another interesting result. Since

$$\begin{aligned}1 - x &= \frac{1}{(1+x)(1+x^2)(1+x^4)(1+x^8)\cdots(1+x^{2^r})\cdots} \\ &= (1 - x + x^2 - x^3 + x^4 - \cdots) \\ &\quad (1 - x^2 + x^4 - x^6 + x^8 - \cdots) \\ &\quad (1 - x^4 + x^8 - x^{12} + x^{16} - \cdots)\cdots \\ &\qquad\qquad (1 - x^{2^r} + x^{2\cdot 2^r} - x^{3\cdot 2^r} + x^{4\cdot 2^r} - \cdots)\cdots\end{aligned}$$

we conclude that to partition any integer n larger than 1 into parts that are powers of 2, namely, $1, 2, 4, 8, \ldots, 2^r, \ldots$, the number of partitions that have an even number of parts is equal to the number of partitions that have an odd number of parts. The series

$$1 - x + x^2 - x^3 + x^4 - \cdots$$

enumerates the number of 1's in a partition, with terms corresponding to an even number of 1's in the partition having $+1$ as the coefficients and terms corresponding to an odd number of 1's in the partition having -1 as the coefficients. Similarly, the series

$$1 - x^2 + x^4 - x^6 + x^8 - \cdots$$

enumerates the number of 2's in a partition, and the series

$$1 - x^4 + x^8 - x^{12} + x^{16} - \cdots$$

enumerates the number of 4's in a partition, with terms corresponding to an even number of 2's (or 4's) having positive coefficients and terms corresponding to an odd number of 2's (or 4's) having negative coefficients. Therefore, in the expansion of the product

$$\begin{aligned}&(1 - x + x^2 - x^3 + x^4 - \cdots)(1 - x^2 + x^4 - x^6 + x^8 - \cdots) \\ &\quad (1 - x^4 + x^8 - x^{12} + x^{16} - \cdots)\cdots \\ &\qquad\qquad (1 - x^{2^r} + x^{2\cdot 2^r} - x^{3\cdot 2^r} + x^{4\cdot 2^r} - \cdots)\cdots\end{aligned}$$

a term $+x^n$ corresponds to a partition of the integer n into an even number of parts, and a term $-x^n$ corresponds to a partition of the integer n into an odd number of parts.

As an example, we see that $4 + 1$, $2 + 1 + 1 + 1$, $2 + 2 + 1$, and $1 + 1 + 1 + 1 + 1$ are the four partitions of the integer 5 into parts that are powers of 2. Two of these partitions have an even number of parts, and the other two have an odd number of parts. ■

★2-6 THE FERRERS GRAPH

A *Ferrers graph* consists of rows of dots. The dots are arranged in such a way that an upper row has at least as many dots as a lower row. A partition of an integer can be represented by a Ferrers graph by making each row in the graph correspond to a part in the partition, with the number of dots in a row specifying the value of the corresponding part. For example, the partition of the integer 14 into $6 + 3 + 3 + 2$ is represented by the Ferrers graph shown in Fig. 2-1.

```
. . . . . .
. . .
. . .
. .
```

Figure 2-1

Using the Ferrers graphs, we can derive some very interesting results:

1. The number of partitions of an integer into exactly m parts is equal to the number of partitions of the integer into parts, the largest of which is m. This comes from the fact that the transposition of a Ferrers graph (the leftmost column becomes the uppermost row and so on) is also a Ferrers graph. It follows that the transposition of the Ferrers graph of a partition having exactly m parts becomes the Ferrers graph of a partition having m as the largest part. For example, there are two partitions of the integer 6 that have exactly four parts each. They are $2 + 2 + 1 + 1$ and $3 + 1 + 1 + 1$. There are also two partitions that have 4 as their largest parts. They are $4 + 2$ and $4 + 1 + 1$. The corresponding Ferrers graphs are shown in Fig. 2-2.

```
. .          . . .
. .          .
.            .
.            .
```

$2 + 2 + 1 + 1$
($4 + 2$, after transposition)

$3 + 1 + 1 + 1$
($4 + 1 + 1$, after transposition)

Figure 2-2

2. Following the same argument as that in (1), we observe that the number of partitions of an integer into at most m parts is equal to the number of partitions of the integer into parts not exceeding m. Therefore, the ordinary generating function of the numbers of partitions of integers into at most m parts is also

$$\frac{1}{(1-x)(1-x^2)\cdots(1-x^m)}$$

Now, since the ordinary generating function of the numbers of partitions of integers into at most $m-1$ parts is

$$\frac{1}{(1-x)(1-x^2)\cdots(1-x^{m-1})}$$

the ordinary generating function of the numbers of partitions of integers into exactly m parts is

$$\frac{1}{(1-x)(1-x^2)\cdots(1-x^m)} - \frac{1}{(1-x)(1-x^2)\cdots(1-x^{m-1})} = \frac{x^m}{(1-x)(1-x^2)\cdots(1-x^m)}$$

3. Using the previous result, we can find the ordinary generating function of the numbers of partitions of integers into exactly m unequal parts. If we add $m-1$ dots to the first row of the Ferrers graph of an m-part partition of the integer $n-[m(m-1)/2]$, $m-2$ dots to the second row, $m-3$ dots to the third row, . . . , and one dot to the $(m-1)$th row, we have the Ferrers graph of a partition of the integer n into m unequal parts. Since this gives a one-to-one correspondence between the m-part partitions of the integer $n-[m(m-1)/2]$ and the partitions of the integer n into m unequal parts, we conclude that the number of partitions of the integer n into exactly m unequal parts equals the number of partitions of the integer $n-[m(m-1)/2]$ into exactly m parts. Therefore, the ordinary generating function of the numbers of the partitions of integers into m distinct parts is

$$x^{m(m-1)/2}\frac{x^m}{(1-x)(1-x^2)\cdots(1-x^m)} = \frac{x^{m(m+1)/2}}{(1-x)(1-x^2)\cdots(1-x^m)}$$

2-7 ELEMENTARY RELATIONS

Using mainly problems in combinations and permutations as examples, we have introduced the notions of ordinary and exponential generating functions. However, we must emphasize that the concept of representing sequences by their generating functions is not limited to the area of combinations and permutations, as we shall see in the following chapters. In this section, we present some elementary relations on the operations of generating functions.

Let $A(x)$, $B(x)$, and $C(x)$ be the ordinary generating functions of the sequences $(a_0,a_1, \ldots ,a_r, \ldots)$, $(b_0,b_1, \ldots ,b_r, \ldots)$, and $(c_0,c_1, \ldots , c_r, \ldots)$, respectively. By definition,

$$C(x) = A(x) + B(x)$$

if and only if the members of the sequences are related as follows:

$$c_0 = a_0 + b_0$$
$$c_1 = a_1 + b_1$$
$$\cdots\cdots$$
$$c_r = a_r + b_r$$
$$\cdots\cdots$$

Similarly, by definition,

$$C(x) = A(x) \times B(x)$$

if and only if the members of the sequences are related as follows:

$$c_0 = a_0b_0$$
$$c_1 = a_1b_0 + a_0b_1$$
$$\cdots\cdots$$
$$c_r = a_rb_0 + a_{r-1}b_1 + a_{r-2}b_2 + \cdots + a_1b_{r-1} + a_0b_r$$
$$\cdots\cdots\cdots\cdots$$

Example 2-17 Let $A(x)$ be the ordinary generating function of the sequence $(a_0,a_1,a_2, \ldots ,a_r, \ldots)$. Since $1/(1-x)$ is the ordinary generating function of the sequence $(1,1,1, \ldots ,1, \ldots)$, $[1/(1-x)]A(x)$ is the ordinary generating function of the sequence $(a_0, a_0 + a_1, a_0 + a_1 + a_2, \ldots , a_0 + a_1 + a_2 + \cdots + a_r, \ldots)$. [$1/(1-x)$ is, therefore, called the summing operator.]

For instance, to find the coefficient of the term x^{37} in $(1 - 3x^2 + 4x^7 + 12x^{21} - 5x^{45})/(1-x)$, we have as the answer $1 - 3 + 4 + 12 = 14$. ■

Example 2-18 Evaluate the sum

$$1^2 + 2^2 + 3^2 + \cdots + r^2$$

Let us first find the ordinary generating function of the sequence $(0^2,1^2,2^2,3^2, \ldots ,r^2, \ldots)$. Differentiating both sides of the identity

$$\frac{1}{1-x} = 1 + x + x^2 + x^3 + x^4 + \cdots + x^r + \cdots$$

we obtain

$$\frac{1}{(1-x)^2} = 1 + 2x + 3x^2 + 4x^3 + \cdots + rx^{r-1} + \cdots$$

It follows that

$$\frac{d}{dx}\frac{x}{(1-x)^2} = 1^2 + 2^2 \times x + 3^2 \times x^2 + 4^2 \times x^3 + \cdots + r^2 \times x^{r-1} + \cdots$$

and

$$x\frac{d}{dx}\frac{x}{(1-x)^2} = 0^2 + 1^2 \times x + 2^2 \times x^2 + 3^2 \times x^3 + 4^2 \times x^4 + \cdots + r^2 \times x^r + \cdots$$

Thus, $x(d/dx)[x/(1-x)^2]$, which is equal to $[x(1+x)]/(1-x)^3$, is the ordinary generating function of the sequence $(0^2,1^2,2^2,3^2, \ldots , r^2, \ldots)$.

According to the result in Example 2-17, $[x(1+x)]/(1-x)^4$ is the ordinary generating function of the sequence $(0^2, 0^2+1^2, 0^2+1^2+2^2, 0^2+1^2+2^2+3^2, \ldots, 0^2+1^2+2^2+3^2+\cdots+r^2, \ldots)$. According to the binomial theorem, the coefficient of x^r in $1/(1-x)^4$ is

$$\frac{(-4)(-4-1)(-4-2)\cdots(-4-r+1)}{r!}(-1)^r = \frac{4 \times 5 \times 6 \times \cdots \times (r+3)}{r!} = \frac{(r+1)(r+2)(r+3)}{1 \times 2 \times 3}$$

Therefore, the coefficient of x^r in the expansion of $[x(1+x)]/(1-x)^4$ is

$$\frac{r(r+1)(r+2)}{1 \times 2 \times 3} + \frac{(r-1)r(r+1)}{1 \times 2 \times 3} = \frac{r(r+1)(2r+1)}{6}$$

that is,

$$1^2 + 2^2 + 3^2 + \cdots + r^2 = \frac{r(r+1)(2r+1)}{6} \qquad \blacksquare$$

Let $A(x)$, $B(x)$, and $C(x)$ be the exponential generating functions of the sequences $(a_0, a_1, \ldots, a_r, \ldots)$, $(b_0, b_1, \ldots, b_r, \ldots)$, and $(c_0, c_1, \ldots, c_r, \ldots)$, respectively. By definition,

$$C(x) = A(x) + B(x)$$

if and only if the members of the sequences are related as follows:

$$\begin{aligned} c_0 &= a_0 + b_0 \\ c_1 &= a_1 + b_1 \\ &\cdots\cdots \\ c_r &= a_r + b_r \\ &\cdots\cdots \end{aligned}$$

Similarly, by definition,

$$C(x) = A(x) \times B(x)$$

if and only if the members of the sequences are related as follows:

$$\begin{aligned} c_0 &= a_0 b_0 \\ c_1 &= a_1 b_0 + a_0 b_1 \\ c_2 &= 2!\left(\frac{a_2 b_0}{2!} + \frac{a_1 b_1}{1!1!} + \frac{a_0 b_2}{2!}\right) \\ &\cdots\cdots\cdots\cdots \\ c_r &= r!\left[\frac{a_r b_0}{r!} + \frac{a_{r-1} b_1}{(r-1)!1!} + \frac{a_{r-2} b_2}{(r-2)!2!} + \cdots + \frac{a_0 b_r}{r!}\right] \\ &= \sum_{i=0}^{r} \binom{r}{i} a_{r-i} b_i \\ &\cdots\cdots\cdots\cdots\cdots\cdots \end{aligned}$$

Example 2-19 Evaluate the sum

$$\sum_{i=0}^{r} \frac{r!}{(r-i+1)!(i+1)!}$$

We observe first that

$$\begin{aligned} \sum_{i=0}^{r} \frac{r!}{(r-i+1)!(i+1)!} &= \sum_{i=0}^{r} \frac{r!}{(r-i)!i!} \frac{1}{r-i+1} \frac{1}{i+1} \\ &= \sum_{i=0}^{r} \binom{r}{i} \frac{1}{r-i+1} \frac{1}{i+1} \end{aligned}$$

Since

$$e^x = 1 + \frac{1}{1!}x + \frac{1}{2!}x^2 + \frac{1}{3!}x^3 + \frac{1}{4!}x^4 + \cdots + \frac{1}{r!}x^r + \cdots$$

and

$$\frac{1}{x}(e^x - 1) = 1 + \frac{1}{2!}x + \frac{1}{3!}x^2 + \frac{1}{4!}x^3 + \cdots + \frac{1}{r!}x^{r-1} + \cdots$$

the exponential generating function of the sequence $\left(1, \frac{1}{2}, \frac{1}{3}, \frac{1}{4}, \ldots, \frac{1}{r}, \ldots\right)$ is $(1/x)(e^x - 1)$. It follows that $(1/x^2)(e^x - 1)^2$ is the exponential generating function of the sequence

$$1 \times 1, \frac{1}{2} \times 1 + 1 \times \frac{1}{2}, \binom{2}{0} \times \frac{1}{3} \times 1 + \binom{2}{1} \times \frac{1}{2} \times \frac{1}{2} + \binom{2}{2} \times 1 \times \frac{1}{3}, \ldots, \sum_{i=0}^{r} \binom{r}{i} \frac{1}{r-i+1} \frac{1}{i+1}, \ldots$$

Since

$$\begin{aligned}\frac{1}{x^2}(e^x - 1)^2 &= \frac{1}{x^2}(e^{2x} - 2e^x + 1) \\ &= \left(\frac{2^2}{2!} - \frac{2}{2!}\right) + \left(\frac{2^3}{3!} - \frac{2}{3!}\right)x + \left(\frac{2^4}{4!} - \frac{2}{4!}\right)x^2 + \cdots \\ &\quad + \left[\frac{2^{r+2}}{(r+2)!} - \frac{2}{(r+2)!}\right]x^r + \cdots\end{aligned}$$

we obtain the result

$$\begin{aligned}\sum_{i=0}^{r} \frac{r!}{(r-i+1)!(i+1)!} &= r!\left[\frac{2^{r+2}}{(r+2)!} - \frac{2}{(r+2)!}\right] \\ &= \frac{2(2^{r+1} - 1)}{(r+2)(r+1)} \quad \blacksquare\end{aligned}$$

2-8 SUMMARY AND REFERENCES

We studied in this chapter the representation of sequences by their generating functions. This is a key concept in our subsequent discussion of enumerative techniques. The generating function representation not only makes the algebraic manipulation of sequences of numbers easier as shall be seen in Chaps. 3 and 4, but such a representation is also very useful for symbolic description of events as shall be seen in Chap. 5.

The concept of generating function is used extensively in probability theory, e.g., the moment-generating function for probability moments. Also, in the study of sampled-data systems and digital filters, generating function is a very useful tool. To many engineers, the ordinary generating function of a sequence of numbers is known as the z-transform.

For the reader who is familiar with the topic of Laplace transformation, the close analogy between the Laplace transform of a continuous function and the ordinary generating function of a sequence should be mentioned. A continuous function $a(r)$ and its Laplace transform $A(x)$ are related by the integral

$$A(x) = \int_0^\infty a(r)e^{-xr}\,dr$$

If $a(r)$ consists of a sequence of delta functions (impulses) at discrete values of r for $r = 0, 1, 2, \ldots$, then

$$A(x) = a(0) + a(1)e^{-x} + a(2)e^{-2x} + a(3)e^{-3x} + \cdots$$

$A(x)$ can be viewed as the ordinary generating function of a discrete sequence of numbers $(a(0),a(1),a(2), \ldots)$ with e^{-ix} for $i = 0, 1, 2, \ldots$ as indicator functions. Such an analogy is more than superficial, as shall be seen in Chap. 3.

MacMahon's book [5] contains a most thorough development of the subject matter, generating functions. Chapters 1 and 2 of Riordan [6] and Chap. 3 of Beckenbach [1] are more compatible with our level of presentation and scope of coverage. The topic of partition of numbers can be found in most books on number theory. See, for example, Chap. 19 of Hardy and Wright [4]. See also Chap. 6 of Riordan [6]. The use of generating functions in probability theory can be found in Chap. 9 of Feller [3] and Chap. 3 of Drake [2]. For an introduction to sampled-data systems, see Tou [7].

1. Beckenbach, E. F. (ed.): "Applied Combinatorial Mathematics," John Wiley & Sons, Inc., New York, 1964.
2. Drake, A. W.: "Fundamentals of Applied Probability Theory," McGraw-Hill Book Company, New York, 1967.
3. Feller, W.: "An Introduction to Probability Theory and Its Applications," vol. I, 2d ed., John Wiley & Sons, Inc., New York, 1957.
4. Hardy, G. H., and E. M. Wright: "An Introduction to the Theory of Numbers," 4th ed., Oxford University Press, London, 1960.
5. MacMahon, P. A.: "Combinatory Analysis," vols. I and II reprinted in one volume (originally published in two volumes in 1915 and 1916), Chelsea Publishing Company, New York, 1960.
6. Riordan, J.: "An Introduction to Combinatorial Analysis," John Wiley & Sons, Inc., New York, 1958.
7. Tou, J. T.: "Digital and Sampled-data Control Systems," McGraw-Hill Book Company, New York, 1959.

PROBLEMS

2-1. Evaluate the following sums in which n, m, and k are nonnegative integers.

(*a*) $\binom{n}{0} + \binom{n}{2} + \binom{n}{4} + \cdots + \binom{n}{q}$ where $q = n$ if n is even, and

$q = n - 1$ if n is odd

(b) $\binom{n}{k} + \binom{n+1}{k} + \binom{n+2}{k} + \cdots + \binom{n+m}{k}$

(c) $\binom{2n}{0} - \binom{2n-1}{1} + \binom{2n-2}{2} - \cdots + (-1)^n \binom{n}{n}$

(d) $\binom{2n}{n} + 2\binom{2n-1}{n} + 2^2\binom{2n-2}{n} + \cdots + 2^n\binom{n}{n}$

(e) $\binom{n}{0}\binom{m}{k} + \binom{n}{1}\binom{m}{k-1} + \binom{n}{2}\binom{m}{k-2} + \cdots + \binom{n}{k}\binom{m}{0}$

$k \leq \min(m,n)$

(f) $\binom{n}{0}\binom{n}{k} - \binom{n}{1}\binom{n-1}{k-1} + \binom{n}{2}\binom{n-2}{k-2} - \cdots$

$+ (-1)^k \binom{n}{k}\binom{n-k}{0}$

(g) $\binom{n}{k} + \binom{n-1}{k-1} + \binom{n-2}{k-2} + \cdots + \binom{n-k+1}{1} + \binom{n-k}{0}$

(h) $\binom{n}{0} + 2\binom{n}{1} + 2^2\binom{n}{2} + \cdots + 2^n\binom{n}{n}$

2-2. (a) Show that

$$\binom{m}{r+1} + \sum_{k=0}^{n-1} \binom{m+k}{r} = \binom{m+n}{r+1}$$

(b) Show that

$$\sum_{k=0}^{n-1} \binom{2+k}{2} = \binom{2+n}{3}$$

and thus

$$1 \times 2 + 2 \times 3 + 3 \times 4 + \cdots + n(n+1) = \tfrac{1}{3}n(n+1)(n+2)$$

2-3. Among the three representatives from each of the 50 states, either none, or one, or two of them will be selected to form a special committee.

(a) In how many ways can the selection be made?

(b) If the committee has exactly 50 members, in how many ways can the selection be made? (The answer may be expressed as a summation.)

2-4. Find the value of a_{50} in the following expansion:

$$\frac{x-3}{x^2-3x+2} = a_0 + a_1x + a_2x^2 + \cdots + a_{50}x^{50} + \cdots$$

2-5. In how many ways can 200 identical chairs be divided among four conference rooms such that each room will have 20 or 40 or 60 or 80 or 100 chairs?

2-6. In how many ways can $3n$ letters be selected from $2n$ A's, $2n$ B's, and $2n$ C's?

2-7. In how many ways can n letters be selected from an unlimited supply of A's, B's, and C's if each selection must include an even number of A's?

2-8. There are p kinds of objects with four of each kind and q kinds of objects with two of each kind (q is an even number). In how many ways can they be divided into two equal piles such that there is an even number of each kind of object in both piles? (0 is taken as an even number.) Evaluate your result for $p = 4$ and $q = 2$.

2-9. (*a*) Let a_r denote the number of ways in which the sum r will show when two *distinct* dice are rolled. Find the ordinary generating function of the sequence $(a_0,a_1,a_2, \ldots)$.

(*b*) Let a_r denote the number of ways in which the sum r can be obtained by rolling a die any number of times. Show that the ordinary generating function of the sequence $(a_0,a_1,a_2, \ldots)$ is $(1 - x - x^2 - x^3 - x^4 - x^5 - x^6)^{-1}$.

2-10. (*a*) Find the ordinary generating function of the sequence $(a_0,a_1,a_2, \ldots)$ where a_r is the number of ways of selecting r objects from a set of six distinct objects, where each object can be selected no more than thrice.

(*b*) From the generating function found in part (*a*), determine the number of outcomes when three indistinguishable dice are rolled.

2-11. Find the ordinary generating function of the sequence $(a_0,a_1,a_2, \ldots)$ where a_r is the number of partitions of the integer r into distinct primes.

2-12. (*a*) Find the ordinary generating function of the sequence $(a_0,a_1,a_2, \ldots)$ where a_r is the number of ways in which the sum r will show when two distinct dice are rolled with the first one showing even and the second one showing odd.

(*b*) Find the ordinary generating function of the sequence $(a_0,a_1,a_2, \ldots)$ where a_r is the number of ways in which the sum r will show when 10 distinct dice are rolled, with five of them showing even and the other five showing odd.

2-13. (*a*) Find the ordinary generating function of the sequence $(a_0,a_1,a_2, \ldots)$ where a_r is the number of ways in which r letters can be selected from the alphabet $\{0,1,2\}$ with unlimited repetitions except that the letter 0 must be selected an even number of times.

(*b*) From the generating function found in part (*a*) find an explicit expression for the number of ways of making such a selection of k letters.

2-14. A "crooked die" is one whose six faces are marked with 1, 2, 3, 4, 5, and 7. Let a_r denote the number of ways in which the sum of the scores in r tosses of the "crooked die" is an even number. Show that the ordinary generating function of the sequence $(a_0,a_1,a_2, \ldots)$ is $\frac{1}{2}[1/(1 - x)^2]\{[1/(1 - x)^4] + [1/(1 + x)^4]\}$.

2-15. Let $A(x)$ denote the ordinary generating function of the sequence of numbers $(a_0,a_1,a_2, \ldots)$. Find the sequence whose ordinary generating function is $A(x)(1 - x)$. ($1 - x$ is therefore called the difference operator.)

2-16. Let $A(x)$ be the ordinary generating function of the sequence $(a_0,a_1,a_2, \ldots)$. Find the ordinary generating function of the sequence $(q_0,q_1,q_2, \ldots)$, where $q_n = \sum_{i=n+1}^{\infty} a_i$. (Assume that all the q's are finite.)

2-17. It is known that the ordinary generating function of the sequence $(1,b,b^2, \ldots, b^r, \ldots)$ is $1/(1 - bx)$. Find the sequence whose ordinary generating function is $b^kx^k/(1 - bx)^{k+1}$.

2-18. Find the exponential generating function of the sequence $(1, 1 \times 4, 1 \times 4 \times 7, \ldots, 1 \times 4 \times 7 \times \cdots \times (3r + 1), \ldots)$.

2-19. Let a_r denote the number of ways of permuting r of the 10 letters A, A, A, A, B, C, C, D, E, E. Find the exponential generating function of the sequence $(a_0,a_1,a_2, \ldots)$.

2-20. Find the number of n-digit words generated from the alphabet $\{0,1,2,3\}$ in each of which the number of 0's is even.

2-21. Find the number of n-digit words generated from the alphabet $\{0,1,2,3,4\}$ in each of which the total number of 0's and 1's is even.

2-22. Find the number of n-digit words generated from the alphabet $\{0,1,2\}$ in each of which none of the digits appears exactly three times.

2-23. (*a*) Evaluate the definite integral $\int_0^\infty e^{-s}s^k\,ds$.

(*b*) Let $A(x)$ and $E(x)$ be the ordinary and exponential generating functions of the sequence $(a_0,a_1,a_2,\ldots)$, respectively. Show that

$$A(x) = \int_0^\infty e^{-s}E(sx)\,ds$$

2-24. (*a*) Let $A(x)$ be the ordinary generating function of the sequence of numbers $(a_0,a_1,a_2,\ldots)$. Find the sequences whose ordinary generating functions are

1. $\dfrac{d^j}{dx^j}A(x)$ where j is a nonnegative integer
2. $x\dfrac{d}{dx}\left(x\dfrac{d}{dx}\left(x\dfrac{d}{dx}A(x)\right)\right)$

(*b*) Let θ denote the operator $x(d/dx)$, and let $P(\theta) = p_0 + p_1\theta + p_2\theta^2 + \cdots + p_k\theta^k$, where $p_0, p_1, p_2, \ldots, p_k$ are real constants. What is the sequence whose ordinary generating function is $P(\theta)A(x)$?

2-25. Let $(p_0,p_1,p_2,\ldots)$ be a sequence of nonnegative numbers which are less than or equal to 1. Let $P(x)$ denote the ordinary generating function of this sequence.

(*a*) The *kth moment* of the sequence $(p_0,p_1,p_2,\ldots)$, m_k, is defined as

$$m_k = \sum_{j=0}^{\infty} j^k p_j$$

Suppose that m_k exists for $k = 0, 1, 2, \ldots$. Show that the exponential generating function of the sequence $(m_0,m_1,m_2,\ldots)$, $M(x)$, is given by

$$M(x) = P(e^x)$$

[$M(x)$ is also called the moment generating function.]

(*b*) The *kth factorial moment* of the sequence $(p_0,p_1,p_2,\ldots)$, n_k, is defined as

$$n_k = \sum_{j=k}^{\infty} \frac{j!}{(j-k)!}p_j$$

Show that the exponential generating function of the sequence $(n_0,n_1,n_2,\ldots)$, $N(x)$, is given by $P(x+1)$.

2-26. Let $A(x)$ and $B(x)$ be the ordinary generating functions of the two sequences of numbers $(a_0,a_1,a_2,\ldots)$ and $(b_0,b_1,b_2,\ldots)$, respectively.

(*a*) Define

$$c_{n,j} = b_0c_{n-1,j} + b_1c_{n-1,j-1} + \cdots + b_{j-1}c_{n-1,1} + b_jc_{n-1,0}$$

and

$$c_{1,j} = b_j$$

Suppose that $c_{n,j}$ exists for $j = 0, 1, 2, \ldots$. Find the ordinary generating function $C_n(x)$ of the sequence $(c_{n,0}, c_{n,1} c_{n,2}, \ldots)$.

(b) Let $d_j = \sum_{k=0}^{\infty} a_k c_{k,j}$. Suppose that d_j exists for $j = 0, 1, 2, \ldots$. Show that the ordinary generating function of the sequence $(d_0, d_1, d_2, \ldots)$, $D(x)$, is equal to $A(B(x))$.

2-27. In this problem we investigate the following representation of an arbitrary positive integer n

$$n = \binom{b_1}{1} + \binom{b_2}{2} + \binom{b_3}{3} + \cdots + \binom{b_k}{k}$$

for a fixed k, where the b's satisfy the condition

$$0 \le b_1 < b_2 < b_3 < \cdots < b_k$$

That such a representation is possible can be seen from the following procedure for determining the b's:

b_k is the greatest integer such that $\binom{b_k}{k} \le n$

b_{k-1} is the greatest integer such that $\binom{b_{k-1}}{k-1} \le n - \binom{b_k}{k}$

. . . .

b_r is the greatest integer such that $\binom{b_r}{r} \le n - \sum_{j=r+1}^{k} \binom{b_j}{j}$

. . . .

b_1 is the greatest integer such that $\binom{b_1}{1} \le n - \sum_{j=2}^{k} \binom{b_j}{j}$

We can show that this representation is unique by proving the following inequality:

$$\binom{n}{k} > \binom{n-1}{k} + \binom{n-2}{k-1} + \binom{n-3}{k-2} + \cdots + \binom{n-k}{1}$$

(a) Explain why.

(b) Show that

$$\binom{n}{k} - 1 = \binom{n-1}{k} + \binom{n-2}{k-1} + \binom{n-3}{k-2} + \cdots + \binom{n-k}{1}$$

2-28. Prove the identity

$$\frac{1}{1-x} = (1 + x + x^2 + \cdots + x^9)(1 + x^{10} + x^{20} + \cdots + x^{90})$$
$$(1 + x^{100} + x^{200} + \cdots + x^{900}) \cdots$$

In terms of the partitions of integers, what is the significance of this identity?

2-29. Show that the number of partitions of the integer $2r + k$ into exactly $r + k$ parts is the same for any nonnegative integer k.

2-30. Prove that the number of partitions of the integer n into m distinct parts is equal to the number of partitions of the integer $n - [m(m + 1)/2]$ into at most m parts $(n > m(m + 1)/2)$.

Hint: Use a Ferrers graph.

2-31. Show that the number of partitions of the integer $2n$ into three parts which are such that the sum of any two parts is greater than the third is equal to the number of partitions of n into exactly three parts.

2-32. (*The Theorems of Euler*)

(*a*) A partition of an integer is said to be a self-conjugate partition if its corresponding Ferrers graph is symmetrical about its main diagonal; i.e., the transposition of the graph is identical to itself. Show that there is a one-to-one correspondence between the self-conjugate partitions and partitions into odd and unequal parts of an integer.

(*b*) Find the ordinary generating function of the sequence $(a_0, a_1, a_2, \ldots, a_r, \ldots)$ where a_r is the number of partitions of the integer $r - N$ into even parts, the largest of which is less than or equal to $2m$. Clearly, a_r is the same as the number of partitions of the integer $\frac{1}{2}(r - N)$ such that the largest part in a partition is less than or equal to m.

(*c*) The largest square of dots at the upper left-hand corner of a Ferrers graph (as illustrated in Fig. 2P-1) is called the *Durfee square.* Let a_r denote the number of

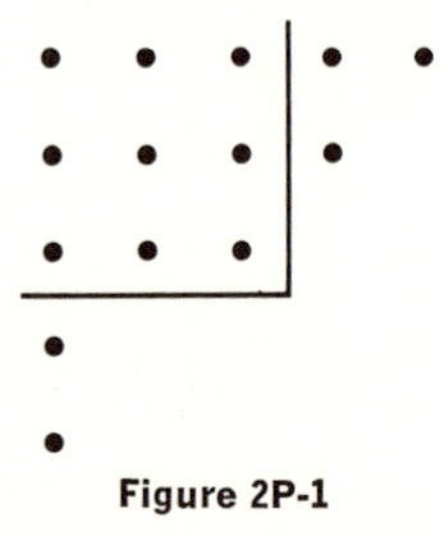

Figure 2P-1

self-conjugate partitions of the integer r whose Ferrers graphs contain an $m \times m$ Durfee square. Show that the ordinary generating function of the sequence $(a_0, a_1, a_2, \ldots, a_r, \ldots)$ is

$$\frac{x^{m^2}}{(1 - x^2)(1 - x^4) \cdots (1 - x^{2m})}$$

(*d*) Use the results of parts (*a*), (*b*), and (*c*) to prove the identity

$$(1 + x)(1 + x^3)(1 + x^5) \cdots = 1 + \frac{x}{1 - x^2} + \frac{x^4}{(1 - x^2)(1 - x^4)} + \frac{x^9}{(1 - x^2)(1 - x^4)(1 - x^6)} + \cdots$$

$$= 1 + \sum_{m=1}^{\infty} \frac{x^{m^2}}{(1 - x^2)(1 - x^4)(1 - x^6) \cdots (1 - x^{2m})}$$

(First theorem of Euler)

(*e*) Prove the identity

$$(1 + x^2)(1 + x^4)(1 + x^6) \cdots = 1 + \frac{x^2}{1 - x^2} + \frac{x^6}{(1 - x^2)(1 - x^4)} + \frac{x^{12}}{(1 - x^2)(1 - x^4)(1 - x^6)} + \cdots$$

$$= 1 + \sum_{m=1}^{\infty} \frac{x^{m(m+1)}}{(1 - x^2)(1 - x^4)(1 - x^6) \cdots (1 - x^{2m})}$$

(Second theorem of Euler)

Hint: Replace x^2 by x. Instead of the Durfee squares in the Ferrers graphs, consider isosceles right triangles of dots at the upper left-hand corner of the Ferrers graphs.

2-33. Let a_r be the number of incongruent triangles with integral sides and perimeter r. Find the ordinary generating function of the sequence $(a_0, a_1, a_2, \ldots)$.

Chapter 3
Recurrence Relations

3-1 INTRODUCTION

Consider the geometric series $(1,3,3^2,3^3, \ldots ,3^n, \ldots)$. Clearly, this sequence of numbers can be described by the expression for the general term

$$a_n = 3^n \qquad n = 0, 1, 2, \ldots$$

An alternative way of describing this sequence of numbers is to express the nth number in terms of the $(n - 1)$st number, together with the specification of the first number in the sequence; that is,

$$a_n = 3a_{n-1} \qquad a_0 = 1$$

For a sequence of numbers $(a_0,a_1,a_2, \ldots ,a_n, \ldots)$ an equation relating a number a_n to some of its predecessors in the sequence, for any n, is called a *recurrence relation*. A recurrence relation is also called a *difference equation*, and these two terms will be used interchangeably. In this example the recurrence relation specifies that the nth number is computed as three times the $(n - 1)$st number in the sequence. To initiate the computation, one must know one (or several) number(s) in the sequence, called the *boundary condition(s)*. In this example the boundary condition is $a_0 = 1$.

As another example, consider the sequence of numbers known as the *Fibonacci numbers*. The sequence starts with the two numbers 1, 1 and contains numbers which are equal to the sum of their two immediate predecessors. A portion of the sequence is

$$1, 1, 2, 3, 5, 8, 13, 21, 34, \ldots$$

It is quite difficult in this case to obtain a general expression for the nth number in the sequence by observation. On the other hand, the sequence can be described by the recurrence relation

$$a_n = a_{n-1} + a_{n-2}$$

together with the boundary conditions

$$a_0 = 1 \qquad a_1 = 1$$

Our immediate interest lies in the solution of a recurrence relation to obtain a general expression for the nth number in a sequence. In most practical cases, the converse problem of obtaining a recurrence relation from a general expression for the nth number is of less interest.

The technique of using recurrence relations is a very powerful one in enumeration problems. Before we discuss the solution of recurrence relations, consider the following example. Let there be n ovals drawn on the plane. If an oval intersects each of the other ovals at exactly two points and no three ovals meet at the same point, into how many regions do these ovals divide the plane? Let a_n denote the number of regions into which the plane is divided by n ovals. It is clear that $a_1 = 2$. By the construction shown in Fig. 3-1 we also find that $a_2 = 4$, $a_3 = 8$,

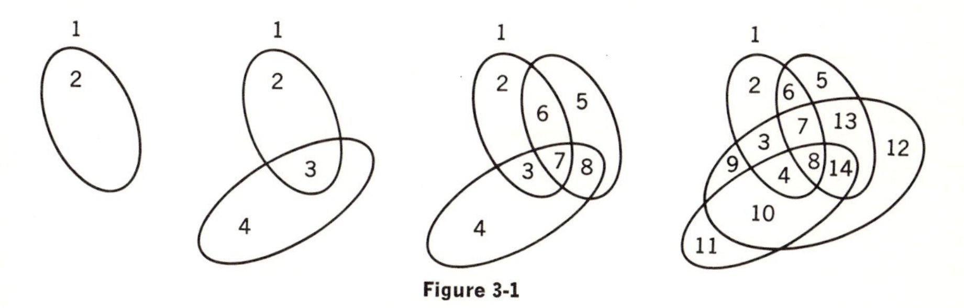

Figure 3-1

and $a_4 = 14$. Beyond this point actual construction becomes complicated, but a general expression for a_n is still not immediately obvious. Suppose that we have drawn $n - 1$ ovals that divide the plane into a_{n-1} regions. The nth oval will intersect these $n - 1$ ovals at $2(n - 1)$ points.

In other words, the nth oval will be divided into $2(n-1)$ arcs. Since each of these arcs will divide one of the a_{n-1} regions in two, we have the recurrence relation

$$a_n = a_{n-1} + 2(n-1)$$

With this relation and the boundary condition $a_1 = 2$, one can compute the value of a_n for any given n simply by repeatedly applying the recurrence relation. For example,

$$a_5 = a_4 + 2 \times (5-1) = 14 + 8 = 22$$

and

$$a_6 = a_5 + 2 \times (6-1) = 22 + 10 = 32$$

Subsequently we shall see how this recurrence relation can be solved to obtain a general expression for a_n. Although such a general expression is most frequently the desired result, in many cases obtaining the recurrence relation is a big step toward the solution of an enumeration problem. This is true because even when a general expression is not readily solvable from the recurrence relation, one can always resort to a step-by-step computation for the value of a desired a_n. As illustrated in this example, without the recurrence relation the possibility of such a computation is not at all obvious.

3-2 LINEAR RECURRENCE RELATIONS WITH CONSTANT COEFFICIENTS

A recurrence relation of the form

$$C_0 a_n + C_1 a_{n-1} + \cdots + C_r a_{n-r} = f(n) \tag{3-1}$$

is called a *linear* recurrence relation (difference equation) with constant coefficients where all the C's are constants. For example,

$$3a_n - 5a_{n-1} + 2a_{n-2} = n^2 + 5$$

is a linear difference equation with constant coefficients.

As was pointed out previously, if the values of r consecutive a's in the sequence, $a_{k-r}, a_{k-r+1}, \ldots, a_{k-1}$, are known for some k, the value of a_k can be calculated by use of (3-1). Also, the values of $a_{k+1}, a_{k+2}, \ldots$ and the values of $a_{k-r-1}, a_{k-r-2}, \ldots$ can then be calculated recursively. It follows that the solution to (3-1) is determined uniquely by the values of r consecutive a's (the boundary conditions). As a matter of fact, as will be seen shortly, the general form of the solution to Eq. (3-1) contains r undetermined constants. These constants can be determined by the values of r consecutive a's in the sequence.

Analogous to the solution of a linear differential equation with constant coefficients, the (total) solution of a linear difference equation with constant coefficients is the sum of two parts—the *homogeneous solution*, which satisfies the difference equation when the right-hand side of the equation is set to 0, and the *particular solution*, which satisfies the difference equation with $f(n)$ at the right-hand side. Let $a_n^{(h)}$ denote the homogeneous solution and $a_n^{(p)}$ denote the particular solution to the difference equation. Since

$$C_0a_n^{(h)} + C_1a_{n-1}^{(h)} + \cdots + C_ra_{n-r}^{(h)} = 0$$

and

$$C_0a_n^{(p)} + C_1a_{n-1}^{(p)} + \cdots + C_ra_{n-r}^{(p)} = f(n)$$

we have

$$C_0(a_n^{(h)} + a_n^{(p)}) + C_1(a_{n-1}^{(h)} + a_{n-1}^{(p)}) + \cdots + C_r(a_{n-r}^{(h)} + a_{n-r}^{(p)}) = f(n)$$

Clearly, the total solution, $a_n = a_n^{(h)} + a_n^{(p)}$, satisfies the difference equation. In Appendix 3-1 it will be shown that a_n is determined uniquely by the boundary conditions.

The homogeneous solution of a linear difference equation is of the form

$$a_n^{(h)} = A\alpha_1^n$$

where α_1 is called a *characteristic root* and A is a constant determined by the boundary conditions. Substituting $A\alpha^n$ for a_n in the difference equation with the right-hand side of the equation set to 0, we obtain

$$C_0A\alpha^n + C_1A\alpha^{n-1} + C_2A\alpha^{n-2} + \cdots + C_rA\alpha^{n-r} = 0$$

This equation can be simplified into the polynomial

$$C_0\alpha^r + C_1\alpha^{r-1} + C_2\alpha^{r-2} + \cdots + C_r = 0$$

which is called the *characteristic equation* of the difference equation. Therefore, if α_1 is one of the roots of the characteristic equation (it is for this reason that α_1 is called a characteristic root), $A\alpha_1^n$ is a homogeneous solution to the difference equation.

A characteristic equation of rth degree has r characteristic roots. Suppose the roots of the characteristic equation are distinct. In this case it is easy to verify that the homogeneous solution is

$$a_n^{(h)} = A_1\alpha_1^n + A_2\alpha_2^n + \cdots + A_r\alpha_r^n$$

where $\alpha_1, \alpha_2, \ldots, \alpha_r$ are the distinct characteristic roots and $A_1, A_2, \ldots, A_r$ are constants which can be determined by the boundary conditions. Let us revisit the example of the Fibonacci sequence of numbers discussed in Sec. 3-1.

Example 3-1 The recurrence relation for the Fibonacci sequence of numbers is

$$a_n = a_{n-1} + a_{n-2}$$

The corresponding characteristic equation is

$$\alpha^2 - \alpha - 1 = 0$$

which has two distinct roots

$$\alpha_1 = \frac{1 + \sqrt{5}}{2} \qquad \alpha_2 = \frac{1 - \sqrt{5}}{2}$$

The homogeneous solution (in this case, also the total solution, since the particular solution is 0) is

$$a_n = a_n^{(h)} = A_1 \left(\frac{1 + \sqrt{5}}{2}\right)^n + A_2 \left(\frac{1 - \sqrt{5}}{2}\right)^n$$

The two constants A_1 and A_2 can be determined from the boundary conditions $a_0 = 1$ and $a_1 = 1$ by solving the two equations

$$a_0 = 1 = A_1 + A_2$$

and

$$a_1 = 1 = A_1 \frac{1 + \sqrt{5}}{2} + A_2 \frac{1 - \sqrt{5}}{2}$$

These equations yield

$$A_1 = \frac{1}{\sqrt{5}} \frac{1 + \sqrt{5}}{2} \qquad \text{and} \qquad A_2 = -\frac{1}{\sqrt{5}} \frac{1 - \sqrt{5}}{2}$$

Thus,

$$a_n = \frac{1}{\sqrt{5}} \left(\frac{1 + \sqrt{5}}{2}\right)^{n+1} - \frac{1}{\sqrt{5}} \left(\frac{1 - \sqrt{5}}{2}\right)^{n+1} \quad \blacksquare$$

When the coefficients of the characteristic equation are real numbers but some of the characteristic roots are complex numbers, the homogeneous solution can be written in a different form. If a polynomial has real coefficients, then the complex conjugate of every root is also a root of the polynomial. Hence, complex roots always appear in pairs. Let $\alpha_1 = \delta + i\omega$ and $\alpha_2 = \delta - i\omega$ be a pair of complex characteristic roots. The corresponding homogeneous solution will be

$$\begin{aligned} A_1(\alpha_1)^n + A_2(\alpha_2)^n &= A_1(\delta + i\omega)^n + A_2(\delta - i\omega)^n \\ &= B_1\rho^n \cos n\theta + B_2\rho^n \sin n\theta \end{aligned}$$

where $\rho = \sqrt{\delta^2 + \omega^2}$, $\theta = \tan^{-1}(\omega/\delta)$, $B_1 = (A_1 + A_2)$, and $B_2 = i(A_1 - A_2)$. Note that B_1 and B_2 are constants determined by the boundary conditions.

Example 3-2 Evaluate the $n \times n$ determinant

$$\begin{vmatrix}
1 & 1 & 0 & 0 & 0 & 0 & \cdots & 0 & 0 & 0 & 0 & 0 \\
1 & 1 & 1 & 0 & 0 & 0 & \cdots & 0 & 0 & 0 & 0 & 0 \\
0 & 1 & 1 & 1 & 0 & 0 & \cdots & 0 & 0 & 0 & 0 & 0 \\
0 & 0 & 1 & 1 & 1 & 0 & \cdots & 0 & 0 & 0 & 0 & 0 \\
\cdots & \cdots & \cdots & \cdots & \cdots & \cdots & \cdots & \cdots & \cdots & \cdots & \cdots & \cdots \\
0 & 0 & 0 & 0 & 0 & 0 & \cdots & 0 & 1 & 1 & 1 & 0 \\
0 & 0 & 0 & 0 & 0 & 0 & \cdots & 0 & 0 & 1 & 1 & 1 \\
0 & 0 & 0 & 0 & 0 & 0 & \cdots & 0 & 0 & 0 & 1 & 1
\end{vmatrix}$$

Let a_k denote the value of the $k \times k$ determinant that is of this form. Expanding the $n \times n$ determinant with respect to the first column, we have

$$\left.\underbrace{\begin{vmatrix}
1 & 1 & 0 & 0 & 0 & 0 & \cdots & 0 & 0 & 0 & 0 & 0 \\
1 & 1 & 1 & 0 & 0 & 0 & \cdots & 0 & 0 & 0 & 0 & 0 \\
0 & 1 & 1 & 1 & 0 & 0 & \cdots & 0 & 0 & 0 & 0 & 0 \\
0 & 0 & 1 & 1 & 1 & 0 & \cdots & 0 & 0 & 0 & 0 & 0 \\
\cdots & \cdots & \cdots & \cdots & \cdots & \cdots & \cdots & \cdots & \cdots & \cdots & \cdots & \cdots \\
0 & 0 & 0 & 0 & 0 & 0 & \cdots & 0 & 1 & 1 & 1 & 0 \\
0 & 0 & 0 & 0 & 0 & 0 & \cdots & 0 & 0 & 1 & 1 & 1 \\
0 & 0 & 0 & 0 & 0 & 0 & \cdots & 0 & 0 & 0 & 1 & 1
\end{vmatrix}}_{n}\right\} n$$

$$= \left.\underbrace{\begin{vmatrix}
1 & 1 & 0 & 0 & 0 & \cdots & 0 & 0 & 0 & 0 & 0 \\
1 & 1 & 1 & 0 & 0 & \cdots & 0 & 0 & 0 & 0 & 0 \\
0 & 1 & 1 & 1 & 0 & \cdots & 0 & 0 & 0 & 0 & 0 \\
\cdots & \cdots & \cdots & \cdots & \cdots & \cdots & \cdots & \cdots & \cdots & \cdots & \cdots \\
0 & 0 & 0 & 0 & 0 & \cdots & 0 & 1 & 1 & 1 & 0 \\
0 & 0 & 0 & 0 & 0 & \cdots & 0 & 0 & 1 & 1 & 1 \\
0 & 0 & 0 & 0 & 0 & \cdots & 0 & 0 & 0 & 1 & 1
\end{vmatrix}}_{n-1}\right\} n-1$$

$$- \left.\underbrace{\begin{vmatrix}
1 & 0 & 0 & 0 & 0 & \cdots & 0 & 0 & 0 & 0 & 0 \\
1 & 1 & 1 & 0 & 0 & \cdots & 0 & 0 & 0 & 0 & 0 \\
0 & 1 & 1 & 1 & 0 & \cdots & 0 & 0 & 0 & 0 & 0 \\
\cdots & \cdots & \cdots & \cdots & \cdots & \cdots & \cdots & \cdots & \cdots & \cdots & \cdots \\
0 & 0 & 0 & 0 & 0 & \cdots & 0 & 1 & 1 & 1 & 0 \\
0 & 0 & 0 & 0 & 0 & \cdots & 0 & 0 & 1 & 1 & 1 \\
0 & 0 & 0 & 0 & 0 & \cdots & 0 & 0 & 0 & 1 & 1
\end{vmatrix}}_{n-1}\right\} n-1$$

Expanding the second determinant on the right-hand side with respect to the first row, we obtain the recurrence relation

$$a_n = a_{n-1} - a_{n-2}$$

The corresponding characteristic equation is

$$\alpha^2 - \alpha + 1 = 0$$

The characteristic roots are

$$\alpha_1 = \frac{1}{2} + i\frac{\sqrt{3}}{2} \qquad \alpha_2 = \frac{1}{2} - i\frac{\sqrt{3}}{2}$$

Thus, we have

$$a_n = B_1 \cos\frac{n\pi}{3} + B_2 \sin\frac{n\pi}{3}$$

since $\rho = \sqrt{\left(\frac{1}{2}\right)^2 + \left(\frac{\sqrt{3}}{2}\right)^2} = 1$ and $\tan^{-1}\left(\frac{\sqrt{3}}{2} \Big/ \frac{1}{2}\right) = \pi/3$. (The particular solution is 0.) From the boundary conditions

$$a_1 = 1 \qquad a_2 = 0$$

the constants B_1 and B_2 are determined as

$$B_1 = 1 \qquad B_2 = \frac{1}{\sqrt{3}}$$

Therefore, the solution of the difference equation is

$$a_n = \cos\frac{n\pi}{3} + \frac{1}{\sqrt{3}}\sin\frac{n\pi}{3}$$

which gives $a_1 = 1$, $a_2 = 0$, $a_3 = -1$, $a_4 = -1$, $a_5 = 0$, . . . , as we can check by directly calculating the values of the corresponding determinants. ∎

So much for the case of distinct characteristic roots. Now suppose some of the roots of the characteristic equation are multiple roots. Let α_1 be a k-multiple root. The corresponding homogeneous solution is

$$(A_1 n^{k-1} + A_2 n^{k-2} + \cdots + A_{k-2} n^2 + A_{k-1} n + A_k)\alpha_1^n$$

where the A's are constants which are determined by the boundary conditions. It is clear that $a_n^{(h)} = A_k \alpha_1^n$ is a homogeneous solution of the difference equation (3-1). To show that $a_n^{(h)} = A_{k-1} n \alpha_1^n$ is also a homogeneous solution, we recall that α_1 not only satisfies the equation

$$C_0\alpha^n + C_1\alpha^{n-1} + C_2\alpha^{n-2} + \cdots + C_r\alpha^{n-r} = 0 \tag{3-2}$$

but also satisfies the derivative of Eq. (3-2), which is

$$C_0 n\alpha^{n-1} + C_1(n-1)\alpha^{n-2} + C_2(n-2)\alpha^{n-3} + \cdots + C_r(n-r)\alpha^{n-r-1} = 0 \tag{3-3}$$

because α_1 is a multiple root of Eq. (3-2). Multiplying Eq. (3-3) by $A_{k-1}\alpha$ and replacing α by α_1, we obtain

$$C_0A_{k-1}n\alpha_1^n + C_1A_{k-1}(n-1)\alpha_1^{n-1} + C_2A_{k-1}(n-2)\alpha_1^{n-2} + \cdots + C_rA_{k-1}(n-r)\alpha_1^{n-r} = 0$$

which shows that $A_{k-1}n\alpha_1^n$ is indeed a homogeneous solution.

The fact that α_1 satisfies the second, third, . . . , $(k-1)$st derivatives of Eq. (3-2) enables us to prove that $A_{k-2}n^2\alpha_1^n$, $A_{k-3}n^3\alpha_1^n$, . . . , $A_1n^{k-1}\alpha_1^n$ are also homogeneous solutions.

Example 3-3 Solve the difference equation

$$a_n + 6a_{n-1} + 12a_{n-2} + 8a_{n-3} = 0$$

with the boundary conditions $a_0 = 1$, $a_1 = -2$, and $a_2 = 8$. The characteristic equation is

$$\alpha^3 + 6\alpha^2 + 12\alpha + 8 = 0$$

The solution is

$$a_n = (A_1n^2 + A_2n + A_3)(-2)^n$$

since -2 is a triple characteristic root. From the boundary conditions, the constants are determined as

$$A_1 = \tfrac{1}{2} \qquad A_2 = -\tfrac{1}{2} \qquad A_3 = 1 \quad \blacksquare$$

Example 3-4 Evaluate the $n \times n$ determinant

$$\begin{vmatrix}
2 & 1 & 0 & 0 & 0 & 0 & \cdots & 0 & 0 & 0 & 0 & 0 \\
1 & 2 & 1 & 0 & 0 & 0 & \cdots & 0 & 0 & 0 & 0 & 0 \\
0 & 1 & 2 & 1 & 0 & 0 & \cdots & 0 & 0 & 0 & 0 & 0 \\
0 & 0 & 1 & 2 & 1 & 0 & \cdots & 0 & 0 & 0 & 0 & 0 \\
\cdots & \cdots & \cdots & \cdots & \cdots & \cdots & \cdots & \cdots & \cdots & \cdots & \cdots & \cdots \\
0 & 0 & 0 & 0 & 0 & 0 & \cdots & 0 & 1 & 2 & 1 & 0 \\
0 & 0 & 0 & 0 & 0 & 0 & \cdots & 0 & 0 & 1 & 2 & 1 \\
0 & 0 & 0 & 0 & 0 & 0 & \cdots & 0 & 0 & 0 & 1 & 2
\end{vmatrix}$$

Let a_k denote the value of the $k \times k$ determinant that is of this form. As in Example 3-2, the recurrence relation is

$$a_n = 2a_{n-1} - a_{n-2}$$

The characteristic roots are found to be 1 and 1, and the solution is

$$a_n = (A_1 n + A_2)(1)^n = A_1 n + A_2$$

From the boundary conditions $a_1 = 2$ and $a_2 = 3$, we obtain

$$A_1 = 1 \qquad A_2 = 1$$

Thus, the solution is

$$a_n = n + 1 \quad \blacksquare$$

As to the particular solution of a difference equation, there is no general way of finding it. However, when the function $f(n)$ is in a relatively simple form, the particular solution can be determined by inspection, as illustrated in the following examples. In Sec. 3-3, it is shown that both the particular solution and the homogeneous solution can be determined at once by the use of generating functions.

Example 3-5 Solve the difference equation

$$a_n + 2a_{n-1} = n + 3$$

with the boundary condition $a_0 = 3$. The homogeneous solution is $A(-2)^n$. To determine the particular solution, we try a solution of the form $a_n^{(p)} = Bn + D$. Substituting this into the difference equation, we have

$$Bn + D + 2[B(n - 1) + D] = n + 3$$

which gives

$$3Bn + 3D - 2B = n + 3$$

Comparing the coefficients of n and the constant terms on the two sides of this equation, we have

$$3B = 1 \qquad \text{and} \qquad 3D - 2B = 3$$

that is,

$$B = 1/3 \qquad D = 11/9$$

and thus,

$$a_n^{(p)} = \frac{n}{3} + \frac{11}{9}$$

The total solution of the difference equation is simply the sum of the homogeneous and particular solutions. Thus,

$$a_n = A(-2)^n + \frac{n}{3} + \frac{11}{9}$$

From the boundary condition, the constant A is determined as $^{16}\!/_9$. ■

Example 3-6 Solve the difference equation

$$a_n + 2a_{n-1} + a_{n-2} = 2^n$$

The homogeneous solution is $(A_1 n + A_2)(-1)^n$. The particular solution is found by trying a solution of the form $B \times 2^n$. Since

$$B \times 2^n + 2 \times B \times 2^{n-1} + B \times 2^{n-2} = 2^n$$

B is determined as $^4\!/_9$. ■

Example 3-7 (*The Tower of Hanoi problem*) n circular rings of tapering size are slipped onto a peg with the largest ring at the bottom as shown in Fig. 3-2. These rings are to be transferred one at a time

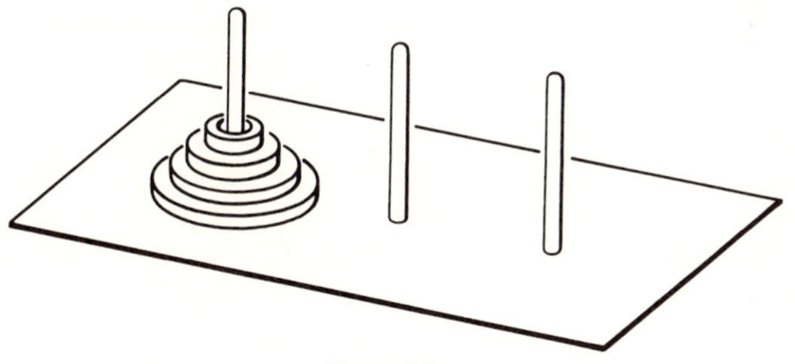

Figure 3-2

onto another peg, and there is a third peg available on which rings can be left temporarily. If, during the course of transferring the rings, no ring may ever be placed on top of a smaller one, in how many moves can these rings be transferred with their relative positions unchanged? We transfer the n rings by first moving the top $n - 1$ rings onto the third peg. Then we place the largest ring onto the second peg and move the $n - 1$ rings from the third peg onto the second peg. If we let a_n denote the number of moves it takes to transfer n rings from one peg to another, we have the recurrence relation

$$a_n = 2a_{n-1} + 1$$

The homogeneous solution is $A \times 2^n$ and the particular solution is -1. Since the boundary condition is $a_1 = 1$, we have the solution

$$a_n = 2^n - 1 \quad ■$$

3-3 SOLUTION BY THE TECHNIQUE OF GENERATING FUNCTIONS

There are many physical problems in which a difference equation of the form

$$C_0a_n + C_1a_{n-1} + \cdots + C_ra_{n-r} = f(n)$$

has physical significance and is valid only for n being larger than or equal to some integer k. We shall only be interested in determining the values of the a_n's for $n \geq k - r$ because these are the only a_n's that are related by the difference equation and may have physical significance. (Among these a_n's, a_{k-r}, a_{k-r+1}, . . . , a_{k-1} are boundary conditions specified by the problem.) Moreover, for problems in which a_n has physical significance only for $n \geq 0$, k is larger than or equal to r. We shall limit our discussion to such a case in which a difference equation is valid for $n \geq k$, with $k \geq r$. (Notice that this is *not* equivalent to saying that the difference equation is valid for $n \geq r$.) Since the values of the a_n's for $n < k - r$ are not constrained by the difference equation, they can be chosen arbitrarily. If we set a_n to 0 for $n < 0$ and choose some arbitrary values† for the a_n's for $0 \leq n < k - r$, we can solve for the generating function of the sequence $(a_0,a_1,a_2, \ldots ,a_n, \ldots)$, instead of solving for a general expression for a_n.

Let $A(x)$ denote the ordinary generating function of the sequence $(a_0,a_1,a_2, \ldots ,a_n, \ldots)$; that is,

$$A(x) = a_0 + a_1x + a_2x^2 + \cdots + a_nx^n + \cdots$$

[Henceforth, unless otherwise specified, the ordinary generating function of a sequence of numbers denoted by lowercase letters will be denoted by the corresponding uppercase letter; e.g., the generating function of the sequence $(g_0,g_1,g_2, \ldots ,g_n, \ldots)$ will be denoted by $G(x)$.] Let

† In many problems, we choose their values such that the range of n for which the difference equation is valid can be extended. That is, the difference equation becomes valid for $n \geq k'$ $(k' < k)$. The advantage of such a choice is to simplify the computation. However, let us emphasize that such an extension of the range of validity does not change the physical problem and will not affect the values of the a_n's that we are interested in $(n \geq k - r)$.

both sides of the difference equation be multiplied by x^n and then summed from $n = k$ to $n = \infty$; that is,

$$\sum_{n=k}^{\infty} (C_0 a_n + C_1 a_{n-1} + \cdots + C_r a_{n-r}) x^n = \sum_{n=k}^{\infty} f(n) x^n$$

Since

$$\sum_{n=k}^{\infty} C_0 a_n x^n = C_0[A(x) - a_0 - a_1 x - a_2 x^2 - \cdots - a_{k-1} x^{k-1}]$$

$$\sum_{n=k}^{\infty} C_1 a_{n-1} x^n = C_1 x[A(x) - a_0 - a_1 x - a_2 x^2 - \cdots - a_{k-2} x^{k-2}]$$

$$\cdots\cdots\cdots\cdots\cdots\cdots\cdots\cdots\cdots$$

$$\sum_{n=k}^{\infty} C_r a_{n-r} x^n = C_r x^r[A(x) - a_0 - a_1 x - a_2 x^2 - \cdots - a_{k-r-1} x^{k-r-1}]$$

we have an algebraic equation by which $A(x)$ can be solved; that is,

$$A(x) = a_0 + a_1 x + \cdots + a_{k-r-1} x^{k-r-1}$$
$$+ \frac{1}{C_0 + C_1 x + \cdots + C_r x^r} \Bigg[\sum_{n=k}^{\infty} f(n) x^n + C_0(a_{k-r} x^{k-r} + \cdots$$
$$+ a_{k-1} x^{k-1}) + C_1(a_{k-r} x^{k-r+1} + \cdots + a_{k-2} x^{k-1})$$
$$+ \cdots + C_{r-1} a_{k-r} x^{k-1} \Bigg]$$

It is clear now that the values of $a_0, a_1, \ldots, a_{k-r-1}$ which were chosen arbitrarily will not affect the values of the a_n's for $n \geq k - r$ as was pointed above. Notice that to determine $A(x)$ we need to know the values of $a_{k-r}, a_{k-r+1}, \ldots, a_{k-1}$ which are the boundary conditions that are used to determine the coefficients in the homogeneous solution in the method of solution in Sec. 3-2.

We shall illustrate the procedure by solving the recurrence relation obtained in Sec. 3-1 for the problem of the ovals. The recurrence relation is

$$a_n = a_{n-1} + 2(n - 1)$$

Since a_i has physical meaning only for $i \geq 1$, the recurrence relation is valid for $n \geq 2$. Because a_0 has no physical significance, we can choose any arbitrary value for a_0†. One choice is to have a value for a_0 such that

† The reader may wonder why such a point never arises in the method of solution discussed in Sec. 3-2. The reason is that in Sec. 3-2, we have tacitly assumed that those a's having no physical significance are always chosen in such a way that the recurrence relation is satisfied for all n (positive as well as negative). For example, we can see that in Example 3-2, the general expression $a_n = \cos(n\pi/3) + 1/\sqrt{3} \sin(n\pi/3)$ yields $a_0 = 1$, $a_{-1} = 0$, $a_{-2} = -1 \ldots$. Although these values have no physical significance, they do satisfy the recurrence relation.

the range of validity of the recurrence relation is extended. We therefore choose a_0 to be equal to 2 because a_1 is equal to 2. The recurrence relation is now valid for $n \geq 1$. Multiplying both sides of the recurrence relation by x^n and summing both sides from $n = 1$ to $n = \infty$, we have

$$\sum_{n=1}^{\infty} a_n x^n = \sum_{n=1}^{\infty} a_{n-1} x^n + 2 \sum_{n=1}^{\infty} (n-1) x^n$$

that is,

$$A(x) - a_0 = xA(x) + \frac{2x^2}{(1-x)^2}$$

or

$$A(x) = \frac{2x^2}{(1-x)^3} + \frac{2}{1-x}$$

It follows that

$$a_n = n(n-1) + 2 \qquad n = 0, 1, 2, \ldots$$

Note that $2x^2/(1-x)^3$ is the ordinary generating function of the sequence $(0, 0, 2, 6, \ldots, n(n-1), \ldots)$ since the coefficient of the term x^{n-2} in $1/(1-x)^3$ is

$$\frac{(-3)(-4) \cdots (-3-n+2+1)}{(n-2)!}(-1)^{n-2}$$
$$= \frac{3 \times 4 \times 5 \cdots \times n}{(n-2)!} = \frac{1}{2} n(n-1)$$

To show that any arbitrary choice of the value of a_0 will give us the correct answer for a_n for $n \geq 1$, we shall solve the recurrence relation with $a_0 = 5$ (a random choice). Multiplying both sides of the recurrence relation by x^n and summing from $n = 2$ to $n = \infty$, we obtain

$$\sum_{n=2}^{\infty} a_n x^n = \sum_{n=2}^{\infty} a_{n-1} x^n + 2 \sum_{n=2}^{\infty} (n-1) x^n$$
$$A(x) - a_1 x - a_0 = x[A(x) - a_0] + \frac{2x^2}{(1-x)^2}$$

Because the boundary condition is $a_1 = 2$, it follows that

$$A(x) = \frac{2x^2}{(1-x)^3} + \frac{2x}{1-x} + 5$$

Thus,

$$a_n = \begin{cases} 5 & n = 0 \\ n(n-1) + 2 & n = 1, 2, 3, \ldots \end{cases}$$

Example 3-8 Among the 4^n n-digit quaternary sequences, how many of them have an even number of 0's? How many of them have an even number of 0's and an even number of 1's? (The reader may recall that we have discussed this problem in Example 1-9. It is instructive to compare the two methods of solution.)

If we let a_{n-1} denote the number of $(n-1)$-digit quaternary sequences that have an even number of 0's, then the number of $(n-1)$-digit quaternary sequences that have an odd number of 0's is $4^{n-1} - a_{n-1}$. To each of the a_{n-1} sequences that have an even number of 0's, the digit 1, 2, or 3 can be appended to yield sequences of length n that contain an even number of 0's. To each of the $4^{n-1} - a_{n-1}$ sequences that have an odd number of 0's, the digit 0 can be appended to yield a sequence of length n that contains an even number of 0's. Therefore, for $n \geq 2$,

$$a_n = 3a_{n-1} + 4^{n-1} - a_{n-1}$$

which simplifies to

$$a_n - 2a_{n-1} = 4^{n-1}$$

Since $a_1 = 3$, we choose $a_0 = 1$ so that the recurrence relation is also valid for $n = 1$. Multiplying both sides of the recurrence relation by x^n and summing from $n = 1$ to $n = \infty$, we obtain

$$\sum_{n=1}^{\infty} a_n x^n - 2 \sum_{n=1}^{\infty} a_{n-1} x^n = \sum_{n=1}^{\infty} 4^{n-1} x^n$$

$$A(x) - 1 - 2xA(x) = \frac{x}{1-4x}$$

$$A(x) = \frac{1}{1-2x}\left(\frac{x}{1-4x} + 1\right) = \frac{\frac{1}{2}}{1-4x} + \frac{\frac{1}{2}}{1-2x}$$

It follows that

$$a_n = \tfrac{1}{2}4^n + \tfrac{1}{2}2^n \qquad n \geq 0$$

Let b_{n-1} denote the number of $(n-1)$-digit quaternary sequences that have an even number of 0's and an even number of 1's. Let c_{n-1} denote the number of $(n-1)$-digit quaternary sequences that have an even number of 0's and an odd number of 1's. Let d_{n-1} denote the number of $(n-1)$-digit quaternary sequences that have an odd number of 0's and an even number of 1's. It follows that there are $4^{n-1} - b_{n-1} - c_{n-1} - d_{n-1}$ $(n-1)$-digit quaternary sequences that have an odd number of 0's and an odd number of 1's. According to the information in the following table,

Sequences of length n − 1 \ *Resultant sequences of length n*	*Digit appended* 0	1	2 *or* 3
Even number of 0's Even number of 1's	Odd number of 0's Even number of 1's	Even number of 0's Odd number of 1's	Even number of 0's Even number of 1's
Even number of 0's Odd number of 1's	Odd number of 0's Odd number of 1's	Even number of 0's Even number of 1's	Even number of 0's Odd number of 1's
Odd number of 0's Even number of 1's	Even number of 0's Even number of 1's	Odd number of 0's Odd number of 1's	Odd number of 0's Even number of 1's
Odd number of 0's Odd number of 1's	Even number of 0's Odd number of 1's	Odd number of 0's Even number of 1's	Odd number of 0's Odd number of 1's

we have the recurrence relations

$$b_n = 2b_{n-1} + c_{n-1} + d_{n-1} \tag{3-4}$$

$$c_n = b_{n-1} + 2c_{n-1} + 4^{n-1} - b_{n-1} - c_{n-1} - d_{n-1} \tag{3-5}$$

$$d_n = b_{n-1} + 2d_{n-1} + 4^{n-1} - b_{n-1} - c_{n-1} - d_{n-1} \tag{3-6}$$

which are valid for $n \geq 2$. After simplification,

$$b_n = 2b_{n-1} + c_{n-1} + d_{n-1}$$

$$c_n = c_{n-1} - d_{n-1} + 4^{n-1}$$

$$d_n = -c_{n-1} + d_{n-1} + 4^{n-1}$$

The values of b_0, c_0, and d_0 can be chosen as

$$b_0 = \tfrac{3}{4} \qquad c_0 = \tfrac{1}{4} \qquad d_0 = \tfrac{1}{4}$$

so that the recurrence relations will be valid for $n \geq 1$. Thus,

$$\sum_{n=1}^{\infty} b_n x^n = 2\sum_{n=1}^{\infty} b_{n-1}x^n + \sum_{n=1}^{\infty} c_{n-1}x^n + \sum_{n=1}^{\infty} d_{n-1}x^n$$

$$\sum_{n=1}^{\infty} c_n x^n = \sum_{n=1}^{\infty} c_{n-1}x^n - \sum_{n=1}^{\infty} d_{n-1}x^n + \sum_{n=1}^{\infty} 4^{n-1}x^n$$

$$\sum_{n=1}^{\infty} d_n x^n = -\sum_{n=1}^{\infty} c_{n-1}x^n + \sum_{n=1}^{\infty} d_{n-1}x^n + \sum_{n=1}^{\infty} 4^{n-1}x^n$$

Using the generating function representation, these relations become

$$B(x) - \tfrac{3}{4} = 2xB(x) + xC(x) + xD(x)$$
$$C(x) - \tfrac{1}{4} = xC(x) - xD(x) + \frac{x}{1-4x}$$
$$D(x) - \tfrac{1}{4} = -xC(x) + xD(x) + \frac{x}{1-4x}$$

Solving these equations, we obtain

$$B(x) = \frac{\tfrac{1}{4}}{1-4x} + \frac{\tfrac{1}{2}}{1-2x}$$
$$C(x) = D(x) = \frac{\tfrac{1}{4}}{1-4x}$$

and

$$b_n = \tfrac{1}{4}4^n + \tfrac{1}{2}2^n \qquad n = 0, 1, 2, \ldots$$
$$c_n = d_n = \tfrac{1}{4}4^n \qquad n = 0, 1, 2, \ldots$$

It should be pointed out that this example also illustrates the use and solution of simultaneous linear difference equations with constant coefficients [Eqs. (3-4) to (3-6)]. With the generating function representation the solution of a set of simultaneous linear difference equations with constant coefficients is reduced to the solution of a set of linear algebraic simultaneous equations for the generating functions of the sequences. Although it is also possible to reduce a given set of simultaneous difference equations to a set of independent difference equations, one for each of the unknown sequences, the generating function approach is usually easier and more straightforward. ∎

★3-4 A SPECIAL CLASS OF NONLINEAR DIFFERENCE EQUATIONS

Thus far we have discussed the solution of linear difference equations with constant coefficients. The solution of nonlinear or variable-coefficient difference equations is a topic that is beyond our scope of discussion. However, there is one very common class of nonlinear difference equations that can be handled elegantly by the use of generating functions. Consider a difference equation of the form

$$a_n = a_{n-r}a_0 + a_{n-r-1}a_1 + \cdots + a_0a_{n-r} \qquad (3\text{-}7)$$

which is valid for $n \geq k$. Again, we shall limit our discussion to the case in which $k \geq r$. Notice that the value of a_n for $n \geq k$ can be computed recursively when the values of $a_0, a_1, \ldots, a_{k-1}$ are known.

These values are the boundary conditions that determine the solution uniquely. Multiplying both sides of Eq. (3-7) by x^n and summing from $n = k$ to $n = \infty$, we obtain

$$\sum_{n=k}^{\infty} a_n x^n = \sum_{n=k}^{\infty} (a_{n-r}a_0 + a_{n-r-1}a_1 + \cdots + a_0 a_{n-r})x^n$$

Recognizing that $(a_{n-r}a_0 + a_{n-r-1}a_1 + \cdots + a_0a_{n-r})$ is the coefficient of x^{n-r} in $A(x)A(x)$, we can write

$$\begin{aligned} A(x) &- a_0 - a_1x - \cdots - a_{k-1}x^{k-1} \\ &= x^r[A(x)A(x) - a_0^2 - (a_1a_0 + a_0a_1)x - \cdots - (a_{k-r-1}a_0 \\ &\qquad + a_{k-r-2}a_1 + \cdots + a_0a_{k-r-1})x^{k-r-1}] \qquad (3\text{-}8) \end{aligned}$$

Equation (3-8) is a second-order algebraic equation in $A(x)$ which can be solved for $A(x)$ by the ordinary algebraic method. (The values of $a_0, a_1, \ldots, a_{k-1}$ are the known boundary conditions.)

Example 3-9 Find the number of ways to parenthesize the expression

$$w_1 + w_2 + \cdots + w_{n-1} + w_n$$

so that only two terms will be added at one time. [For instance, the expression $w_1 + w_2 + w_3 + w_4$ can be parenthesized as $((w_1 + w_2) + (w_3 + w_4))$, $(w_1 + ((w_2 + w_3) + w_4))$, and so on.] Let a_i denote the number of ways of parenthesizing an expression with i terms. Consider the two subexpressions

$$w_1 + w_2 + \cdots + w_{n-r} \qquad w_{n-r+1} + w_{n-r+2} + \cdots + w_n$$

There are a_{n-r} ways to parenthesize the first expression and a_r ways to parenthesize the second expression. It follows that there are $a_{n-r}a_r$ ways to parenthesize the overall expression in which the last pair of parentheses added joins these two subexpressions. Letting r range from 1 to $n - 1$, we obtain the difference equation

$$a_n = a_{n-1}a_1 + a_{n-2}a_2 + \cdots + a_2a_{n-2} + a_1a_{n-1}$$

This equation is valid for $n \geq 2$. (Notice that $a_1 = 1$.) Since a_0 is not constrained by the difference equation, it can be chosen in an arbitrary manner. Letting $a_0 = 0$, we rewrite the difference equation as

$$a_n = a_na_0 + a_{n-1}a_1 + \cdots + a_1a_{n-1} + a_0a_n \qquad n \geq 2$$

It follows that

$$\sum_{n=2}^{\infty} a_n x^n = \sum_{n=2}^{\infty} (a_n a_0 + a_{n-1} a_1 + \cdots + a_1 a_{n-1} + a_0 a_n) x^n$$

$$A(x) - a_1 x - a_0 = [A(x)]^2 - a_0^2 - (a_1 a_0 + a_0 a_1) x$$

$$[A(x)]^2 - A(x) + x = 0$$

$$A(x) = \frac{1 \pm \sqrt{1 - 4x}}{2}$$

Although there are two solutions for $A(x)$, only the one that generates a sequence of positive numbers will be chosen. Since the general term in $\sqrt{1 - 4x}$ is

$$\frac{(\frac{1}{2})(\frac{1}{2} - 1)(\frac{1}{2} - 2) \cdots (\frac{1}{2} - n + 1)}{n!} (-4x)^n$$

$$= -\frac{1 \times 1 \times 3 \times 5 \times \cdots \times (2n - 3)}{n!} 2^n x^n$$

$$= -\frac{2}{n} \binom{2n - 2}{n - 1}$$

the solution

$$A(x) = \tfrac{1}{2} - \tfrac{1}{2} \sqrt{1 - 4x}$$

should be chosen. It follows that

$$a_n = \begin{cases} 0 & n = 0 \\ \dfrac{1}{n} \dbinom{2n - 2}{n - 1} & n = 1, 2, 3, \ldots \end{cases} \quad \blacksquare$$

The extension of this case leads us to the solution of another large class of problems. Consider a difference equation of the form

$$b_n = a_{n-r} b_0 + a_{n-r-1} b_1 + \cdots + a_0 b_{n-r} \qquad n \geq k$$

where $k \geq r$. Multiplying both sides by x^n and summing from $n = k$ to $n = \infty$, we obtain

$$\sum_{n=k}^{\infty} b_n x^n = \sum_{n=k}^{\infty} (a_{n-r} b_0 + a_{n-r-1} b_1 + \cdots + a_0 b_{n-r}) x^n$$

$$B(x) - b_0 - b_1 x - \cdots - b_{k-1} x^{k-1}$$
$$= x^r [A(x)B(x) - a_0 b_0 - (a_1 b_0 + a_0 b_1) x - \cdots$$
$$- (a_{k-r-1} b_0 + a_{k-r-2} b_1 + \cdots + a_0 b_{k-r-1}) x^{k-r-1}]$$

If either $A(x)$ or $B(x)$ together with the appropriate boundary conditions are known, then the other can be obtained.

We now discuss several examples concerning the occurrence of patterns in a binary sequence. A *pattern* consists of one or more consecutive binary digits like 01 and 1011. A pattern is said to occur at the kth digit of a sequence if, in scanning the sequence from left to right, the pattern appears after the kth digit is scanned. After a pattern occurs, scanning starts all over again to search for the second occurrence of the pattern that just occurred or for the occurrence of other patterns. For example, the pattern 010 occurs at the fifth and the ninth digits in the sequence $11\underline{010}1\underline{010}101$, but not at the seventh and eleventh digits.

Example 3-10 Find the number of n-digit binary sequences that have the pattern 010 occurring at the nth digit.

Let b_n denote the number of such sequences. Among all the n-digit binary sequences, there are 2^{n-3} sequences that have 010 as the last three digits. These sequences can be divided into two groups: those that have the pattern 010 occurring at the nth digit and those that do not have the pattern 010 occurring at the nth digit. There are b_n sequences in the former group. The sequences in the latter group must have the pattern 010 occurring at the $(n-2)$nd digit (for example, a sequence such as $\cdots 0\underline{010}10$) since this is the only reason that the last three digits in these n-digit sequences were not accepted as a 010 pattern. It follows that there are b_{n-2} sequences in the latter group, and thus

$$2^{n-3} = b_n + b_{n-2}$$

Clearly, this difference equation is valid for $n \geq 5$. Since the values of b_0, b_1, and b_2 are not constrained by the difference equation, they can be chosen in an arbitrary manner. For a reason that we shall discuss later, we set

$$b_0 = 1 \qquad b_1 = b_2 = 0$$

For such a choice of the unconstrained values, the difference equation is valid for $n \geq 3$. The solution of the difference equation is straightforward, namely,

$$\sum_{n=3}^{\infty} 2^{n-3}x^n = \sum_{n=3}^{\infty} b_n x^n + \sum_{n=3}^{\infty} b_{n-2}x^n$$

$$\frac{x^3}{1-2x} = B(x) - 1 + x^2[B(x) - 1]$$

and

$$B(x) = \frac{1 - 2x + x^2 - x^3}{1 - 2x + x^2 - 2x^3} = 1 + x^3 + 2x^4 + 3x^5 + 6x^6 + \cdots$$

■

Example 3-11 Find the number of n-digit binary sequences that have the pattern 010 occurring *for the first time* at the nth digit.

Let a_n denote the number of such sequences. There are 2^{n-3} n-digit binary sequences that have 010 as the last three digits. These sequences can be classified according to the digit at which the pattern 010 occurs for the first time. There are a_n sequences in which the first occurrence of the pattern is at the nth digit and a_{n-2} sequences in which the first occurrence of the pattern is at the $(n-2)$nd digit. [Notice that there can be no sequence among these 2^{n-3} sequences in which the first occurrence of the pattern is at the $(n-1)$st digit.] For $3 \leq r \leq n-3$, there are $a_r 2^{n-r-3}$ sequences in which the first occurrence of the pattern is at the rth digit, because to each of the a_r r-digit sequences that have the pattern 010 occurring for the first time at the rth digit, $n-r-3$ digits can be appended arbitrarily. Therefore,

$$2^{n-3} = a_n + a_{n-2} + a_{n-3}2^0 + a_{n-4}2^1 + \cdots + a_3 2^{n-6}$$

The difference equation is valid for $n \geq 6$. Because a_0, a_1, a_2 are not constrained by the difference equation, they can be chosen arbitrarily. Let $a_0 = a_1 = a_2 = 0$. Let

$$\begin{aligned} B(x) &= b_0 + b_1 x + b_2 x^2 + b_3 x^3 + \cdots + b_n x^n + \cdots \\ &= 1 + x^2 + 2^0 x^3 + 2^1 x^4 + \cdots + 2^{n-3} x^n + \cdots \\ &= 1 + x^2 + \frac{x^3}{1-2x} \end{aligned}$$

We can rewrite the difference equation as

$$2^{n-3} = a_n b_0 + a_{n-1} b_1 + a_{n-2} b_2 + \cdots + a_2 b_{n-2} + a_1 b_{n-1} + a_0 b_n$$

Notice that the difference equation is now valid for $n \geq 3$. It follows that

$$\sum_{n=3}^{\infty} 2^{n-3} x^n = \sum_{n=3}^{\infty} (a_n b_0 + a_{n-1} b_1 + a_{n-2} b_2 + \cdots + a_2 b_{n-2} + a_1 b_{n-1} + a_0 b_n) x^n$$

$$B(x) - 1 - x^2 = A(x)B(x) - a_0 b_0 - (a_1 b_0 + a_0 b_1)x - (a_2 b_0 + a_1 b_1 + a_0 b_2)x^2$$

and

$$A(x) = \frac{x^3}{1 - 2x + x^2 - x^3} = x^3 + 2x^4 + 3x^5 + 5x^6 + 9x^7 + \cdots$$

Notice that in this example $B(x)$ is chosen in such a way that the simple form $A(x)B(x)$ appears in the algebraic equation relating the generating functions from which $A(x)$ can be solved. However, this is not the only way in which the function $B(x)$ can be chosen. For example, by considering all the n-digit binary sequences, the last five digits of which are either 00010 or 10010 or 11010 (there are $3 \times 2^{n-5}$ such sequences), we obtain the difference equation

$$3 \times 2^{n-5} = a_n + a_{n-3} + a_{n-4} + a_{n-5}(3 \times 2^0) + a_{n-6}(3 \times 2^1) + \cdots + a_3(3 \times 2^{n-8}) \qquad n \geq 8$$

where, as before, a_n is the number of sequences in which the pattern 010 occurs for the first time at the nth digit. Again, let $a_0 = a_1 = a_2 = 0$. Let

$$\begin{aligned} B(x) &= 1 + x^3 + x^4 + b_5x^5 + b_6x^6 + \cdots + b_nx^n + \cdots \\ &= 1 + x^3 + x^4 + 3(2^0x^5 + 2x^6 + \cdots + 2^{n-5}x^n + \cdots) \end{aligned}$$

The difference equation can be rewritten as

$$3 \times 2^{n-5} = a_nb_0 + a_{n-1}b_1 + a_{n-2}b_2 + \cdots + a_2b_{n-2} + a_1b_{n-1} + a_0b_n \qquad n \geq 5$$

Thus,

$$\begin{aligned} B(x) - 1 - x^3 - x^4 &= A(x)B(x) - a_3x^3 - a_4x^4 \\ &= A(x)B(x) - x^3 - 2x^4 \end{aligned}$$

Solving for $A(x)$, we obtain

$$A(x) = \frac{x^3}{1 - 2x + x^2 - x^3}$$

As another possibility, we let b_n be the number of sequences of length n in which the pattern 010 occurs at the nth digit. In this case, we have the difference equation

$$b_n = a_n + a_{n-3}b_3 + a_{n-4}b_4 + \cdots + a_3b_{n-3} \qquad n \geq 6$$

Because a_0, a_1, a_2, b_0, b_1, b_2 are not constrained by the difference equation, we let $a_0 = a_1 = a_2 = 0$ and $b_0 = 1$, $b_1 = b_2 = 0$. We can rewrite the difference equation as

$$b_n = a_nb_0 + a_{n-1}b_1 + a_{n-2}b_2 + a_{n-3}b_3 + \cdots + a_3b_{n-3} + a_2b_{n-2} + a_1b_{n-1} + a_0b_n \qquad n \geq 3$$

Multiplying both sides of the difference equation by x^n and summing from $n = 3$ to $n = \infty$, we have

$$B(x) - 1 = A(x)B(x)$$

Since $B(x)$ has been found to be $(1 - 2x + x^2 - x^3)/(1 - 2x + x^2 - 2x^3)$ in Example 3-10, solving for $A(x)$, we obtain

$$A(x) = 1 - \frac{1}{B(x)} = 1 - \frac{1 - 2x + x^2 - 2x^3}{1 - 2x + x^2 - x^3} = \frac{x^3}{1 - 2x + x^2 - x^3}$$

■

The last choice of $B(x)$ in the preceding example suggests a useful formula for the solution of the first-occurrence problems. Let a_n be the number of n-digit sequences in which a particular pattern of p digits occurs for the first time at the nth digit. Let b_n be the number of n-digit sequences in which the pattern occurs at the nth digit. By choosing the unconstrained values as

$$a_0 = a_1 = a_2 = \cdots = a_{p-1} = 0$$
$$b_0 = 1 \qquad b_1 = b_2 = \cdots = b_{p-1} = 0$$

we see that the difference equation always leads to

$$B(x) - b_0 = A(x)B(x)$$

and

$$A(x) = 1 - \frac{1}{B(x)} \tag{3-9}$$

Example 3-12 Find the number of n-digit binary sequences in which an occurrence of the pattern 010 is followed by an occurrence of the pattern 110. Let c_n be the number of such sequences. Let a_n be the number of n-digit binary sequences in which the pattern 010 occurs for the first time at the nth digit, and let b_n be the number of n-digit binary sequences in which the pattern 110 occurs at least once.

Clearly,

$$c_n = a_3 b_{n-3} + a_4 b_{n-4} + a_5 b_{n-5} + \cdots + a_{n-3} b_3 \qquad n \geq 6$$

Let $a_0 = a_1 = a_2 = 0$, $b_0 = b_1 = b_2 = 0$, and $c_0 = c_1 = c_2 = \cdots = c_5 = 0$, since they are not constrained by the difference equation. It follows that

$$\sum_{n=6}^{\infty} c_n x^n = \sum_{n=6}^{\infty} (a_3 b_{n-3} + a_4 b_{n-4} + a_5 b_{n-5} + \cdots + a_{n-3} b_3) x^n$$

and

$$C(x) = A(x)B(x)$$

According to the result in Example 3-11,

$$A(x) = \frac{x^3}{1 - 2x + x^2 - x^3}$$

To find $B(x)$, let us define d_n as the number of n-digit sequences in which the pattern 110 occurs at the nth digit for the first time, and let $d_0 = d_1 = d_2 = 0$. Then

$$b_n = d_3 \times 2^{n-3} + d_4 \times 2^{n-4} + \cdots + d_{n-1} \times 2 + d_n \qquad n \geq 3$$

Consequently,

$$B(x) = D(x) \frac{1}{1 - 2x}$$

Using Eq. (3-9), we find $D(x)$ to be

$$D(x) = 1 - \frac{1}{1 + x^3/(1 - 2x)} = \frac{x^3}{1 - 2x + x^3}$$

Therefore,

$$\begin{aligned} C(x) &= \frac{x^3}{1 - 2x + x^2 - x^3} \frac{x^3}{1 - 2x + x^3} \frac{1}{1 - 2x} \\ &= \frac{x^6}{1 - 6x + 13x^2 - 12x^3 + 4x^4 + x^5 - 3x^6 + 2x^7} \\ &= x^6 + 6x^7 + 23x^8 + \cdots \quad \blacksquare \end{aligned}$$

3-5 RECURRENCE RELATIONS WITH TWO INDICES

For the combinations of distinct objects, we have derived in Example 1-2 the relation

$$C(n,r) = C(n - 1, r - 1) + C(n - 1, r) \tag{3-10}$$

This is an example of a recurrence relation with two indices. With the boundary conditions $C(n,0) = 1$ and $C(0,r) = 0$ for $r > 0$, the recurrence relation is valid for $n \geq 1$ and $r \geq 1$. The value of $C(n,r)$ can be computed recursively. Thus,

$$C(0,0) = 1$$

$$C(1,0) = 1 \qquad C(1,1) = C(0,0) + C(0,1) = 1$$

$$C(2,0) = 1 \qquad C(2,1) = C(1,0) + C(1,1) = 1 + 1 = 2 \quad \cdots$$

The reader may recall that the construction of the famous *Pascal triangle* (Fig. 3-3) for the binomial coefficients is based on this recurrence relation.

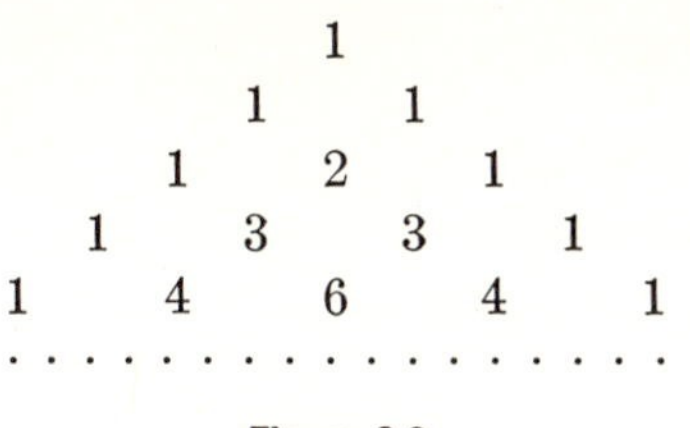

Figure 3-3

The general form of a linear recurrence relation with constant coefficients that has two indices is

$$\begin{aligned} &C_0 a_{n,r} + C_1 a_{n,r-1} + C_2 a_{n,r-2} + \cdots \\ &\quad + D_0 a_{n-1,r} + D_1 a_{n-1,r-1} + D_2 a_{n-1,r-2} + \cdots \\ &\quad + \cdots\cdots\cdots\cdots\cdots\cdots \\ &\quad + G_0 a_{n-k,r} + G_1 a_{n-k,r-1} + G_2 a_{n-k,r-2} + \cdots = f(n,r) \end{aligned}$$

where the C's, D's, . . . , G's are constants. As pointed out at the beginning of this chapter, although it may be tedious, we can always evaluate $a_{n,r}$ using the recurrence relation and starting with the known boundary conditions.

To solve a recurrence relation with two indices by the generating function technique, we first define a sequence of generating functions with one function for each value of one of the two indices; that is,

$$\begin{aligned} A_0(x) &= a_{0,0} + a_{0,1}x + a_{0,2}x^2 + \cdots + a_{0,r}x^r + \cdots \\ A_1(x) &= a_{1,0} + a_{1,1}x + a_{1,2}x^2 + \cdots + a_{1,r}x^r + \cdots \\ &\cdots\cdots\cdots\cdots\cdots\cdots \\ A_n(x) &= a_{n,0} + a_{n,1}x + a_{n,2}x^2 + \cdots + a_{n,r}x^r + \cdots \\ &\cdots\cdots\cdots\cdots\cdots\cdots \end{aligned}$$

Now, we can also define a generating function of the sequence $(A_0(x), A_1(x), A_2(x), \ldots)$, that is, using powers of y as indicator functions, we define

$$\begin{aligned} \mathcal{A}(y,x) &= A_0(x) + A_1(x)y + A_2(x)y^2 + \cdots + A_n(x)y^n + \cdots \\ &= [a_{0,0} + a_{0,1}x + a_{0,2}x^2 + \cdots + a_{0,r}x^r + \cdots] \\ &\quad + [a_{1,0} + a_{1,1}x + a_{1,2}x^2 + \cdots + a_{1,r}x^r + \cdots]y \\ &\quad + [a_{2,0} + a_{2,1}x + a_{2,2}x^2 + \cdots + a_{2,r}x^r + \cdots]y^2 \\ &\quad + \cdots\cdots\cdots\cdots\cdots\cdots \\ &\quad + [a_{n,0} + a_{n,1}x + a_{n,2}x^2 + \cdots + a_{n,r}x^r + \cdots]y^n \\ &\quad + \cdots\cdots\cdots\cdots\cdots\cdots \end{aligned}$$

Multiplying out, we have

$$\begin{aligned}\mathcal{A}(y,x) = {} & a_{0,0} + a_{0,1}x + a_{0,2}x^2 + \cdots + a_{0,r}x^r + \cdots \\ & + a_{1,0}y + a_{1,1}yx + a_{1,2}yx^2 + \cdots + a_{1,r}yx^r + \cdots \\ & + a_{2,0}y^2 + a_{2,1}y^2x + a_{2,2}y^2x^2 + \cdots + a_{2,r}y^2x^r + \cdots \\ & + \cdots\cdots\cdots\cdots\cdots\cdots\cdots\cdots \\ & + a_{n,0}y^n + a_{n,1}y^nx + a_{n,2}y^nx^2 + \cdots + a_{n,r}y^nx^r + \cdots \\ & + \cdots\cdots\cdots\cdots\cdots\cdots\cdots\cdots\end{aligned}$$

An alternative point of view is that for a sequence of two indices like $(a_{0,0},a_{0,1},a_{0,2}, \ldots ,a_{0,r}, \ldots,a_{1,0},a_{1,1}, \ldots ,a_{1,r}, \ldots)$, we use x and y as formal variables and define the generating function of two variables, $\mathcal{A}(y,x)$, as a function of x and y in which the coefficient of y^ix^j is $a_{i,j}$.

Let us consider the following illustrative examples.

Example 3-13 Find the generating function of the $C(n,r)$'s,

$$F_n(x) = C(n,0) + C(n,1)x + C(n,2)x^2 + \cdots + C(n,r)x^r + \cdots$$

From the recurrence relation in (3-10) we have

$$\begin{aligned}\sum_{r=1}^{\infty} C(n,r)x^r &= \sum_{r=1}^{\infty} C(n-1, r-1)x^r + \sum_{r=1}^{\infty} C(n-1, r)x^r \\ F_n(x) - C(n,0) &= xF_{n-1}(x) + F_{n-1}(x) - C(n-1, 0) \\ F_n(x) &= (1+x)F_{n-1}(x)\end{aligned}$$

It follows that

$$\begin{aligned}F_n(x) &= (1+x)^2F_{n-2}(x) \\ &= (1+x)^3F_{n-3}(x) \\ &= \cdots \\ &= (1+x)^nF_0(x) \\ &= (1+x)^nC(0,0) \\ &= (1+x)^n\end{aligned}$$

which was derived in Chap. 2. ■

Example 3-14 Find the number of r-combinations of n distinct objects with unlimited repetitions. Denote this number by $f(n,r)$. Let one of the n objects be labeled as a special one. There are $f(n, r-1)$ r-combinations in which this special object is selected at least once. There are $f(n-1, r)$ r-combinations in which this special object is not selected. Therefore,

$$f(n,r) = f(n, r-1) + f(n-1, r) \qquad n \geq 1, r \geq 1$$

Let

$$F_n(x) = f(n,0) + f(n,1)x + f(n,2)x^2 + \cdots + f(n,r)x^r + \cdots$$

for every $n \geq 0$. Thus,

$$\sum_{r=1}^{\infty} f(n,r)x^r = \sum_{r=1}^{\infty} f(n, r-1)x^r + \sum_{r=1}^{\infty} f(n-1, r)x^r$$

$$F_n(x) - f(n,0) = xF_n(x) + F_{n-1}(x) - f(n-1, 0)$$

With the boundary conditions $f(n,0) = 1$ for $n \geq 0$ and $f(0,r) = 0$ for $r > 0$,

$$F_n(x) = (1-x)^{-1}F_{n-1}(x) = (1-x)^{-n}F_0(x) = (1-x)^{-n}$$

which again is a result derived in Chap. 2. ■

Example 3-15 Find the number of n-digit binary sequences that have exactly r pairs of adjacent 1's and no adjacent 0's. Notice that every two successive 1's are counted as a pair. For instance, there are two pairs of adjacent 1's in the sequence 111. Let $a_{n,r}$ denote the number of such sequences. Also let $b_{n,r}$ denote the number of such sequences that have a 1 as the nth digit, and let $c_{n,r}$ denote the number of such sequences that have a 0 as the nth digit. Clearly,

$$a_{n,r} = b_{n,r} + c_{n,r} \tag{3-11}$$

Since an n-digit sequence that has r pairs of 1's, no adjacent 0's, and a 1 as the nth digit can be formed by appending a 1 either to an $(n-1)$-digit sequence that has $r-1$ pairs of 1's, no adjacent 0's, and a 1 as the $(n-1)$st digit or to an $(n-1)$-digit sequence that has r pairs of 1's, no adjacent 0's, and a 0 as the $(n-1)$st digit, we have the relation

$$b_{n,r} = b_{n-1,r-1} + c_{n-1,r} \tag{3-12}$$

Similarly, an n-digit sequence that has r pairs of 1's, no adjacent 0's, and a 0 as the nth digit can be formed by appending a 0 to an $(n-1)$-digit sequence that has r pairs of 1's, no adjacent 0's, and a 1 as the $(n-1)$st digit. Hence,

$$c_{n,r} = b_{n-1,r} \tag{3-13}$$

Combining Eqs. (3-12) and (3-13), we obtain

$$b_{n,r} = b_{n-1,r-1} + b_{n-2,r} \tag{3-14}$$

Since the value of $b_{i,j}$ has physical significance only for $i \geq 1$ and $j \geq 0$, we see that Eq. (3-14) is valid for $n \geq 3$ and $r \geq 1$. As to the boundary conditions, we have

$$b_{n,0} = 1 \qquad \text{for } n \geq 1$$

because there is exactly one n-digit sequence that contains neither adjacent 0's nor adjacent 1's and has a 1 as the nth digit, namely, the sequence that consists of alternating 0's and 1's. We also have

$$b_{i,j} = 0 \qquad \text{for } i \leq j$$

The value of $b_{0,0}$ is not constrained by the difference equation. Let it be chosen as 1. Let

$$B_n(x) = b_{n,0} + b_{n,1}x + b_{n,2}x^2 + \cdots + b_{n,r}x^r + \cdots$$

Multiplying both sides of Eq. (3-14) by x^r and summing from $r = 1$ to $r = \infty$, we obtain

$$\sum_{r=1}^{\infty} b_{n,r}x^r = \sum_{r=1}^{\infty} b_{n-1,r-1}x^r + \sum_{r=1}^{\infty} b_{n-2,r}x^r$$

which yields

$$B_n(x) - b_{n,0} = xB_{n-1}(x) + B_{n-2}(x) - b_{n-2,0} \qquad n \geq 3$$

that is,

$$B_n(x) = xB_{n-1}(x) + B_{n-2}(x) \qquad n \geq 3 \tag{3-15}$$

To solve Eq. (3-15), we observe first the following boundary conditions:

$$B_0(x) = b_{0,0} = 1$$
$$B_1(x) = b_{1,0} + b_{1,1}x = 1$$
$$B_2(x) = b_{2,0} + b_{2,1}x + b_{2,2}x = 1 + x$$

Equation (3-15) yields

$$\sum_{n=3}^{\infty} B_n(x)y^n = \sum_{n=3}^{\infty} xB_{n-1}(x)y^n + \sum_{n=3}^{\infty} B_{n-2}(x)y^n$$

$$\mathcal{B}(y,x) - B_2(x)y^2 - B_1(x)y - B_0(x) = xy[\mathcal{B}(y,x) - B_1(x)y - B_0(x)] + y^2[\mathcal{B}(y,x) - B_0(x)]$$

that is,

$$\mathcal{B}(y,x) - (1 + x)y^2 - y - 1 = xy[\mathcal{B}(y,x) - y - 1] + y^2[\mathcal{B}(y,x) - 1]$$

$$\begin{aligned}\mathcal{B}(y,x) &= \frac{1 + (1 - x)y}{1 - xy - y^2} \\ &= 1 + y + (1 + x)y^2 + (1 + x + x^2)y^3 \\ &\qquad + (1 + 2x + x^2 + x^3)y^4 \\ &\qquad + (1 + 2x + 3x^2 + x^3 + x^4)y^5 + \cdots\end{aligned}$$

According to Eq. (3-13),

$$\mathcal{C}(y,x) = y\mathcal{B}(y,x)$$

by choosing $c_{0,0}$ to be 0. Therefore,

$$\begin{aligned}\mathcal{A}(y,x) &= \mathcal{B}(y,x) + \mathcal{C}(y,x) \\ &= (1 + y)\mathcal{B}(x,y) \\ &= 1 + 2y + (2 + x)y^2 + (2 + 2x + x^2)y^3 \\ &\qquad + (2 + 3x + 2x^2 + x^3)y^4 \\ &\qquad + (2 + 4x + 4x^2 + 2x^3 + x^4)y^5 + \cdots\end{aligned}$$

or

$$\begin{aligned}A_0(x) &= 1 \\ A_1(x) &= 2 \\ A_2(x) &= 2 + x \\ A_3(x) &= 2 + 2x + x^2 \\ A_4(x) &= 2 + 3x + 2x^2 + x^3 \\ A_5(x) &= 2 + 4x + 4x^2 + 2x^3 + x^4\end{aligned}$$

. . . . ■

Example 3-16 The number of r-permutations of n distinct objects, $P(n,r)$, satisfies the recurrence relation

$$P(n,r) = P(n - 1, r) + rP(n - 1, r - 1) \qquad n \geq 1, r \geq 1 \quad (3\text{-}16)$$

$P(n - 1, r)$ is the number of r-permutations in which a special object does not appear and $rP(n - 1, r - 1)$ is the number of r-permutations in which the special object appears. Let

$$F_n(x) = P(n,0) + P(n,1)x + P(n,2)x^2 + \cdots + P(n,r)x^r + \cdots$$

Then

$$\sum_{r=1}^{\infty} P(n,r)x^r = \sum_{r=1}^{\infty} P(n-1, r)x^r + \sum_{r=1}^{\infty} rP(n-1, r-1)x^r$$

$$F_n(x) - P(n,0) = F_{n-1}(x) - P(n-1, 0) + x\frac{d}{dx}[xF_{n-1}(x)]$$

that is,

$$F_n(x) = (1+x)F_{n-1}(x) + x^2\frac{d}{dx}F_{n-1}(x)$$

With the boundary condition

$$F_0(x) = 1$$

this recurrence relation enables us to carry out a step-by-step computation to yield

$$F_1(x) = 1 + x$$

$$F_2(x) = (1+x)(1+x) + x^2\frac{d}{dx}(1+x) = 1 + 2x + 2x^2$$

$$F_3(x) = (1+x)(1+2x+2x^2) + x^2\frac{d}{dx}(1+2x+2x^2)$$
$$= 1 + 3x + 6x^2 + 6x^3$$

. . . .

That there is no closed-form expression for $F_n(x)$ was pointed out in Sec. 2-3. (It was exactly for this reason that we introduced the notion of exponential generating function.) Let us define

$$G_n(x) = P(n,0) + \frac{P(n,1)}{1!}x + \frac{P(n,2)}{2!}x^2 + \cdots + \frac{P(n,r)}{r!}x^r + \cdots$$

for every $n \geq 0$. Multiplying both sides of Eq. (3-16) by $(1/r!)x^r$ and summing both sides from $r = 1$ to $r = \infty$, we obtain

$$\sum_{r=1}^{\infty} \frac{P(n,r)}{r!}x^r = \sum_{r=1}^{\infty} \frac{P(n-1, r)}{r!}x^r + \sum_{r=1}^{\infty} \frac{rP(n-1, r-1)}{r!}x^r$$

that is,

$$G_n(x) - P(n,0) = G_{n-1}(x) - P(n-1, 0) + xG_{n-1}(x)$$

which can be simplified to

$$G_n(x) = (1+x)G_{n-1}(x)$$

It follows that

$$G_n(x) = (1+x)^n G_0(x) = (1+x)^n$$

the result obtained previously in Chap. 2. ■

3-6 SUMMARY AND REFERENCES

We have seen in this chapter the solution of combinatorial problems using the technique of recurrence relations. As was pointed out earlier, such an approach is particularly attractive when a high-speed digital computer is available for step-by-step computations. On several occasions, we have observed the application of the enumerative techniques developed in Chap. 2 and in this chapter to the solution of problems which were solved in Chap. 1 through some tricky and less obvious argument. It is hoped that such observations will help to develop the reader's perspective on various methods of solution. We also wish to point out that a sequence can be represented either by its generating function or by the corresponding recurrence relation. Instead of being two disparate techniques, what we studied in Chaps. 2 and 3 are simply different points of view of the same subject, the effective representation and manipulation of sequences of events.

The subject of finite difference equations is by no means limited to the study of enumerative theory. As a matter of fact, it is one of the most important subjects in numerical analysis. (See, for example, Ralston [5].) Corresponding to a differential equation, the solution of which is a continuous function, there is a difference equation, the solution of which is a discrete function. When a digital computer, which can handle only discrete data, is used to find the approximate solution of a differential equation, the differential equation is approximated by its corresponding difference equation. It is therefore not a coincidence that a linear difference equation with constant coefficients has a homogeneous solution and a particular solution just as does a linear differential equation with constant coefficients. Also, we observe that corresponding to homogeneous solutions of the form $Ae^{\alpha x}$ for linear differential equations with constant coefficients, the homogeneous solutions for linear difference equations with constant coefficients are of the form $A\alpha^n$.

For those readers who are familiar with the technique of using Laplace transformation to solve linear differential equations with constant coefficients, the analogy between the Laplace transform of a continuous function and the ordinary generating function of a discrete function is brought out more clearly in this chapter.

Chapters 1 and 2 of Riodan [6] and Chap. 3 of Ryser [7] give a brief account of the use of recurrence relations in solving combinatorial problems. There are many books on finite difference equations. For example, see Levy and Lessman [3] and Milne-Thomson [4]. Also, see Chap. 3 of Hildebrand [2]. In probability theory, recurrence relations are used in the study of random walk, recurrence events, and ruin problems. See Chaps. 9, 13, and 14 of Feller [1].

1. Feller, W.: "An Introduction to Probability Theory and Its Applications," vol. I, 2d ed., John Wiley & Sons, Inc., 1957.
2. Hildebrand, F. B.: "Methods of Applied Mathematics," Prentice-Hall, Inc., Englewood Cliffs, N.J., 1952.
3. Levy, H., and F. Lessman: "Finite Difference Equations," The Macmillan Company, New York, 1961.
4. Milne-Thomson, L. M.: "The Calculus of Finite Differences," The Macmillan Company, New York, 1933.
5. Ralston, A.: "A First Course in Numerical Analysis," McGraw-Hill Book Company, New York, 1965.
6. Riordan, J.: "An Introduction to Combinatorial Analysis," John Wiley & Sons, Inc., New York, 1958.
7. Ryser, H. J.: "Combinatorial Mathematics," published by the Mathematical Association of America, distributed by John Wiley & Sons, Inc., New York, 1963.

APPENDIX 3-1 UNIQUENESS OF THE SOLUTION TO A DIFFERENCE EQUATION

In Sec. 3-2, we show that the total solution $a_n = a_n^{(h)} + a_n^{(p)}$ satisfies the linear difference equation with constant coefficients

$$C_0 a_n + C_1 a_{n-1} + \cdots + C_r a_{n-r} = f(n) \tag{3A-1}$$

We show next that the boundary conditions (the values of $a_0, a_1, \ldots, a_{r-1}$) determine the total solution uniquely. Suppose that both $\bar{a}_n$ and $\bar{\bar{a}}_n$ are solutions that satisfy the difference equation and the boundary conditions. Because

$$C_0 \bar{a}_n + C_1 \bar{a}_{n-1} + \cdots + C_r \bar{a}_{n-r} = f(n)$$

and

$$C_0 \bar{\bar{a}}_n + C_1 \bar{\bar{a}}_{n-1} + \cdots + C_r \bar{\bar{a}}_{n-r} = f(n)$$

we have

$$C_0(\bar{a}_n - \bar{\bar{a}}_n) + C_1(\bar{a}_{n-1} - \bar{\bar{a}}_{n-1}) + \cdots + C_r(\bar{a}_{n-r} - \bar{\bar{a}}_{n-r}) = 0$$

Let $b_n = \bar{a}_n - \bar{\bar{a}}_n$. Clearly, b_n is a solution to the difference equation

$$C_0 b_n + C_1 b_{n-1} + \cdots + C_r b_{n-r} = 0 \tag{3A-2}$$

with the boundary conditions being $b_0 = b_1 = b_2 = \cdots = b_{r-1} = 0$. It follows that, in a step-by-step computation based on the difference equation (3A-2) and the boundary conditions, $b_n = 0$ for all n. Therefore, we conclude that $\bar{a}_n = \bar{\bar{a}}_n$ for all n, and the solution to the difference equation (3A-1) is unique.

We show next that the undetermined coefficients in the homogeneous solution can be determined uniquely by the values of $a_0, a_1, \ldots, a_{r-1}$. There are two cases to be examined.

Case 1 The characteristic roots are all distinct. In this case, the homogeneous solution is

$$A_1\alpha_1^n + A_2\alpha_2^n + \cdots + A_r\alpha_r^n$$

To determine the coefficients $A_1, A_2, \ldots, A_r$, we solve the set of simultaneous equations

$$\begin{aligned}
&A_1 + A_2 + \cdots + A_r = a_0 - a_0^{(p)}\\
&A_1\alpha_1 + A_2\alpha_2 + \cdots + A_r\alpha_r = a_1 - a_1^{(p)}\\
&A_1\alpha_1^2 + A_2\alpha_2^2 + \cdots + A_r\alpha_r^2 = a_2 - a_2^{(p)}\\
&\cdots\cdots\cdots\cdots\cdots\cdots\cdots\cdots\\
&A_1\alpha_1^{r-1} + A_2\alpha_2^{r-1} + \cdots + A_r\alpha_r^{r-1} = a_{r-1} - a_{r-1}^{(p)}
\end{aligned}$$

According to the theory of simultaneous linear equations, $A_1, A_2, \ldots, A_r$ can be solved uniquely if and only if the value of the determinant

$$\begin{vmatrix}
1 & 1 & \cdots & 1\\
\alpha_1 & \alpha_2 & \cdots & \alpha_r\\
\alpha_1^2 & \alpha_2^2 & \cdots & \alpha_r^2\\
\cdots & \cdots & \cdots & \cdots\\
\alpha_1^{r-1} & \alpha_2^{r-1} & \cdots & \alpha_r^{r-1}
\end{vmatrix}$$

is nonzero. This determinant is the famous *Vandermonde determinant*, the value of which is equal to

$$(\alpha_1 - \alpha_2)(\alpha_1 - \alpha_3) \cdots (\alpha_1 - \alpha_r)(\alpha_2 - \alpha_3)(\alpha_2 - \alpha_4) \cdots (\alpha_2 - \alpha_r) \cdots (\alpha_{r-1} - \alpha_r) = \prod_{\substack{i<j\\ 1\le i<r\\ 1<j\le r}} (\alpha_i - \alpha_j)$$

Since all the α's are distinct, the value of the determinant is nonzero. It follows that the coefficients $A_1, A_2, \ldots, A_r$ can be determined uniquely.

Case 2 The characteristic roots are not all distinct. In this case, the homogeneous solution is

$$\begin{aligned}
&(A_1n^{k_1-1} + A_2n^{k_1-2} + \cdots + A_{k_1-1}n + A_{k_1})\alpha_1^n\\
&+ (A_{k_1+1}n^{k_2-1} + A_{k_1+2}n^{k_2-2} + \cdots + A_{k_1+k_2-1}n + A_{k_1+k_2})\alpha_2^n\\
&\quad + \cdots + (A_{r-k_t+1}n^{k_t-1} + A_{r-k_t+2}n^{k_t-2} + \cdots + A_{r-1}n + A_r)\alpha_t^n
\end{aligned}$$

where $k_1, k_2, \ldots, k_t$ are the multiplicities of the characteristic roots $\alpha_1, \alpha_2, \ldots \alpha_t$, respectively. To determine the A's, we shall solve the set of equations

$$\begin{aligned}
&0 + 0 + \cdots + 0 + A_{k_1} + \cdots + 0 + 0 + \cdots + 0 + A_r \\
&\qquad = a_0 - a_0^{(p)} \\
&(A_1 + A_2 + \cdots + A_{k_1-1} + A_{k_1})\alpha_1 + \cdots \\
&\qquad + (A_{r-k_t+1} + A_{r-k_t+2} + \cdots + A_{r-1} + A_r)\alpha_t = a_1 - a_1^{(p)} \\
&(A_1 2^{k_1-1} + A_2 2^{k_1-2} + \cdots A_{k_1-1}2 + A_{k_1})\alpha_1^2 + \cdots \\
&\qquad + (A_{r-k_t+1}2^{k_t-1} + A_{r-k_t+2}2^{k_t-2} + \cdots + A_{r-1}2 + A_r)\alpha_t^2 \\
&\qquad = a_2 - a_2^{(p)} \\
&\cdots\cdots\cdots\cdots\cdots\cdots\cdots\cdots\cdots\cdots \\
&[A_1(r-1)^{k_1-1} + A_2(r-1)^{k_1-2} + \cdots + A_{k_1-1}(r-1) + A_{k_1}]\alpha_1^{r-1} \\
&\qquad + \cdots + [A_{r-k_t+1}(r-1)^{k_t-1} + A_{r-k_t+2}(r-1)^{k_t-2} + \cdots \\
&\qquad + A_{r-1}(r-1) + A_r]\alpha_t^{r-1} = a_{r-1} - a_{r-1}^{(p)}
\end{aligned}$$

The value of the determinant

$$\begin{vmatrix}
0 & 0 & \cdots & 1 & \cdots & 0 & 0 & \cdots & 1 \\
\alpha_1 & \alpha_1 & \cdots & \alpha_1 & \cdots & \alpha_t & \alpha_t & \cdots & \alpha_t \\
2^{k_1-1}\alpha_1^2 & 2^{k_1-2}\alpha_1^2 & \cdots & \alpha_1^2 & \cdots & 2^{k_t-1}\alpha_t^2 & 2^{k_t-2}\alpha_t^2 & \cdots & \alpha_t^2 \\
\cdots & \cdots & \cdots & \cdots & \cdots & \cdots & \cdots & \cdots & \cdots \\
(r-1)^{k_1-1}\alpha_1^{r-1} & (r-1)^{k_1-2}\alpha_1^{r-1} & \cdots & \alpha_1^{r-1} & \cdots & (r-1)^{k_t-1}\alpha_t^{r-1} & (r-1)^{k_t-2}\alpha_t^{r-1} & \cdots & \alpha_t^{r-1}
\end{vmatrix}$$

is equal to

$$\left[\prod_{1 \le i \le t} (\alpha_i)^{\binom{k_i}{2}}\right]\left[\prod_{\substack{i<j \\ 1 \le i < t \\ 1 < j \le t}} (\alpha_i - \alpha_j)^{k_i k_j}\right]$$

Since the α's are nonzero and assume distinct values, the coefficients A_1, $A_2, \ldots, A_r$ can be solved uniquely.

It should be pointed out that the values of r *nonconsecutive* a's might not determine the coefficients A's uniquely. For instance, in Example 3-2, that $a_2 = 0$ and $a_5 = 0$ is not sufficient to determine the coefficients in the homogeneous solution.

PROBLEMS

3-1. (*a*) Find the ordinary generating functions of the sequences

1. $0, 0, 2 \times 1, 3 \times 2, \ldots, k(k-1)$
2. $0, 1^2, 2^2, 3^2, \ldots, k^2$
3. $0, 1 \times 3, 2 \times 4, 3 \times 5, \ldots, k(k+2)$
4. $0, 1 \times 2 \times 3, 2 \times 3 \times 4, \ldots, k(k+1)(k+2)$

(*b*) Evaluate the sums

1. $s_m = \sum_{k=0}^{m} k(k-1)$
2. $s_m = \sum_{k=0}^{m} k^2$
3. $s_m = \sum_{k=0}^{m} k(k+2)$
4. $s_m = \sum_{k=0}^{m} k(k+1)(k+2)$

3-2. A circle and n straight lines are drawn on a plane. Each of these lines intersects all the other lines inside the circle. If no three or more lines meet at one point, into how many regions do these lines divide the circle?

3-3. A circle is drawn on a plane, and n straight lines are then drawn one by one. The first line passes through the circle. The rth line intersects only one of the other $r - 1$ lines inside the circle if r is odd and greater than 1. The rth line intersects all the other $r - 1$ lines inside the circle if r is even. If no three or more lines meet at one point inside the circle, into how many regions will n lines divide the circle, when n is even? when n is odd?

3-4. Let a_n be the number of incongruent triangles with integral sides and perimeter n.

(*a*) Show that

$$a_n = a_{n-3} \qquad \text{if } n \text{ is even}$$
$$a_n = a_{n-3} + \frac{n + (-1)^{(n+1)/2}}{4} \qquad \text{if } n \text{ is odd}$$

(*b*) Find the ordinary generating function of the sequence $(a_0, a_1, a_2, \ldots)$.

3-5. (*a*) Show that the number of incongruent triangles whose sides are integral numbers with the longest one equal to l is

$$\tfrac{1}{4}(l+1)^2 \qquad \text{if } l \text{ is odd}$$
$$\frac{l}{4}(l+2) \qquad \text{if } l \text{ is even}$$

(*b*) Let f_n be the number of triangles every side of which does not exceed $2n$, and let g_n be the number of triangles every side of which does not exceed $2n + 1$. Find the expressions for f_n and g_n.

3-6. In the linear resistance network shown in Fig. 3P-1, find the voltage v_0.
Hint: Assume first that $v_0 = 1$.

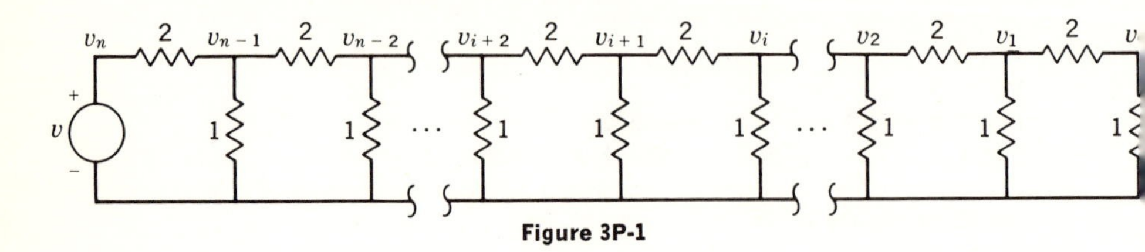

Figure 3P-1

3-7. (*a*) How many n-digit binary sequences that have no adjacent 0's are there?
(*b*) How many n-digit ternary sequences that have no adjacent 0's are there?

3-8. The Fibonacci numbers are defined by the recurrence relation

$$f_n = f_{n-1} + f_{n-2} \qquad n \geq 2$$

together with the boundary conditions

$$f_0 = f_1 = 1$$

(*a*) Find the ordinary generating function $F(x) = \sum_{n=0}^{\infty} f_n x^n$ of the sequence $(f_0, f_1, f_2, \ldots)$.

(*b*) Let

$$P(x) = \sum_{n=0}^{\infty} f_{2n} x^n$$

$$Q(x) = \sum_{n=1}^{\infty} f_{2n-1} x^n$$

Find $P(x)$ and $Q(x)$.

3-9. Define

$$a_n = \sum_{k=0}^{n} \binom{n+k}{2k} \qquad n \geq 1$$

$$b_n = \sum_{k=0}^{n-1} \binom{n+k}{2k+1} \qquad n \geq 1$$

and $a_0 = 1$ and $b_0 = 0$.

(*a*) Show that a_n and b_n satisfy the recurrence relations

$$a_{n+1} = a_n + b_{n+1} \qquad n \geq 0$$

$$b_{n+1} = a_n + b_n \qquad n \geq 0$$

(*b*) Find the ordinary generating functions of the sequences $(a_0, a_1, a_2, \ldots)$ and $(b_0, b_1, b_2, \ldots)$.

(*c*) Express a_n and b_n in terms of the Fibonacci numbers.

3-10. Let $f(n,k)$ denote the number of ways of selecting k numbers from the n numbers 1, 2, 3, . . . , n so that no consecutive numbers will be selected.

(a) Find a recurrence relation for $f(n,k)$.

(b) With the boundary conditions $f(n,1) = n$ and $f(n,n) = 0$, show that $f(n,k) = \binom{n-k+1}{k}$ by induction.

(c) Let $g(n,k)$ denote the number of ways of selecting k numbers from the n numbers 1, 2, 3, . . . , n so that no consecutive numbers will be selected. Here, n and 1 are also taken as consecutive numbers. Express $g(n,k)$ in terms of the $f(n,k)$'s and use the result in part (b) to find $g(n,k)$.

3-11. Let f_n denote the number of nonoverlapping regions into which the interior of a convex n-gon is divided by its diagonals. Suppose that no three diagonals meet at one point.

(a) Show that

$$f_n - f_{n-1} = \frac{(n-1)(n-2)(n-3)}{6} + n - 2 \qquad n \geq 3$$

(b) Let $F(x) = \sum_{n=0}^{\infty} f_n x^n$, and let $f_0 = f_1 = f_2 = 0$. Find $F(x)$ and thus an expression for f_n.

3-12. The Bernoulli numbers $b_0, b_1, b_2, \ldots$ are defined by the recurrence relation

$$b_n = \sum_{k=0}^{n} \binom{n}{k} b_{n-k} \qquad n \geq 1$$

together with the boundary condition

$$b_0 = 1$$

(a) Compute b_0, b_1, b_2, b_3, b_4, and b_5.

(b) Show that the exponential generating function of the sequence $(b_0, b_1, b_2, \ldots)$ is $x/(e^x - 1)$.

3-13. In how many ways can a convex n-gon be divided into triangles by nonintersecting diagonals?

3-14. When a coin is tossed n times all possible outcomes can be represented by sequences of heads (H) and tails (T) of length n. A sequence of H's and T's contains as many patterns of $HHTHH$ as there are *nonoverlapping* sequences of $HHTHH$. The pattern $HHTHH$ is said to occur at the nth toss, if the nth toss adds a new $HHTHH$ pattern to the sequence. A sequence of length n is acceptable if the $HHTHH$ pattern occurs at the nth toss. Let u_n be the number of acceptable sequences of length n.

(a) Find a recurrence relation for u_n.

(b) Let $u_0 = 1$, and let $U(x) = \sum_{n=0}^{\infty} u_n x^n$. Find $U(x)$.

(c) Let f_n be the number of sequences in which the pattern $HHTHH$ occurs for the first time at the nth toss. If $f_0 = 0$, find the ordinary generating function $F(x)$ of the sequence $(f_0, f_1, f_2, \ldots)$.

3-15. Find the number of n-digit binary sequences in which the pattern 111 occurs exactly twice with the second occurrence at the suffix. For example, the first and the fourth sequences in the following satisfy the condition, whereas the second and the third sequences do not.

0011111010111
00 111 111 01 111
0011011011111
0110110111 111

3-16. Two players, A and B, gamble by tossing a coin which has probabilities p and q to turn up head and tail, respectively. $(p + q = 1.)$ Initially, A has d dollars and B has $t - d$ dollars. At each toss, A wins a dollar if the head shows and loses a dollar if the tail shows. The game lasts as long as neither of the players is broke. Let $f(d,n)$ be the probability that A goes broke as a result of the nth toss.

(*a*) Find a recurrence relation for $f(d,n)$. Also, determine the boundary conditions $f(0,n)$, $f(t,n)$, and $f(d,0)$.

(*b*) Let $F_d(x) = \sum_{n=0}^{\infty} f(d,n)x^n$. Find $F_0(x)$, $F_t(x)$, and a recurrence relation for $F_d(x)$.

(*c*) Solve the recurrence relation found in part (*b*) for $F_d(x)$.

3-17. (*a*) Let u_n be the number of binary sequences of length n in which a run of r 1's occurs at the nth digit. Find a recurrence relation for u_n.

(*b*) Find the ordinary generating function $U(x)$ of the sequence $(u_0,u_1,u_2, \ldots)$.

(*c*) Let v_n be the number of binary sequences of length n in which a run of p 0's occurs at the nth digit. Find the ordinary generating function $V(x)$ of the sequence $(v_0,v_1,v_2, \ldots)$.

(*d*) Let w_n be the number of binary sequences of length n in which either a run of r 1's or a run of p 0's occurs at the nth digit. Find the ordinary generating function $W(x)$ of the sequence $(w_0,w_1,w_2, \ldots)$.

(*e*) Let t_n be the number of binary sequences of length n in which a run of p 0's occurs before a run of r 1's. Find the ordinary generating function $T(x)$ of the sequence $(t_0,t_1,t_2, \ldots)$.

3-18. A coin is tossed $2n$ times. How many of the 2^{2n} outcomes have the number of heads and tails equalized for the first time after $2n$ tosses?

3-19. (*a*) Find the number of binary sequences of length n that have the pattern 0101 occurring at the nth digit and have an even number of 0's. Also find the number of those that have an odd number of 0's.

(*b*) Find the number of binary sequences of length n that have the first occurrence of the pattern 0101 at the nth digit and have an even number of 0's. Also find the number of those that have an odd number of 0's.

3-20. A sequence of binary digits is fed to a counter at the rate of 1 digit/sec. The counter is designed to register 1's in the input sequence. However, it is so slow that it is locked for exactly p sec following each registration, during which time input digits are ignored.

(*a*) How many input sequences of length n would cause the counter to register exactly r 1's, finishing in the unlocked state?

(*b*) Let f_n be the number of binary sequences of length n at the end of which the counter is unlocked. Find a recurrence relation which f_n satisfies, and find the ordinary generating function of the sequence $(f_0,f_1,f_2, \ldots)$.

3-21. Find the number of binary sequences of length n that have exactly one pair of consecutive 0's.

3-22. (*a*) Recall that there are $S(r,n)$ ways to distribute r distinct objects into n nondistinct cells with no cell left empty, where $S(r,n)$ is the Stirling number of the second kind. Using combinatorial arguments, show that $S(r,n)$ satisfies the following recurrence relation:

$$S(r+1,\, n) = S(r,\, n-1) + nS(r,n) \qquad n \geq 1$$

(*b*) Let $G_n(x)$ be the ordinary generating function of the sequence $(S(0,n), S(1,n), S(2,n), \ldots)$; that is, $G_n(x) = \sum_{r=0}^{\infty} S(r,n)x^r$. Find a recurrence relation satisfied by $G_n(x)$. The boundary conditions are

$$S(r,n) = \begin{cases} 1 & r = 0,\, n = 0 \\ 0 & r = 0,\, n > 0 \quad \text{or} \quad r > 0,\, n = 0 \end{cases}$$

(*c*) From the recurrence relation in part (*b*) find $G_n(x)$. [Do not solve the recurrence relation; try to obtain the general form for $G_n(x)$ by observation.]

(*d*) Define

$$E_n(x) = \sum_{r=0}^{\infty} S(r,n)\frac{x^r}{r!}$$

which is the exponential generating function of the sequence $(S(0,n), S(1,n), S(2,n), \ldots)$. Show that

$$\frac{d}{dx}E_n(x) - nE_n(x) = E_{n-1}(x) \qquad n \geq 1$$

3-23. Let $s(r,n)$ denote the Stirling number of the first kind which is defined by the following equation where x is a formal variable:

$$\sum_{n=0}^{r} s(r,n)x^n = x(x-1)\cdots(x-r+1)$$

(*a*) Show that $s(r,n)$ satisfies the recurrence relation

$$s(r+1,\, n) = s(r,\, n-1) - r \times s(r,n)$$

(*b*) If $E_n(x) = \sum_{r=0}^{\infty} s(r,n)\frac{x^r}{r!}$ is the exponential generating function of the sequence $(s(0,n), s(1,n), s(2,n), \ldots)$, show that $E_n(x)$ satisfies the differential equation

$$(1+x)\frac{d}{dx}E_n(x) = E_{n-1}(x)$$

(*c*) Find $E_n(x)$. The boundary conditions are $s(0,0) = 1$ and $s(r,0) = 0$ for $r \neq 0$.

Chapter 4
The Principle of Inclusion and Exclusion

4-1 INTRODUCTION

Let us motivate the subject of this chapter with a simple illustrative example. In a group of ten girls, six have blond hair, five have blue eyes, and three have blond hair and blue eyes. How many girls are there in the group who have neither blond hair nor blue eyes? Clearly the answer is

$$10 - 6 - 5 + 3 = 2$$

Since the three blondes with blue eyes are included in the count of the six blondes and are again included in the count of the five with blue eyes, they are subtracted twice in the expression $10 - 6 - 5$. Therefore, 3 should be added to the expression $10 - 6 - 5$ to give the correct count of girls who have neither blond hair nor blue eyes.

The graphical representation in Fig. 4-1 shows very clearly the same argument. The area inside the large circle represents the total number of girls. The areas inside the two small circles represent, respectively, the number of girls who have blond hair and the number of girls who have

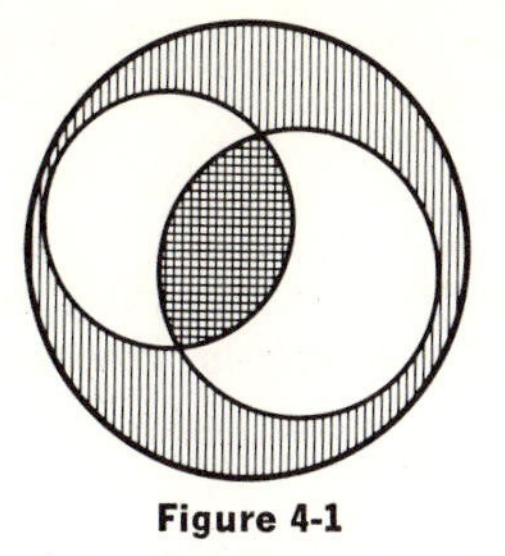

Figure 4-1

blue eyes. The crosshatched area represents the number of girls that have both blond hair and blue eyes. This area is subtracted twice when the areas of the two small circles are subtracted from the area of the large circle. To find the area marked with vertical lines which represents the number of girls who neither are blondes nor have blue eyes, we should, therefore, compensate the oversubtraction by adding back the crosshatched area.

The extension of the logical reasoning in this example leads to a very important counting theorem that is studied in this chapter. To count the number of a certain class of objects, we exclude those that should not be included in the count and, in turn, compensate the count by including those that have been excluded incorrectly. The counting theorem is called the *principle of inclusion and exclusion.*

4-2 THE PRINCIPLE OF INCLUSION AND EXCLUSION

Consider a set of N objects. Let $a_1, a_2, \ldots, a_r$ be a set of properties that these objects may have. In general, these properties are not mutually exclusive; that is, an object can have one or more of these properties. (The case in which these properties are mutually exclusive proves to be an uninteresting special case, as will be seen.) Let $N(a_1)$ denote the number of objects that have the property a_1, let $N(a_2)$ denote the number of objects that have the property $a_2, \ldots,$ and let $N(a_r)$ denote the number of objects that have the property a_r. Notice that an object having the property a_i is included in the count $N(a_i)$ regardless of the other properties it may have. Thus, if an object has both the properties a_i and a_j, it will contribute a count in $N(a_i)$ as well as a count in $N(a_j)$.

Let $N(a_1')$ denote the number of objects that do not have the property a_1, let $N(a_2')$ denote the number of objects that do not have the property $a_2, \ldots,$ and let $N(a_r')$ denote the number of objects that do not have the property a_r. Let $N(a_ia_j)$ denote the number of objects that have *both* the properties a_i and a_j, let $N(a_i'a_j')$ denote the number of objects that have neither the property a_i nor the property a_j, and let $N(a_i'a_j)$

denote the number of objects that have the property a_j but not the property a_i. Logically, we see that

$$N(a_i') = N - N(a_i) \tag{4-1}$$

because each of the N objects either has the property a_i [accounted for in $N(a_i)$] or does not have the property a_i [accounted for in $N(a_i')$]. Also,

$$N(a_i'a_j) = N(a_j) - N(a_ia_j)$$

because for each of the $N(a_j)$ objects that have the property a_j, it either has the property a_i [accounted for in $N(a_ia_j)$] or does not have the property a_i [accounted for in $N(a_i'a_j)$]. Using a similar argument, we have

$$N(a_i'a_j') = N - N(a_ia_j') - N(a_i'a_j) - N(a_ia_j)$$

which can be rewritten as

$$\begin{aligned} N(a_i'a_j') &= N - [N(a_ia_j') + N(a_ia_j)] - [N(a_i'a_j) + N(a_ia_j)] + N(a_ia_j) \\ &= N - N(a_i) - N(a_j) + N(a_ia_j) \end{aligned} \tag{4-2}$$

We now prove the following extension of Eqs. (4-1) and (4-2):

$$\begin{aligned} N(a_1'a_2' \cdots a_r') \\ &= N - N(a_1) - N(a_2) - \cdots - N(a_r) \\ &\quad + N(a_1a_2) + N(a_1a_3) + \cdots + N(a_{r-1}a_r) \\ &\quad - N(a_1a_2a_3) - N(a_1a_2a_4) - \cdots - N(a_{r-2}a_{r-1}a_r) \\ &\quad + \cdots\cdots\cdots\cdots\cdots\cdots \\ &\quad + (-1)^r N(a_1a_2 \cdots a_r) \\ &= N - \sum_i N(a_i) + \sum_{i,j;i\neq j} N(a_ia_j) - \sum_{i,j,k;i\neq j\neq k} N(a_ia_ja_k) \\ &\quad + \cdots + (-1)^r N(a_1a_2 \cdots a_r) \end{aligned} \tag{4-3}$$

This identity, known as the *principle of inclusion and exclusion,* will be proved by induction on the total number of properties the objects may have. As the basis of induction, we have already shown that

$$N(a_1') = N - N(a_1)$$

As the induction hypothesis, we assume that the identity is true for objects having up to $r - 1$ properties; that is,

$$\begin{aligned} N(a_1'a_2' \cdots a_{r-1}') &= N - N(a_1) - N(a_2) - \cdots - N(a_{r-1}) \\ &\quad + N(a_1a_2) + N(a_1a_3) + \cdots + N(a_{r-2}a_{r-1}) \\ &\quad - \cdots\cdots\cdots\cdots\cdots\cdots \\ &\quad + (-1)^{r-1} N(a_1a_2 \cdots a_{r-1}) \end{aligned} \tag{4-4}$$

Now, for a set of N objects having up to r properties, $a_1, a_2, \ldots, a_r$, we consider the set of $N(a_r)$ objects that have the property a_r. Since this set of objects may have any of the $r-1$ properties $a_1, a_2, \ldots, a_{r-1}$, according to the induction hypothesis,

$$\begin{aligned} N(a_1'a_2' \cdots a_{r-1}'a_r) &= N(a_r) - N(a_1a_r) - N(a_2a_r) - \cdots - N(a_{r-1}a_r) \\ &\quad + N(a_1a_2a_r) + N(a_1a_3a_r) + \cdots + N(a_{r-2}a_{r-1}a_r) \\ &\quad - \cdots\cdots\cdots\cdots\cdots\cdots\cdots\cdots \\ &\quad + (-1)^{r-1}N(a_1a_2 \cdots a_{r-1}a_r) \end{aligned} \tag{4-5}$$

Subtracting Eq. (4-5) from Eq. (4-4), we obtain

$$\begin{aligned} N(a_1'a_2' \cdots a_{r-1}') - N(a_1'a_2' \cdots a_{r-1}'a_r) &= N - N(a_1) - N(a_2) - \cdots - N(a_{r-1}) - N(a_r) \\ &\quad + N(a_1a_2) + N(a_1a_3) + \cdots + N(a_1a_r) + \cdots \\ &\quad + N(a_{r-1}a_r) \\ &\quad - \cdots\cdots \\ &\quad + (-1)^r N(a_1a_2 \cdots a_{r-1}a_r) \end{aligned}$$

Since

$$N(a_1'a_2' \cdots a_{r-1}') - N(a_1'a_2' \cdots a_{r-1}'a_r) = N(a_1'a_2' \cdots a_{r-1}'a_r')$$

we have proved Eq. (4-3).

Example 4-1 is an analysis example illustrating the application of the principle of inclusion and exclusion.

Example 4-1 Twelve balls are painted in the following way:

> Two are unpainted.
> Two are painted red, one is painted blue, and one is painted white.
> Two are painted red and blue, and one is painted red and white.
> Three are painted red, blue, and white.

Let a_1, a_2, and a_3 denote the properties that a ball is painted red, blue, and white, respectively; then

$$\begin{array}{lll} N(a_1) = 8 & N(a_2) = 6 & N(a_3) = 5 \\ N(a_1a_2) = 5 & N(a_1a_3) = 4 & N(a_2a_3) = 3 \\ N(a_1a_2a_3) = 3 & & \end{array}$$

It follows that

$$N(a_1'a_2'a_3') = 12 - 8 - 6 - 5 + 5 + 4 + 3 - 3 = 2 \quad \blacksquare$$

Example 4-2 Find the number of integers between 1 and 250 that are not divisible by any of the integers 2, 3, 5, and 7. Let a_1, a_2, a_3, and a_4 denote the properties that a number is divisible by 2, divisible by 3, divisible by 5, and divisible by 7, respectively. Among the integers 1 through 250 there are 125 ($= {}^{250}\!/_2$) integers that are divisible by 2, because every other integer is a multiple of 2. Similarly, there are 83 (= the integral part of ${}^{250}\!/_3$) integers that are multiples of 3 and 50 ($= {}^{250}\!/_5$) integers that are multiples of 5 and so on. Letting $[x]$ denote the integral part of the number x,

$$N(a_1) = \left[\frac{250}{2}\right] = 125 \qquad N(a_2) = \left[\frac{250}{3}\right] = 83$$

$$N(a_3) = \left[\frac{250}{5}\right] = 50 \qquad N(a_4) = \left[\frac{250}{7}\right] = 35$$

$$N(a_1a_2) = \left[\frac{250}{2 \times 3}\right] = 41 \qquad N(a_1a_3) = \left[\frac{250}{2 \times 5}\right] = 25$$

$$N(a_1a_4) = \left[\frac{250}{2 \times 7}\right] = 17 \qquad N(a_2a_3) = \left[\frac{250}{3 \times 5}\right] = 16$$

$$N(a_2a_4) = \left[\frac{250}{3 \times 7}\right] = 11 \qquad N(a_3a_4) = \left[\frac{250}{5 \times 7}\right] = 7$$

$$N(a_1a_2a_3) = \left[\frac{250}{2 \times 3 \times 5}\right] = 8 \qquad N(a_1a_2a_4) = \left[\frac{250}{2 \times 3 \times 7}\right] = 5$$

$$N(a_1a_3a_4) = \left[\frac{250}{2 \times 5 \times 7}\right] = 3 \qquad N(a_2a_3a_4) = \left[\frac{250}{3 \times 5 \times 7}\right] = 2$$

$$N(a_1a_2a_3a_4) = \left[\frac{250}{2 \times 3 \times 5 \times 7}\right] = 1$$

Therefore, the number of integers that are not divisible by any of the integers 2, 3, 5, and 7 is

$$N(a_1'a_2'a_3'a_4') = 250 - (125 + 83 + 50 + 35) + (41 + 25 + 17 + 16 + 11 + 7) - (8 + 5 + 3 + 2) + 1 = 57$$

Similarly, the number of integers that are not divisible by 2 nor by 7 but are divisible by 5 is

$$\begin{aligned} N(a_1'a_3a_4') &= N(a_3) - N(a_1a_3) - N(a_3a_4) + N(a_1a_3a_4) \\ &= 50 - 25 - 7 + 3 \\ &= 21 \quad \blacksquare \end{aligned}$$

Example 4-3 Find the number of r-digit quaternary sequences in which each of the three digits 1, 2, and 3 appears at least once. Let a_1, a_2, and a_3 be the properties that the digits 1, 2, and 3 *do not* appear in a sequence, respectively. Because

$$N(a_1) = N(a_2) = N(a_3) = 3^r$$
$$N(a_1a_2) = N(a_1a_3) = N(a_2a_3) = 2^r$$
$$N(a_1a_2a_3) = 1$$

we have

$$N(a_1'a_2'a_3') = 4^r - 3 \times 3^r + 3 \times 2^r - 1$$

This problem was solved in Example 2-12, and the reader is encouraged to compare the two methods of solution.

Notice that this problem is the same as that of distributing r distinct objects into four distinct cells with three of them never left empty. As a matter of fact, using the generating function technique, we derived a formula for the number of ways of distributing r distinct objects into n distinct cells with no cell left empty in Chap. 2. This formula can also be derived by the use of the principle of inclusion and exclusion as follows: Let a_1, a_2, . . . , a_n be the properties that the 1st, 2d, . . . , nth cell is left empty in the distributions of the r objects, respectively. Then,

$$\begin{aligned} N(a_1'a_2' \cdots a_n') &= n^r - \binom{n}{1}(n-1)^r + \binom{n}{2}(n-2)^r - \cdots \\ &\quad + (-1)^{n-1}\binom{n}{n-1}1^r + (-1)^n\binom{n}{n}0^r \\ &= \sum_{i=0}^{n}(-1)^i\binom{n}{i}(n-i)^r \end{aligned}$$

which is exactly the result obtained in Sec. 2-4. ■

Example 4-4 Consider a single ball that is painted with n colors. Let a_1, a_2, . . . , a_n denote the properties that a ball is painted with the 1st, 2d, . . . , nth color, respectively. Since

$$N(a_1) = N(a_2) = \cdots = N(a_n) = 1$$
$$N(a_1a_2) = N(a_1a_3) = \cdots = N(a_{n-1}a_n) = 1$$
$$\cdots\cdots\cdots\cdots\cdots\cdots\cdots\cdots$$
$$N(a_1a_2 \cdots a_n) = 1$$

we have

$$N(a_1'a_2' \cdots a_n') = 1 - \binom{n}{1} + \binom{n}{2} - \cdots + (-1)^n \binom{n}{n}$$

However,

$$N(a_1'a_2' \cdots a_n') = 0$$

because there is no unpainted ball. Therefore, we have the identity

$$1 - \binom{n}{1} + \binom{n}{2} - \cdots + (-1)^n \binom{n}{n} = 0$$

which was proved in Example 2-1. ■

4-3 THE GENERAL FORMULA

In a set of N objects with properties $a_1, a_2, \ldots, a_r$, the number of objects that do not have any of these properties, $N(a_1'a_2' \cdots a_r')$, is given by Eq. (4-3). In this section, we derive a more general formula for the number of objects that have exactly m of the r properties for $m = 0, 1, \ldots, r$. Let us introduce the notations

$$s_0 = N$$

$$s_1 = N(a_1) + N(a_2) + \cdots + N(a_r) = \sum_i N(a_i)$$

$$s_2 = N(a_1a_2) + N(a_1a_3) + \cdots + N(a_{r-1}a_r) = \sum_{i,j;i \neq j} N(a_ia_j)$$

$$s_3 = N(a_1a_2a_3) + N(a_1a_2a_4) + \cdots + N(a_{r-2}a_{r-1}a_r)$$
$$= \sum_{i,j,k;i \neq j \neq k} N(a_ia_ja_k)$$

$$\cdots\cdots\cdots\cdots$$

$$s_r = N(a_1a_2 \cdots a_r)$$

Also,

$$e_0 = N(a_1'a_2' \cdots a_r')$$

$$e_1 = N(a_1a_2'a_3' \cdots a_r') + N(a_1'a_2a_3' \cdots a_r') + \cdots + N(a_1'a_2'a_3' \cdots a_r)$$

$$e_2 = N(a_1a_2a_3' \cdots a_r') + N(a_1a_2'a_3 \cdots a_r') + \cdots + N(a_1'a_2'a_3' \cdots a_{r-1}a_r)$$

$$e_3 = N(a_1a_2a_3 \cdots a_r') + N(a_1a_2a_3'a_4 \cdots a_r') + \cdots + N(a_1'a_2'a_3' \cdots a_{r-2}a_{r-1}a_r)$$

$$\cdots\cdots\cdots\cdots$$

$$e_r = N(a_1a_2 \cdots a_r)$$

In other words, e_i is the number of objects that have exactly i properties. With this notation Eq. (4-3) can be rewritten as

$$e_0 = s_0 - s_1 + s_2 - \cdots + (-1)^r s_r$$

We shall now prove the formula

$$e_m = s_m - \binom{m+1}{1} s_{m+1} + \binom{m+2}{2} s_{m+2} - \cdots + (-1)^{r-m} \binom{r}{r-m} s_r \qquad (4\text{-}6)$$

for $m = 0, 1, 2, \ldots, r$. When $m = 0$, Eq. (4-6) reduces to Eq. (4-3).

An object having less than m properties should not be included in the count e_m. Indeed, it contributes no count to the expression on the right-hand side of Eq. (4-6).

On the other hand, an object having exactly m properties should be included in the count e_m. Indeed, it contributes a count of 1 to the expression on the right-hand side of Eq. (4-6), since it is counted exactly once in s_m and is not included in the counts $s_{m+1}, s_{m+2}, \ldots, s_r$.

An object having $m + j$ properties with $0 < j \leq r - m$ should not be included in the count e_m either. Since it contributes $\binom{m+j}{m}$ counts to s_m, $\binom{m+j}{m+1}$ counts to $s_{m+1}, \ldots,$ and $\binom{m+j}{m+j}$ counts to s_{m+j}, the total count it contributes to the expression on the right-hand side of Eq. (4-6) is

$$\binom{m+j}{m} - \binom{m+1}{1}\binom{m+j}{m+1} + \binom{m+2}{2}\binom{m+j}{m+2} - \cdots + (-1)^j \binom{m+j}{j}\binom{m+j}{m+j}$$

Notice that

$$\begin{aligned}\binom{m+k}{k}\binom{m+j}{m+k} &= \frac{(m+k)!}{m!k!}\,\frac{(m+j)!}{(m+k)!(j-k)!} \\ &= \frac{(m+j)!}{m!k!(j-k)!} \\ &= \frac{(m+j)!}{m!j!}\,\frac{j!}{k!(j-k)!} \\ &= \binom{m+j}{m}\binom{j}{k}\end{aligned}$$

Thus, the total count is

$$\binom{m+j}{m} - \binom{m+j}{m}\binom{j}{1} + \binom{m+j}{m}\binom{j}{2} - \cdots + (-1)^j\binom{m+j}{m}\binom{j}{j}$$

$$= \binom{m+j}{m}\left[\binom{j}{0} - \binom{j}{1} + \binom{j}{2} - \cdots + (-1)^j\binom{j}{j}\right]$$

$$= 0$$

Therefore, an object having more than m properties is not included in the count e_m, and Eq. (4-6) is proved.

Example 4-5 In Example 4-1, $s_1 = 19$, $s_2 = 12$, and $s_3 = 3$. Therefore,

$$e_1 = 19 - \binom{2}{1} \times 12 + \binom{3}{2} \times 3 = 19 - 24 + 9 = 4$$

$$e_2 = 12 - \binom{3}{1} \times 3 = 12 - 9 = 3$$

$$e_3 = 3 \quad \blacksquare$$

Let $E(x)$ be the ordinary generating function of the sequence $(e_0, e_1, e_2, \ldots, e_m, \ldots, e_r)$. According to Eq. (4-6),

$$\begin{aligned}
E(x) &= e_0 + e_1x + e_2x^2 + \cdots + e_mx^m + \cdots + e_rx^r \\
&= [s_0 - s_1 + s_2 - \cdots + (-1)^r s_r] \\
&\quad + \left[s_1 - \binom{2}{1}s_2 + \binom{3}{2}s_3 - \cdots + (-1)^{r-1}\binom{r}{r-1}s_r\right]x \\
&\quad + \left[s_2 - \binom{3}{1}s_3 + \binom{4}{2}s_4 - \cdots + (-1)^{r-2}\binom{r}{r-2}s_r\right]x^2 \\
&\quad + \cdots\cdots\cdots\cdots\cdots\cdots\cdots\cdots\cdots \\
&\quad + \left[s_m - \binom{m+1}{1}s_{m+1} + \binom{m+2}{2}s_{m+2} - \cdots + (-1)^{r-m}\binom{r}{r-m}s_r\right]x^m \\
&\quad + \cdots\cdots\cdots\cdots\cdots\cdots\cdots\cdots\cdots \\
&\quad + s_rx^r
\end{aligned}$$

$$
\begin{aligned}
&= s_0 \\
&\quad + s_1[x - 1] \\
&\quad + s_2\left[x^2 - \binom{2}{1}x + 1\right] \\
&\quad + s_3\left[x^3 - \binom{3}{1}x^2 + \binom{3}{2}x - 1\right] \\
&\quad + \cdots\cdots\cdots\cdots\cdots \\
&\quad + s_m\left[x^m - \binom{m}{1}x^{m-1} + \binom{m}{2}x^{m-2} + \cdots \right. \\
&\qquad\qquad \left. + (-1)^{m-1}\binom{m}{m-1}x + (-1)^m\right] \\
&\quad + \cdots\cdots\cdots\cdots\cdots\cdots\cdots\cdots \\
&\quad + s_r\left[x^r - \binom{r}{1}x^{r-1} + \binom{r}{2}x^{r-2} + \cdots + (-1)^r\right] \\
&= \sum_{j=0}^{r} s_j(x-1)^j \qquad (4\text{-}7)
\end{aligned}
$$

Setting $x = 1$, we obtain

$$E(1) = e_0 + e_1 + e_2 + \cdots + e_r = s_0$$

This is, of course, a known result: The sum of the numbers of objects that have no property, one property, . . . , r properties, is equal to N, which is equal to s_0, the total number of objects. Also, observe that

$$\tfrac{1}{2}[E(1) + E(-1)] = e_0 + e_2 + e_4 + \cdots = \tfrac{1}{2}\left[s_0 + \sum_{j=0}^{r}(-2)^j s_j\right]$$

gives the number of objects having an even number of properties, and

$$\tfrac{1}{2}[E(1) - E(-1)] = e_1 + e_3 + e_5 + \cdots = \tfrac{1}{2}\left[s_0 - \sum_{j=0}^{r}(-2)^j s_j\right]$$

gives the number of objects having an odd number of properties.

Example 4-6 Find the number of n-digit ternary sequences that have an even number of 0's. Let a_i be the property that the ith digit of a sequence is 0, $i = 1, 2, \ldots, n$. Let e_j and s_j be defined as above with $j = 0, 1, \ldots, n$. Then since $s_j = \binom{n}{j}3^{n-j}$ with $j = 0, 1,$

. . . , n, it follows that

$$e_0 + e_2 + e_4 + \cdots = \tfrac{1}{2}\left[3^n + \sum_{j=0}^{n} (-2)^j \binom{n}{j} 3^{n-j}\right]$$
$$= \tfrac{1}{2}[3^n + (3-2)^n] = \tfrac{1}{2}(3^n + 1) \quad \blacksquare$$

4-4 DERANGEMENTS

The rest of this chapter is devoted to the discussion of the applications of the principle of inclusion and exclusion to several interesting problems. Consider the permutations of the integers 1, 2, . . . , n. A permutation of these integers is said to be a *derangement* of the integers if no integer appears in its natural position; that is, 1 does not appear in the first position, 2 does not appear in the second position, . . . , and n does not appear in the nth position. In general, when each of a set of objects has a position that it is forbidden to occupy and no two objects have the same forbidden position, a derangement of these objects is a permutation of them such that no object is in its forbidden position. To find the number of derangements of n objects, the principle of inclusion and exclusion can be used. Let a_i be the property of a permutation in which the ith object is placed in its forbidden position with $i = 1, 2, \ldots, n$. It follows that

$$N(a_i) = (n-1)! \qquad i = 1, 2, \ldots, n$$
$$N(a_ia_j) = (n-2)! \qquad i, j = 1, 2, \ldots, n;\ i \neq j$$
$$N(a_ia_ja_k) = (n-3)! \qquad i, j, k = 1, 2, \ldots, n;\ i \neq j \neq k$$
$$\cdots\cdots\cdots\cdots\cdots\cdots\cdots\cdots$$
$$N(a_1a_2 \cdots a_n) = 1$$

and

$$s_1 = \binom{n}{1}(n-1)!$$
$$s_2 = \binom{n}{2}(n-2)!$$
$$s_3 = \binom{n}{3}(n-3)!$$
$$\cdots\cdots\cdots$$
$$s_n = \binom{n}{n}(n-n)!$$

Therefore, the number of derangements of n objects, which will be denoted by d_n, is

$$\begin{aligned} d_n &= N(a_1'a_2' \cdots a_n') \\ &= n! - \binom{n}{1}(n-1)! + \binom{n}{2}(n-2)! - \cdots \\ &\qquad\qquad\qquad\qquad + (-1)^n \binom{n}{n}(n-n)! \\ &= n!\left[1 - \frac{1}{1!} + \frac{1}{2!} - \cdots + (-1)^n \frac{1}{n!}\right] \end{aligned}$$

Observe that the expression in the square brackets is the truncated series of the Taylor expansion of e^{-1}. The value of this expression can therefore be approximated very accurately by the value of e^{-1}, even for a relatively small value of n. For example, when $n = 6$, the exact value of the expression is 0.36806, and the approximated value is 0.36788.

The derangement of integers can be rephrased in many interesting ways. For example, 10 gentlemen check their hats at the coatroom, and later on the hats are returned to them randomly. In how many ways can the hats be returned to them such that no gentleman will get his own hat back? This is exactly the problem of permuting 10 objects (the hats) so that none of them will be in its forbidden position (the owner). Therefore, the number of ways of returning the hats is

$$d_{10} = 1{,}334{,}961 \approx 10! \times e^{-1}$$

For those readers who are familiar with the notion of probability, it can be seen that the probability that none of the gentlemen will have his own hat back is

$$\frac{d_{10}}{10!} \approx e^{-1}$$

Note that this probability is essentially the same for 10 gentlemen as well as for 10,000 gentlemen. Also, when there are 10 gentlemen, the probability is slightly higher than that when there are 9 or 11 gentlemen because of the alternating signs in the expression for d_n.

The number of derangements of integers can also be obtained by the solution of a recurrence relation. Consider the derangements of the integers $1, 2, \ldots, n$ in which the first position is occupied by the integer k ($k \neq 1$). If the integer 1, in turn, occupies the kth position, then there are d_{n-2} ways to derange the $n - 2$ integers $2, 3, \ldots, k-1, k+1, \ldots, n$. If the integer 1 does not occupy the kth position, then there are d_{n-1} ways to derange the integers $1, 2, \ldots, k-1, k+1, \ldots, n$, because in this case we can consider the kth position as the forbidden position for the integer 1. Since k can assume the $n-1$ values 2, 3, . . . , n, we have the recurrence relation

$$d_n = (n-1)(d_{n-1} + d_{n-2})$$

Clearly, the boundary conditions are $d_2 = 1$ and $d_1 = 0$. The difference equation is valid for $n \geq 2$ if we choose the value of d_0 to be 1. The difference equation can be rewritten as

$$\begin{aligned} d_n - nd_{n-1} &= -[d_{n-1} - (n-1)d_{n-2}] \\ &= -[-d_{n-2} + (n-2)d_{n-3}] \\ &\quad \cdots\cdots\cdots\cdots\cdots \\ &= (-1)^{n-2}[d_2 - 2d_1] \\ &= (-1)^{n-2} \\ &= (-1)^n \end{aligned}$$

that is,

$$d_n - nd_{n-1} = (-1)^n \tag{4-8}$$

To solve this equation, let

$$D(x) = d_0 + \frac{d_1}{1!}x + \frac{d_2}{2!}x^2 + \frac{d_3}{3!}x^3 + \cdots + \frac{d_r}{r!}x^r + \cdots$$

be the exponential generating function of the sequence $(d_0, d_1, d_2, \ldots, d_r, \ldots)$. Multiplying both sides of Eq. (4-8) by $x^n/n!$ and summing from $n = 2$ to $n = \infty$, we obtain

$$\sum_{n=2}^{\infty} \frac{d_n}{n!}x^n - \sum_{n=2}^{\infty} \frac{nd_{n-1}}{n!}x^n = \sum_{n=2}^{\infty} \frac{(-1)^n x^n}{n!}$$

that is,

$$D(x) - d_1x - d_0 - x[D(x) - d_0] = e^{-x} - (1-x)$$

or

$$D(x) = \frac{e^{-x}}{1-x}$$

Recalling that $1/(1-x)$ is the summing operator, we have

$$d_n = n!\left[1 - \frac{1}{1!} + \frac{1}{2!} - \cdots + (-1)^n \frac{1}{n!}\right]$$

Let us now study the following illustrative examples.

Example 4-7 Let n books be distributed to n children. The books are returned and distributed to the children again later on. In how many ways can the books be distributed so that no child will get the same book twice? For the first time, the books can be distributed in $n!$ ways. For the second time, the books can be distributed in d_n ways. Therefore, the total number of ways is given by

$$(n!)^2\left[1 - \frac{1}{1!} + \frac{1}{2!} - \cdots + (-1)^n \frac{1}{n!}\right] \approx (n!)^2 e^{-1} \quad \blacksquare$$

Example 4-8 In how many ways can the integers 1, 2, 3, 4, 5, 6, 7, 8, and 9 be permuted such that no odd integer will be in its natural position? Applying the principle of inclusion and exclusion, we have

$$9! - \binom{5}{1} 8! + \binom{5}{2} 7! - \binom{5}{3} 6! + \binom{5}{4} 5! - \binom{5}{5} 4! = 205{,}056$$

As a matter of fact, for a set of n objects, the number of permutations in which a subset of r objects are deranged can be computed by the formula

$$n! - \binom{r}{1}(n-1)! + \binom{r}{2}(n-2)! - \cdots + (-1)^r \binom{r}{r}(n-r)!$$

The number of permutations in which all even integers are in their natural positions and none of the odd integers are in their natural positions is equal to

$$d_5 = 5!\left(1 - \frac{1}{1!} + \frac{1}{2!} - \frac{1}{3!} + \frac{1}{4!} - \frac{1}{5!}\right) = 44$$

Similarly, the number of permutations in which exactly four of the nine integers are in their natural positions (exactly five integers are deranged) is

$$C(9,5) \times d_5 = 5{,}544$$

Consequently, the number of permutations in which five or more integers are deranged is equal to

$$C(9,5) \times d_5 + C(9,6) \times d_6 + C(9,7) \times d_7 + C(9,8) \times d_8 + C(9,9) \times d_9 \quad \blacksquare$$

4-5 PERMUTATIONS WITH RESTRICTIONS ON RELATIVE POSITIONS

In our discussion about the derangement of objects, the forbidden positions are absolute positions in the permutations. Moreover, each object has only one forbidden position and no two objects have the same forbidden position. In this section, we study the case in which the restrictions are on the relative positions of the objects, whereas in Sec. 4-7 we study the case in which an object may have any number of forbidden positions and several objects may have the same forbidden position.

Consider the permutations of the n integers $1, 2, \ldots, n$. We wish to find the number of permutations in which no two adjacent integers are consecutive integers. In other words, the $n-1$ patterns $12, 23, 34, \ldots, (n-1)n$ should not appear in the permutations. Let a_i be the property that the pattern $i(i+1)$ appears in a permutation, with $i = 1, 2, \ldots, n-1$. Since

$$N(a_1) = N(a_2) = \cdots = N(a_{n-1}) = (n-1)!$$

it follows that

$$s_1 = \binom{n-1}{1}(n-1)!$$

Observe that

$$N(a_1a_2) = (n-2)!$$

because $N(a_1a_2)$ is equal to the number of permutations of the $n-2$ "objects" $123, 4,5, \ldots, n$, where 123 is considered to be bound as one object. Similarly, for $1 \leq i < n-1$,

$$N(a_ia_{i+1}) = (n-2)!$$

Also observe that

$$N(a_1a_3) = (n-2)!$$

because $N(a_1a_3)$ is equal to the number of permutations of the $n-2$ "objects" $12, 34, 5, 6, \ldots, n$, where 12 and 34 are considered to be bound as two objects. Similarly, for $1 \leq i < n-2$ and $i+1 < j \leq n-1$,

$$N(a_ia_j) = (n-2)!$$

Therefore,

$$N(a_1a_2) = N(a_1a_3) = \cdots = N(a_{n-2}a_{n-1}) = (n-2)!$$

$$s_2 = \binom{n-1}{2}(n-2)!$$

Furthermore, it can be shown that

$$s_j = \binom{n-1}{j}(n-j)! \qquad j = 0, 1, 2, \ldots, n-1$$

from which we see that

$$N(a_1', a_2', \ldots, a_{n-1}') = n! - \binom{n-1}{1}(n-1)! + \binom{n-1}{2}(n-2)! - \cdots + (-1)^{n-1}\binom{n-1}{n-1}1! \quad (4\text{-}9)$$

Extension to the general case is immediate. Since restrictions on the relative positions of the objects are equivalent to restrictions on the appearance of a set of patterns, the enumeration of permutations with restricted relative positions is the same as the enumeration of permutations in which none of a certain set of patterns can appear.

Example 4-9 Find the number of permutations of the letters a, b, c, d, e, and f in which neither the pattern *ace* nor the pattern *fd* appears. Let a_1 be the property that the pattern *ace* appears in a permutation, and let a_2 be the property that the pattern *fd* appears in a permutation. According to the principle of inclusion and exclusion,

$$\begin{aligned} N(a_1'a_2') &= N - N(a_1) - N(a_2) + N(a_1a_2) \\ &= 6! - 4! - 5! + 3! = 582 \quad \blacksquare \end{aligned}$$

Example 4-10 In how many ways can the letters α, α, α, α, β, β, β, γ, and γ be arranged so that all the letters of the same kind are not in a single block? For the permutations of these letters, let a_1 be the property that the four α's are in one block, let a_2 be the property that the three β's are in one block, and let a_3 be the property that the two γ's are in one block. Then,

$$\begin{aligned} N(a_1'a_2'a_3') &= \frac{9!}{4!3!2!} - \left(\frac{6!}{3!2!} + \frac{7!}{4!2!} + \frac{8!}{4!3!}\right) + \left(\frac{4!}{2!} + \frac{5!}{3!} + \frac{6!}{4!}\right) - 3! \\ &= 871 \quad \blacksquare \end{aligned}$$

★4-6 THE ROOK POLYNOMIALS

In this section, we discuss a seemingly unrelated topic—the problem of "nontaking rooks." It will be shown in Sec. 4-7 that the enumeration of the ways of placing nontaking rooks on a chessboard is useful in counting the number of permutations of objects when there are arbitrary restrictions on the positions they can occupy. A *rook* is a chessboard piece which "captures" on both rows and columns. The problem of nontaking rooks is to enumerate the number of ways of placing k rooks on a chessboard such that no rook will be captured by any other rook. For example, on a regular 8×8 chessboard, there are (trivially) 64 ways to place one nontaking rook. There are $\binom{8}{2}P(8,2) = 1{,}568$ ways to place two nontaking rooks, since there are $\binom{8}{2}$ ways to choose two rows and then $P(8,2)$ ways to choose two cells from the two rows for the rooks. Obviously, we can put at the most eight nontaking rooks on the board and there are 8! ways to do so.

Here, we generalize the problem in that we are interested in placing nontaking rooks not only on a regular 8×8 chessboard, but also on chessboards of arbitrary shapes and sizes. For example, Fig. 4-2 shows a so-called "staircase" chessboard. Clearly, for such a chessboard, there are four ways to place one nontaking rook, three ways to place two nontaking rooks, and no way to place three or more nontaking rooks.

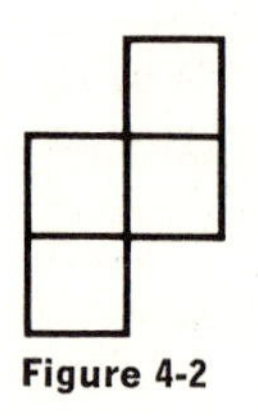

Figure 4-2

For a given chessboard, let r_k denote the number of ways of placing k nontaking rooks on the board, and let

$$R(x) = \sum_{k=0}^{n} r_k x^k$$

be the ordinary generating function of the sequence $(r_0, r_1, r_2, \ldots, r_k, \ldots)$. $R(x)$ is the *rook polynomial* of the given chessboard. Notice that $R(x)$ is a finite polynomial whose degree is at most n, where n is the number of cells of the chessboard, because it is never possible to place more than n rooks on a chessboard of n cells. For the staircase chessboard in Fig. 4-2, the rook polynomial is

$$R(x) = 1 + 4x + 3x^2$$

When there are several chessboards C_1, C_2, C_3, . . . under consideration, let $r_k(C_1)$, $r_k(C_2)$, $r_k(C_3)$, . . . denote the numbers of ways of placing k nontaking rooks on the boards C_1, C_2, C_3, . . . , and let $R(x,C_1)$, $R(x,C_2)$, $R(x,C_3)$, . . . denote the rook polynomials of the boards C_1, C_2, C_3, . . . , respectively.

Suppose that on a given chessboard C, a cell is selected and marked as a special cell. Let C_i denote the chessboard obtained from C by deleting the row and the column that contain the special cell, and let C_e

denote the chessboard obtained from C by deleting the special cell. To find the value of $r_k(C)$, we observe that the ways of placing k nontaking rooks on C can be divided into two classes, those that have a rook in the special cell and those that do not have a rook in the special cell. The number of ways in the first class is equal to $r_{k-1}(C_i)$, and the number of ways in the second class is equal to $r_k(C_e)$. We have then the relation

$$r_k(C) = r_{k-1}(C_i) + r_k(C_e)$$

Correspondingly, we have

$$R(x,C) = xR(x,C_i) + R(x,C_e) \tag{4-10}$$

Equation (4-10) is called the *expansion formula*. The rook polynomial of a chessboard of arbitrary shape and size can be found by the repeated applications of the expansion formula.

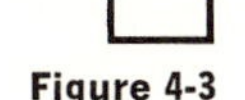

Figure 4-3

Let a pair of parentheses around a chessboard be used to denote the rook polynomial of the board. Thus, the expansion formula applied to the chessboard in Fig. 4-3 can be written as

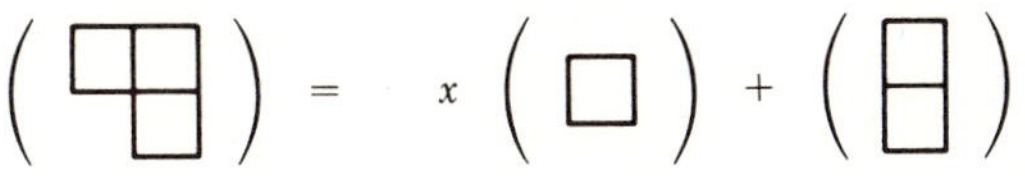

$$(\quad) = x(\quad) + (\quad)$$

where the expansion is carried out with respect to the cell in the upper left corner. Also

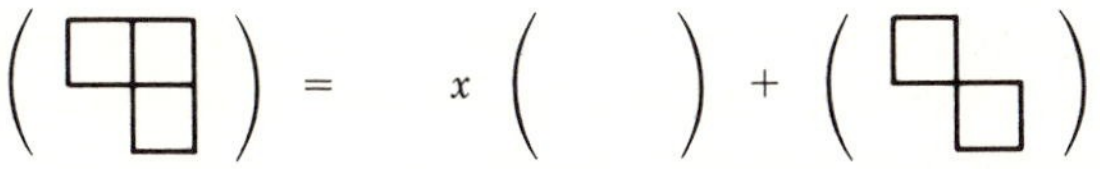

$$(\quad) = x(\quad) + (\quad)$$

where the expansion is carried out with respect to the cell in the upper right corner. Because

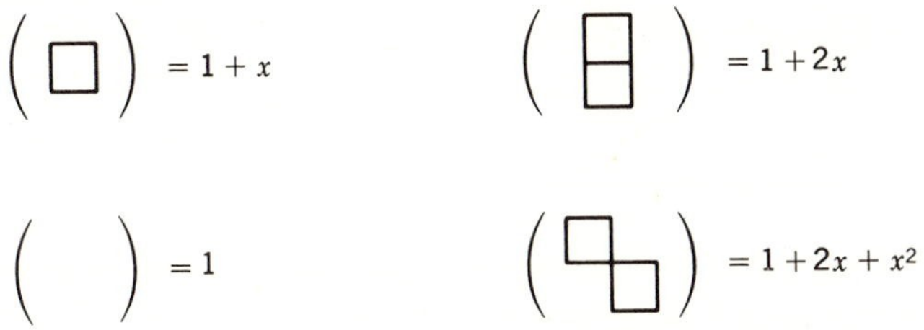

$$(\quad) = 1 + x \qquad (\quad) = 1 + 2x$$

$$(\quad) = 1 \qquad (\quad) = 1 + 2x + x^2$$

it follows that

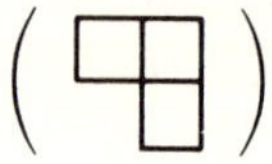

$$= 1 + 3x + x^2$$

As another example of the application of the expansion formula, observe that

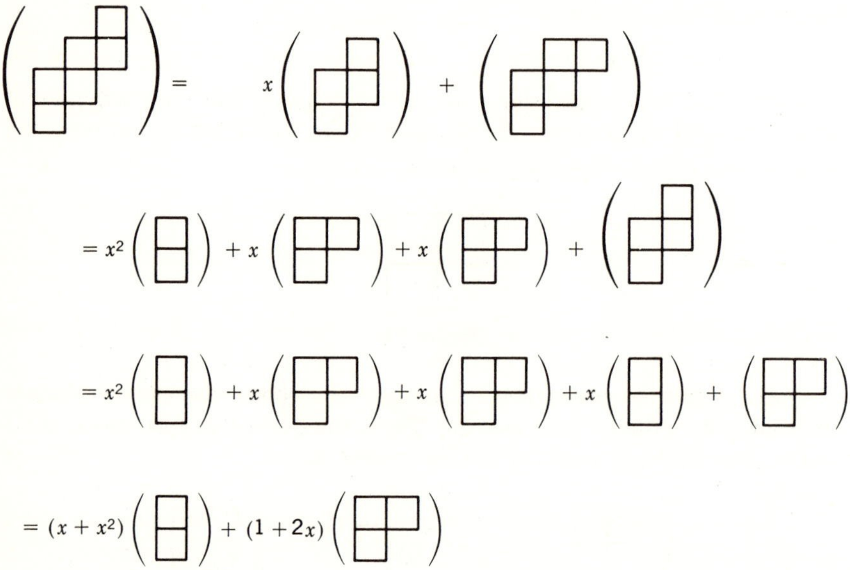

$$= \quad x \quad + $$

$$= x^2 \quad + x \quad + x \quad + $$

$$= x^2 \quad + x \quad + x \quad + x \quad + $$

$$= (x + x^2) \quad + (1 + 2x)$$

$$= (x + x^2)(1 + 2x) + (1 + 2x)(1 + 3x + x^2)$$

$$= 1 + 6x + 10x^2 + 4x^3$$

If a chessboard C consists of two subboards C_1 and C_2 in which no cell of one subboard is in the same row or in the same column of any cell of the other subboard (C_1 and C_2 are said to be *disjunct*), then

$$R(x,C) = R(x,C_1)R(x,C_2) \tag{4-11}$$

This result comes from the observation that the way rooks are placed on C_1 is completely independent of the way rooks are placed on C_2. Therefore,

$$r_k(C) = \sum_{j=0}^{k} r_j(C_1)r_{k-j}(C_2)$$

and Eq. (4-11) follows. (Recall the definition of the product of two ordinary generating functions in Sec. 2-7.)

★4-7 PERMUTATIONS WITH FORBIDDEN POSITIONS

Consider the distribution of four distinct objects, labeled *a*, *b*, *c*, and *d*, into four distinct positions, labeled 1, 2, 3, and 4, with no two objects occupying the same position. A distribution can be represented in the form of a matrix as illustrated in Fig. 4-4, where the rows correspond to the

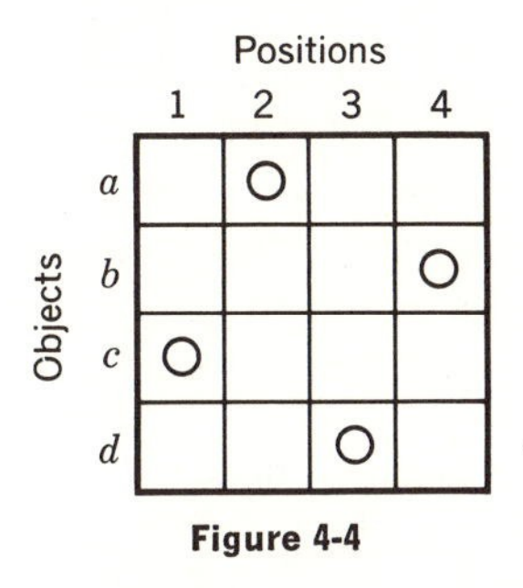

Figure 4-4

objects, and the columns correspond to the positions. A circle in a cell indicates that the object in the row containing the cell occupies the position in the column containing the cell. Thus, the distribution shown in Fig. 4-4 is as follows: *a* is placed in the second position, *b* is placed in the fourth position, *c* is placed in the first position, and *d* is placed in the third position.

Since an object cannot be placed in more than one position and a position cannot hold more than one object, in the matrix representation of an acceptable distribution there will never be more than one circle in a row or column. This is equivalent to placing nontaking rooks on a chessboard, and therefore the problem of enumerating the number of ways of distributing distinct objects into distinct positions is the same as that of enumerating the number of ways of placing nontaking rooks on a chessboard.

The notion of placing nontaking rooks on a chessboard can be extended to the case where these are forbidden positions for the objects. For example, for the derangement of four objects, the forbidden positions are shown as dark cells in the chessboard in Fig. 4-5. It follows that the problem of enumerating the number of derangements of four objects is equivalent to the problem of finding the value of r_4 for the *white* chessboard in Fig. 4-5. In general, the restrictions on the positions that the objects may occupy can be quite arbitrary. For example, consider the problem of painting four houses *a*, *b*, *c*, and *d* with four different colors, green, blue, gray, and yellow, under the restriction that house *a* cannot be painted with yellow, house *b* cannot be painted with gray or

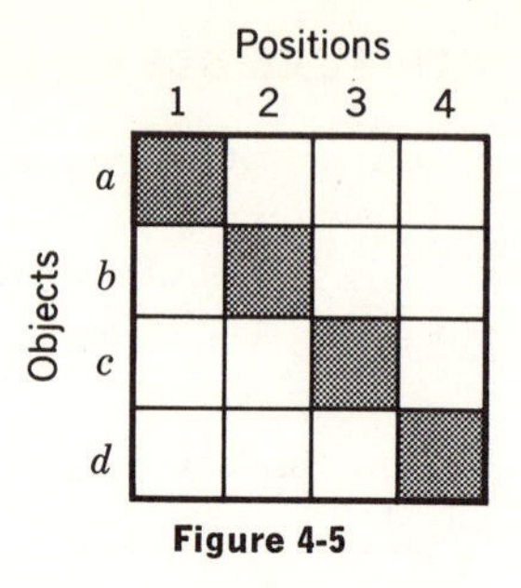

Figure 4-5

yellow, house c cannot be painted with blue or gray, and house d cannot be painted blue. This is a problem of permutations with forbidden positions. The forbidden positions are shown in Fig. 4-6 as dark cells in

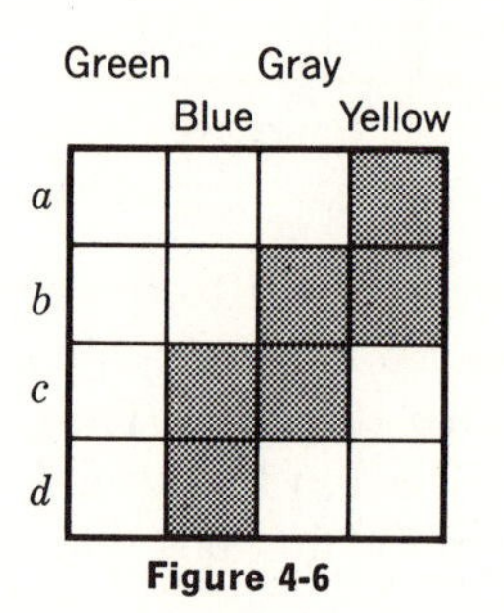

Figure 4-6

the chessboard. Again, the number of ways of painting the houses is equal to the value of r_4 for the white chessboard.

Once we have seen the equivalence between the permutation of objects with restrictions on their positions and the placement of nontaking rooks on a chessboard, it seems that this is the end of the discussion since the problem of placing nontaking rooks on a chessboard has already been studied in Sec. 4-6. However, there is an alternative viewpoint that turns out to be more useful. Consider the permutation of n objects with restrictions on their positions. Let a_i denote the property of a permutation in which the ith object is in a forbidden position ($i = 1, 2, \ldots, n$). Then the number of permutations in which no object is in a forbidden position is

$$\begin{aligned} N(a_1'a_2' \cdots a_n') = e_0 &= n! - r_1 \times (n-1)! + r_2 \times (n-2)! \\ &\quad + \cdots \\ &\quad + (-1)^{n-1} \times r_{n-1} \times 1! + (-1)^n \times r_n \times 0! \\ &= \sum_{j=0}^{n} (-1)^j \times r_j \times (n-j)! \end{aligned} \tag{4-12}$$

where r_k is the number of ways of placing k nontaking rooks on the *dark* chessboard. Equation (4-12) is just the result of another direct application of the principle of inclusion and exclusion, that is, since j of the n objects can be placed in the forbidden positions in r_j ways and the $n - j$ remaining objects can be placed in the $n - j$ remaining positions arbitrarily in $(n - j)!$ ways, we have

$$s_j = r_j \times (n - j)!$$

Therefore, Eq. (4-12) follows from Eq. (4-6).

For the problem of painting four houses with four colors mentioned above, since the rook polynomial for the *board of forbidden positions* is

$$R(x) = 1 + 6x + 10x^2 + 4x^3$$

Eq. (4-12) gives

$$e_0 = 4! - 6 \times 3! + 10 \times 2! - 4 \times 1! = 4$$

We can also compute the number of permutations in which exactly m of the objects are in forbidden positions; that is,

$$e_m = r_m \times (n - m)! - \binom{m+1}{1} \times r_{m+1} \times (n - m - 1)! + \cdots + (-1)^{n-m} \binom{n}{n-m} \times r_n \times 0!$$

Also, according to Eq. (4-7),

$$E(x) = \sum_{j=0}^{n} s_j(x - 1)^j = \sum_{j=0}^{n} r_j \times (n - j)! \times (x - 1)^j$$

Because an object in a forbidden position is said to be a "hit," $E(x)$ is also called the *hit polynomial*. For the problem of painting four houses with four colors, the hit polynomial is

$$\begin{aligned} E(x) &= 4! + 6 \times 3! \times (x - 1) + 10 \times 2! \times (x - 1)^2 \\ &\qquad + 4 \times 1! \times (x - 1)^3 + 0 \times 0! \times (x - 1)^4 \\ &= 4 + 8x + 8x^2 + 4x^3 \end{aligned}$$

Thus, there are four ways to paint the houses so that none of the houses will be painted with forbidden colors, eight ways to paint the houses so that exactly one of the houses will be painted with forbidden colors, and so on. Notice that there is no way to paint all four houses with forbidden colors.

Example 4-11 Find the number of permutations of the letters α, α, β, β, γ, and γ so that no α appears in the first and second positions, no

β appears in the third position, and no γ appears in the fifth and sixth positions. Imagine that the α's, β's, and γ's are marked so that they become distinguishable. The forbidden positions are shown as dark cells in the chessboard in Fig. 4-7. The rook poly-

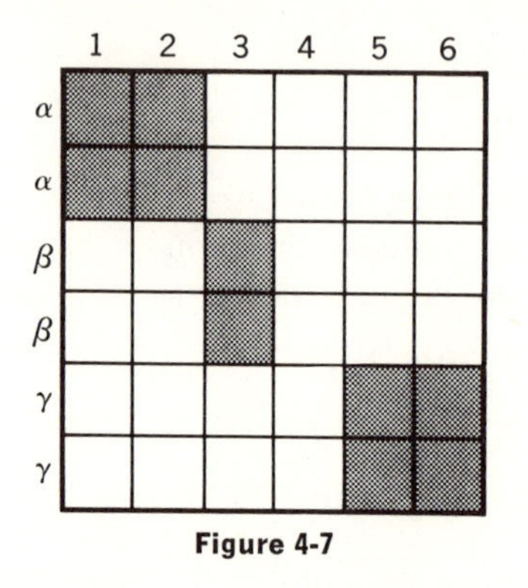

Figure 4-7

nomial for a 2×2 square chessboard is

$$1 + 4x + 2x^2$$

The rook polynomial for a 2×1 rectangular chessboard is

$$1 + 2x$$

The rook polynomial for the board of the forbidden positions is

$$(1 + 4x + 2x^2)^2(1 + 2x) = 1 + 10x + 36x^2 + 56x^3 + 36x^4 + 8x^5$$

It follows that

$$\begin{aligned} e_0 &= 6! - 10 \times 5! + 36 \times 4! - 56 \times 3! + 36 \times 2! - 8 \times 1! \\ &= 112 \end{aligned}$$

The fact that the objects are not all distinct merely introduces a division factor $2! \times 2! \times 2!$. Thus, we have

$$\frac{112}{2! \times 2! \times 2!} = 14$$

as the number of ways of distributing the objects with none of them in a forbidden position. ■

4-8 SUMMARY AND REFERENCES

In a sense, the principle of inclusion and exclusion offers an indirect approach to the solution of enumeration problems. Instead of computing the exact count directly, we "adjust" the count by alternately adding and subtracting appropriate terms according to the principle of inclusion and exclusion. As illustrated by the examples in this chapter, such an approach is often very powerful.

Discussion of the principle of inclusion and exclusion can also be found in Chap. 3 of Riordan [3] and Chap. 2 of Ryser [5]. For further details on rook polynomials and permutations with forbidden positions see Chaps. 7 and 8 of Riordan [3]. Derangement of objects is also discussed in Chap. 4 of Whitworth [6]. For an application in the theory of numbers that leads to a formula known as the Möbius inversion formula, see Chap. 16 of Hardy and Wright [2]. A significant extension of the concept of the Möbius inversion formula was carried out by Rota [4]. See also Chap. 2 of Hall [1].

1. Hall, M., Jr.: "Combinatorial Theory," Blaisdell Co., Waltham, Mass., 1967.
2. Hardy, G. H., and E. M. Wright: "An Introduction to the Theory of Numbers," 4th ed., Oxford University Press, London, 1960.
3. Riordan, J.: "An Introduction to Combinatorial Analysis," John Wiley & Sons, Inc., New York, 1958.
4. Rota, G. C.: On the Foundations of Combinatorial Theory, I. Theory of Möbius Functions, *Z. Wahrscheinlichkeitstheorie und Verw. Gebiete,* **2**:340–368 (1964).
5. Ryser, H. J.: "Combinatorial Mathematics," published by the Mathematical Association of America, distributed by John Wiley & Sons, Inc., New York, 1963.
6. Whitworth, W. A.: "Choice and Chance," reprint of the 5th ed. (1901), Hafner Publishing Company, Inc., New York, 1965.

PROBLEMS

4-1. In how many ways can three 0's, three 1's, and three 2's be arranged so that no three adjacent digits are the same in an arrangement?

4-2. A man has six friends. He has met each of them at dinner 12 times, every two of them six times, every three of them four times, every four of them three times, every five twice, and all six only once. He has dined out eight times without meeting any of them. How many times has he dined out altogether?

4-3. A symmetric expression in three variables x, y, and z contains nine terms. Four terms contain the variable x. Two terms contain the variables x, y, and z. One term is a constant. How many terms contain the variables x and y?

4-4. Find the number of binary sequences of length 5 in which every 1 is adjacent to another 1.

4-5. With three differently colored paints, in how many ways can the walls of a rectangular room be painted so that color changes occur at (and only at) each corner? With two colors?

4-6. Among the numbers 1, 2, . . . , 500, how many of them are not divisible by 7 but are divisible by 3 *or* 5?

4-7. Find the number of permutations of the letters α, α, α, β, β, β, γ, γ, and γ which are such that no identical letters are adjacent.

4-8. Find the number of permutations of the 26 letters $A, B, \ldots, X, Y, Z$ that do not contain the patterns *JOHN, PAUL,* and *SMITH.*

4-9. When two indistinguishable dice are rolled n times:

(*a*) In how many ways can a pair of each of the six numbers show at least once?

(*b*) In how many ways can each of the pairs of 3's, 4's, 5's, and 6's show at least once and either a pair of aces *or* a pair of deuces show at least once?

4-10. (*a*) $5n$ men, standing in a line, are from n countries (five men from each country). Show that the number of ways the line can be formed so that every man is standing next to a countryman of his is equal to

$$120^n \left[(2n)! - \binom{n}{1}(2n-1)! + \binom{n}{2}(2n-2)! - \cdots + (-1)^n \binom{n}{n} n! \right]$$

(*b*) Repeat part (*a*) with $4n$ men from n countries (four men from each country).

4-11. The Euler ϕ-function of a positive integer n, $\phi(n)$, is defined as the number of positive integers less than or equal to n that are prime to n. Show that

$$\phi(n) = n \prod_{i=1}^{m} \left(1 - \frac{1}{p_i}\right)$$

where $p_1, p_2, \ldots, p_m$ are all the distinct prime divisors of n.

4-12. Use a combinatorial argument to prove the identity

$$\binom{r-1}{n-m-1} = \sum_{k=0}^{n-m} (-1)^k \binom{n-m}{k} \binom{n-m-k+r-1}{r}$$

Hint: the number of ways to distribute r nondistinct objects into n distinct cells leaving exactly m cells empty is equal to $\binom{n}{m}\binom{r-1}{n-m-1}$.

4-13. Use a combinatorial argument to prove the identity

$$\binom{n-m}{n-r} = \sum_{k=0}^{m} (-1)^k \binom{m}{k} \binom{n-k}{r} \qquad n \geq r \geq m$$

Hint: $\binom{n-m}{n-r}$ is the number of ways to select r objects from n distinct ones such that m special objects are always included in the selections.

4-14. Show that

$$s_m = \sum_{j=m}^{r} \binom{j}{m} e_j$$

where s_m and e_j are defined as in Sec. 4-3. What is the combinatorial significance of this identity?

4-15. (*a*) Prove the identity

$$\binom{m+r}{m} - \binom{m+r}{m+1} + \binom{m+r}{m+2} - \cdots + (-1)^r \binom{m+r}{m+r} = \binom{m+r-1}{m-1}$$

(*b*) Let $e_{\geq m}$ denote the number of objects having m or more of r properties; that is,

$$e_{\geq m} = e_m + e_{m+1} + e_{m+2} + \cdots + e_r$$

Prove that

$$e_{\geq m} = s_m - \binom{m}{m-1} s_{m+1} + \binom{m+1}{m-1} s_{m+2} - \binom{m+2}{m-1} s_{m+3} + \cdots + (-1)^{r-m} \binom{r-1}{m-1} s_r$$

4-16. Two professors in two different subjects are giving oral examinations to 12 students in the same hour. Each student shall be examined individually for 5 minutes in each subject. In how many ways can a schedule be made up so that no student will have to see both professors at the same time? Give an approximated value to the answer.

4-17. (*The Game of Rencontres*) There are n balls, numbered 1, 2, . . . , n, in an urn. When the balls are drawn one by one without replacement from the urn, a rencontre is said to occur if the ball numbered r appears in the rth drawing. Find the probability of at least one rencontre and the probability of exactly m rencontres. Give approximated values of the probabilities.

4-18. The n integers 1, 2, . . . , n are arranged around a circle Let g_n denote the number of such arrangements in which no two adjacent integers are consecutive in clockwise order (n and 1 are considered to be consecutive). Find an expression for g_n, and show that it satisfies the recurrence equation

$$g_n + g_{n+1} = d_n \qquad n \geq 1$$

4-19. (*a*) Show that the number of permutations of the n integers 1, 2, . . . , n in which no two adjacent integers are consecutive in left-to-right order is equal to $d_n + d_{n-1}$.

(*b*) If n and 1 are also considered as "consecutive numbers," show that the number of permutations of the n integers 1, 2, . . . , n in which no two adjacent integers are "consecutive" is equal to nd_{n-1}.

4-20. (*The Permanent*) Let $A = [a_{ij}]$ be an $m \times n$ matrix with $m \leq n$. Let A_r denote a matrix obtained from A by replacing r of the n columns of A by 0's. Let $S(A_r)$ be the product of the row sums of A_r and $\Sigma S(A_r)$ be the sum of the $S(A_r)$'s over all A_r's. Define the *permanent* of A, denoted by Per (A), as

$$\text{Per}\,(A) = \Sigma a_{1i_1} a_{2i_2} \cdots a_{mi_m}$$

where the summation extends over all m-permutations $(i_1, i_2, \ldots, i_m)$ of the n integers 1, 2, . . . , n.

(*a*) Show that

$$\text{Per}\,(A) = \sum S(A_{n-m}) - \binom{n-m+1}{1} \sum S(A_{n-m+1}) + \binom{n-m+2}{2} \sum S(A_{n-m+2}) - \cdots + (-1)^{m-1} \binom{n-1}{m-1} \sum S(A_{n-1})$$

Hint: Consider as the jth property of a term $a_{1i_1} a_{2i_2} \cdots a_{mi_m}$ that $i_p \neq j$ $(p = 1, 2, \ldots, m)$.

(*b*) Let I be the $n \times n$ identity matrix, and let K be the $n \times n$ matrix whose entries are all 1's. Show that
1. Per $(K) = n!$
2. Per $(K - I) = d_n$, where d_n is the number of derangements of n objects.

4-21. (*a*) A computer matching service has five male subscribers A, B, C, D, and E and four female subscribers a, b, c, and d. After analyzing their personalities and interests, the computer decides that a should not be matched with C and D, b should not be matched with A and E, c should not be matched with B and C, and d would quarrel with E. Find the number of ways in which the subscribers can be matched.

(*b*) Since there are five boys and only four girls, one of the boys will be left with no date. However, B is a good friend of the programmer who programs the computer and sees to it that B is matched. In how many ways can the subscribers be matched?

4-22. Find the rook polynomial $R_{m,l}(x)$ for the staircase chessboard shown in Fig. 4P-1.

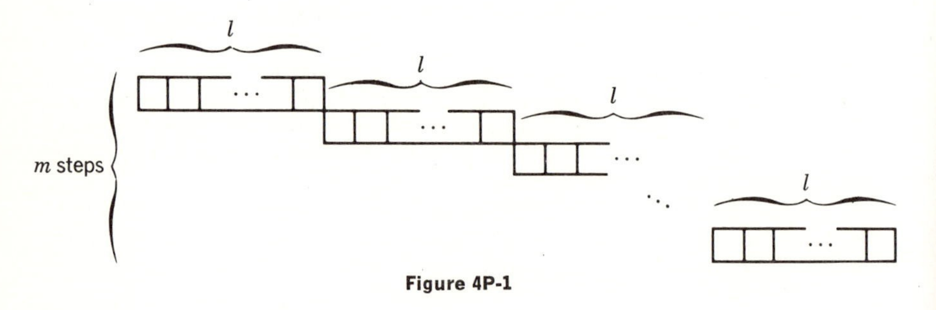

Figure 4P-1

4-23. A pair of distinct dice are rolled six times. It is known that the combinations (1,2), (2,2), (3,2), (2,3), (3,3), and (4,4) did not appear. Find the probability that all faces of the two dice have shown.

4-24. An arrangement of integers is said to contain a *descent* when an integer is immediately followed by a smaller integer in the arrangement. For example, the arrangement 52143 contains three descents.

(*a*) Find the chessboard for the restrictions in the enumeration of the number of arrangements of the integers 1, 2, 3, 4, and 5 that contain no descent. The restriction implied by the cell (i,j) in this chessboard is now interpreted as the integer i being to the immediate left of the integer j in an arrangement, for $i \neq j$.

(*b*) Find the rook polynomial of the chessboard in part (*a*). Show that its corresponding hit polynomial is the ordinary generating function that enumerates the number of arrangements containing zero, one, two, three, and four descents, respectively.

4-25. (*a*) Use the expansion formula to show that the rook polynomial of an $n \times m$ rectangular chessboard, $R_{n,m}(x)$, satisfies the recurrence relation

$$R_{n,m}(x) = R_{n-1,m}(x) + mxR_{n-1,m-1}(x)$$

(*b*) Show that

$$\frac{d}{dx} R_{n,m}(x) = nmR_{n-1,m-1}(x)$$

(*c*) If the cells in an $n \times m$ rectangular chessboard are forbidden positions for the permutations of n elements $(n \geq m)$, find the corresponding hit polynomial $E(x)$.

4-26. (*Equivalence*) (*a*) Let C_1 be a chessboard that has l columns and at the most n rows. Let C be an $n \times m$ rectangular chessboard. The chessboard $C_1 + C$ is defined as a board containing all the cells in C_1 and C arranged as in Fig. 4P-2. Let

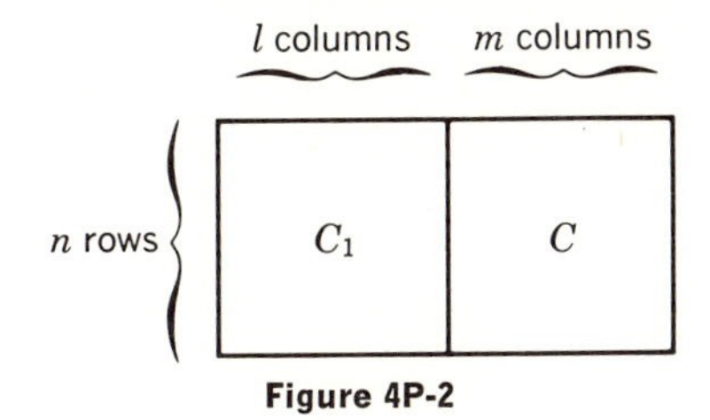

Figure 4P-2

$R(x,C_1) = \sum_k r_k x^k$, and let $R(x, C_1 + C) = \sum_k q_k x^k$. Find q_k, and express it in terms of the r_k's.

(*b*) Two chessboards C_1 and C_2 are said to be *equivalent* if $R(x,C_1) = R(x,C_2)$. Show that if C_1 and C_2 are equivalent, $C_1 + C$ and $C_2 + C$ are also equivalent.

4-27. (*The Complement of a Chessboard*) The complement C' of a chessboard C with respect to an $n \times m$ rectangular chessboard $C_{n,m}$ is defined as the chessboard that contains all the cells that are in $C_{n,m}$ but not in C. (Assume that C has at the most n rows and m columns.)

(*a*) Let r_k and q_k be the number of ways to place k nontaking rooks on a chessboard C and its complement C' with respect to $C_{n,m}$, respectively. Show that

$$q_k = \binom{n}{k} P(m,k) - r_1 \binom{n-1}{k-1} P(m-1, k-1) + r_2 \binom{n-2}{k-2} P(m-2, k-2) - \cdots + (-1)^k r_k \binom{n-k}{0} P(m-k, 0)$$

(*b*) Let $R_{i,j}(x)$ denote the rook polynomial of an $i \times j$ rectangular chessboard. Show that the rook polynomial of C', $Q(x)$, can be expressed as

$$Q(x) = \sum_{k=0}^{\min[n,m]} q_k x^k = \sum_{k=0}^{\min[n,m]} (-1)^k r_k x^k R_{n-k,m-k}(x)$$

(*c*) Let C and C' be complementary chessboards with respect to an $n \times n$ chessboard, and let $E(x,C)$ and $E(x,C')$ be their hit polynomials. Show that

$$E(x,C) = x^n E\left(\frac{1}{x}, C'\right)$$

$$E(x,C') = x^n E\left(\frac{1}{x}, C\right)$$

4-28. (*Triangular Chessboard*) Let $T_n(x)$ be the rook polynomial of an $n \times n$ triangular chessboard as shown in Fig. 4P-3. We shall prove by mathematical induction that

$$T_n(x) = \sum_{k=0}^{n} S(n+1, n+1-k) x^k \qquad \text{(4P-1)}$$

where $S(i,j)$ is the Stirling number of the second kind.

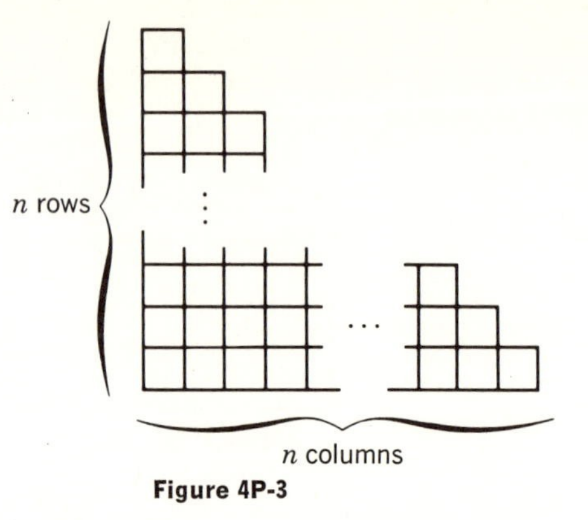

Figure 4P-3

(*a*) Basis of induction: Show that Eq. (4P-1) is satisfied for $n = 1$ and $n = 2$.

(*b*) Induction step: Assume that $T_m(x) = \sum_{k=0}^{m} S(m+1, m+1-k)x^k$ for $m \leq n - 1$. Show that Eq. (4P-1) is true.

$$\textit{Hint: } S(n+1, k) = \sum_{j=0}^{n} \binom{n}{j} S(j, k-1)$$

4-29. (*The Ménage Problem*) In the problem of enumerating the permutations of the integers 1, 2, . . . , n which are such that the integer i is not in positions i and $i + 1$ for $i = 1, 2, \ldots, n - 1$ and the integer n is not in positions n and 1, the chessboard for the forbidden positions is shown in Fig. 4P-4. The rook polynomial of this chessboard is called the Ménage polynomial, which is denoted by $M_n(x)$.

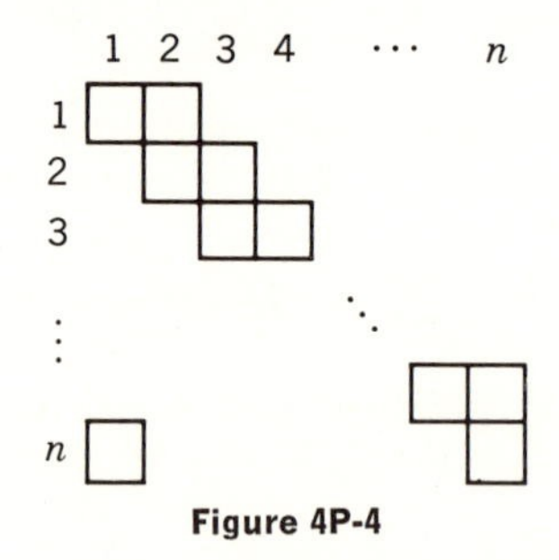

Figure 4P-4

(*a*) Let $A_{2n-1}(x)$ be the rook polynomial of a staircase chessboard containing an odd number of cells, $2n - 1$, as shown in Fig. 4P-5*a*. Let $B_{2n}(x)$ be the rook polynomial of a staircase chessboard containing an even number of cells, $2n$, as shown in Fig. 4P-5*b*. Show that

$$A_{2n-1}(x) = xA_{2n-3}(x) + B_{2n-2}(x)$$
$$B_{2n}(x) = xB_{2n-2}(x) + A_{2n-1}(x)$$

(*b*) With the boundary conditions

$$A_1(x) = 1 + x \qquad A_3(x) = 1 + 3x + x^2$$
$$B_0(x) = 1 \qquad B_2(x) = 1 + 2x$$

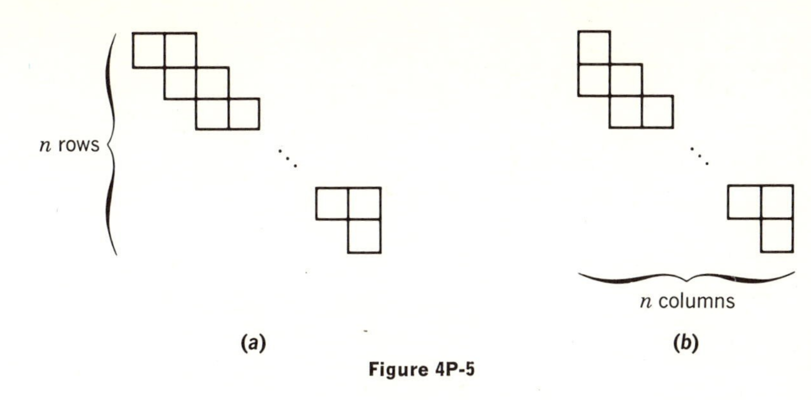

Figure 4P-5

show by mathematical induction that

$$A_{2n-1}(x) = \sum_{k=0}^{n} \binom{2n-k}{k} x^k$$

$$B_{2n}(x) = \sum_{k=0}^{n} \binom{2n-k+1}{k} x^k$$

(*c*) Show that

$$M_n(x) = xA_{2n-3}(x) + A_{2n-1}(x)$$

(*d*) Find $M_n(x)$ and the corresponding hit polynomial.

Chapter 5
Pólya's Theory of Counting

5-1 INTRODUCTION

Consider the problem of counting the number of 2×2 chessboards that contain black and white cells. Clearly, there are 2^4 such chessboards as shown in Fig. 5-1. However, if the four sides of a chessboard are not marked and one side cannot be distinguished from another, then there are chessboards that become indistinguishable from other chessboards after being rotated by 90, 180, or 270°. The reader can check that in this case, chessboards C_2, C_3, C_4, and C_5; chessboards C_6, C_8, C_9, and C_{11}; chessboards C_7 and C_{10}; and chessboards C_{12}, C_{13}, C_{14}, and C_{15} are indistinguishable under rotations. In a sense that will be defined precisely later, chessboards C_2, C_3, C_4, and C_5 are said to be "equivalent," as are chessboards C_6, C_8, C_9, and C_{11}; chessboards C_7 and C_{10}; and chessboards C_{12}, C_{13}, C_{14}, and C_{15}. Therefore, among the 16 chessboards shown in Fig. 5-1, there are only six "nonequivalent" ones.

Suppose that we are not interested in the black-and-white patterns of the chessboards but are only interested in the contrast patterns;

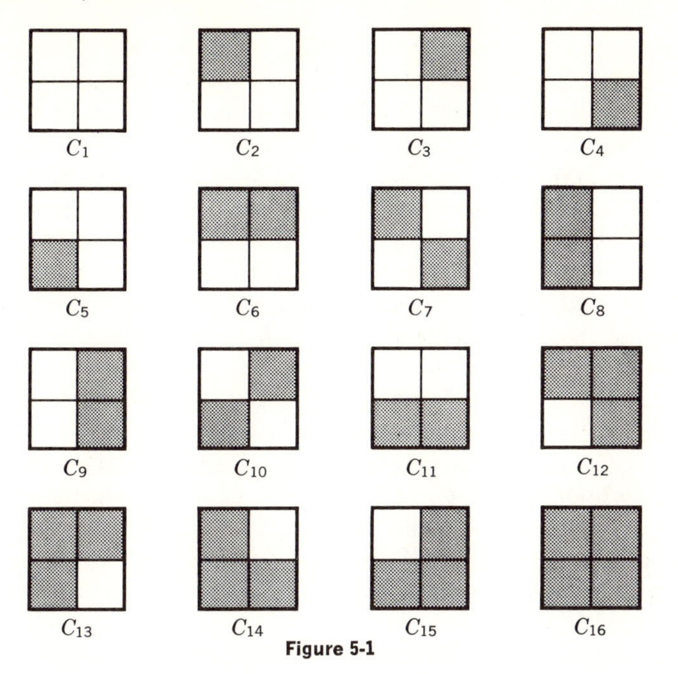

Figure 5-1

that is, chessboards C_1 and C_{16} have the same contrast pattern, chessboards C_2 and C_{15} have the same contrast pattern, and so on. With rotations of the chessboards also allowed, the reader can check that there are only four nonequivalent contrast patterns. In this chapter we shall study the theory of enumerating nonequivalent objects from a point of view first developed by Pólya in 1938.

●5-2 SETS, RELATIONS, AND GROUPS

In this section some of the basic notions and terminologies in algebra that are used throughout this chapter are introduced.

A *set* is a collection of distinct *elements* (objects). We shall use an uppercase Roman letter to denote a set and the notation $S = \{a,b,c,x,z\}$ to denote a set S that contains the elements a, b, c, x, and z. Notice that there is no ordering among the elements in a set. Thus, $\{a,b,c\}$ and $\{c,b,a\}$ denote identically the same set. Also, since the elements in a set are all distinct, $\{a,a,b,c\}$ is a redundant representation of the set $\{a,b,c\}$. The *empty* or *null* set, denoted by ϕ, is a set that contains no elements. A set T is said to be a *subset* of another set S if every element in T is also an element in S. For example, $\{a,b,x\}$ is a subset of $\{a,b,c,x,z\}$, but $\{a,b,y\}$ is not. Notice that every set is, trivially, a subset of itself. A set T is said to be a *proper subset* of S if T is a subset of S but there is at least one element in S that is not in T. We write $a \in S$ to mean that a is an element in the set S, and $T \subseteq S$ to mean that T is a subset of S. Also, $|S|$ denotes the number of elements in the set S. A set is said to be a k-set if it contains k elements.

Let A and B be two sets. The *union* of A and B, denoted by $A \cup B$, is the set that contains the elements in A and the elements in B. For example, $\{a,b,c,d\} \cup \{a,d,e,j\} = \{a,b,c,d,e,j\}$. The *intersection* of A and B, denoted by $A \cap B$, is the set that contains the elements that are in both A and B. For example, $\{a,b,c,d\} \cap \{a,d,e,j\} = \{a,d\}$. The *difference* of A and B, denoted by $A - B$, is a set that contains the elements that are in A but not in B. For example, $\{a,b,c,d\} - \{a,d,e,j\} = \{b,c\}$. The *ring sum* of A and B, denoted by $A \oplus B$, is the set that contains the elements in A and the elements in B which are not in the intersection of A and B. For example, $\{a,b,c,d\} \oplus \{a,d,e,j\} = \{b,c,e,j\}$.

A *partition* on a set is a subdivision of all the elements in the set into *disjoint* subsets. In other words, a partition on a set is a collection of subsets of the set such that every element in the set is in exactly one of the subsets. For example, $\{\{a,b,x\},\{d\},\{c,z\}\}$ is a partition on the set $\{a,b,c,d,x,z\}$.

An *ordered pair* is an ordered arrangement of two (not necessarily distinct) elements. We use the notation (a,b) for an ordered pair that contains the elements a and b, arranged in that order. Thus, (a,b) and (b,a) are two different ordered pairs. The *cartesian product* of two sets S and T, denoted by $S \times T$, is the set of all ordered pairs (x,y) in which x is in S and y is in T. For example,

$$\{a,b,c\} \times \{1,2\} = \{(a,1),(a,2),(b,1),(b,2),(c,1),(c,2)\}$$

A *binary relation* between two sets S and T is a subset of the ordered pairs in the cartesian product $S \times T$. For example, $\{(a,1),(a,2),(c,2)\}$ is a binary relation between the sets $\{a,b,c\}$ and $\{1,2\}$. For an ordered pair like $(a,2)$ in the relation, we say that the element a is *related* to the element 2. A binary relation between two sets can be represented in the form of a matrix. For example, Fig. 5-2 shows a representation of the relation $\{(a,1),(a,3),(b,4),(d,2),(d,4)\}$ between the sets $\{a,b,c,d\}$ and

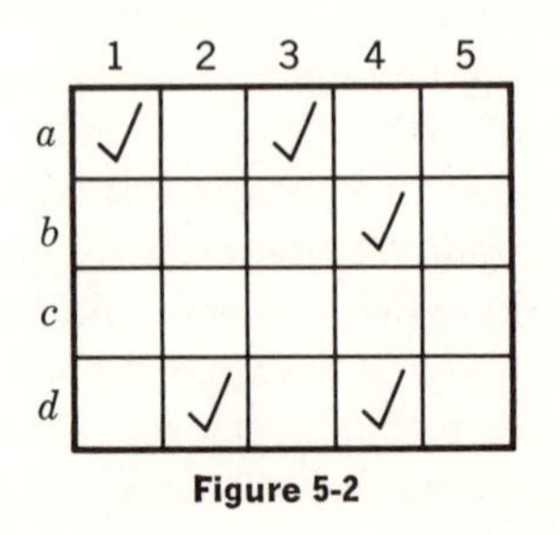

Figure 5-2

$\{1,2,3,4,5\}$. A check mark in a cell indicates that the element identifying the row that contains the cell and the element identifying the column that contains the cell are related. A binary relation on a set S is a binary relation between the set S and itself. For example, $\{(a,a),(a,c),(b,a),(b,c),(c,b)\}$ is a binary relation on the set $\{a,b,c\}$.

A binary relation on a set is called an *equivalence relation* if the following conditions are satisfied:

1. Every element in the set is related to itself (reflexive law).
2. For any two elements a and b in the set, if a is related to b, then b is also related to a (symmetric law).
3. For any three elements a, b, and c in the set, if a is related to b and b is related to c, then a is also related to c (transitive law).

For example, the binary relation shown in Fig. 5-3a is an equivalence relation, whereas the binary relation shown in Fig. 5-3b is not. As

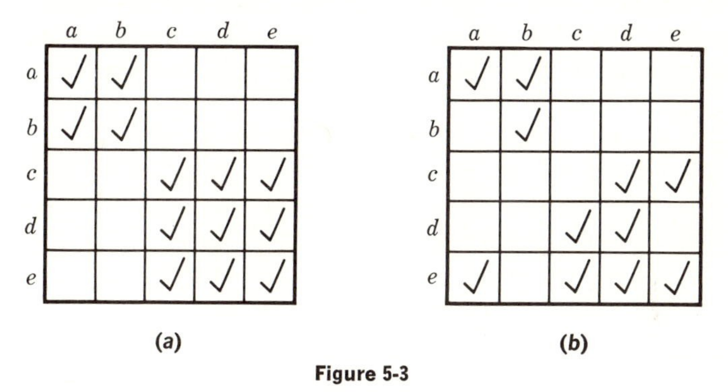

Figure 5-3

another example, we define a binary relation on a group of people which is such that for two persons a and b, a is related to b if and only if a is from the same family as b is. The reader can verify that this relation is an equivalence relation.

Given an equivalence relation on a set S, we can divide the elements of S into classes such that two elements are in the same class if and only if they are related. These classes of elements are called the *equivalence classes* into which the set S is divided by the equivalence relation. Notice that every element is in one of the equivalence classes because it can at least be in a class by itself, according to the reflexive law. The symmetric law ensures that there is no ambiguity regarding membership in the equivalence classes. (If the relation is not symmetric, we might encounter the difficult situation where a is related to b but b is not related to a.) Finally, because of the transitive law, no element can be in more than one equivalence class. Therefore, we say that an equivalence relation on a set induces a partition on the set in which the disjoint subsets are the equivalence classes. For example, the partition induced by the equivalence relation on the set $\{a,b,c,d,e\}$ shown in Fig. 5-3*a* is $\{\{a,b\},\{c,d,e\}\}$. Two elements are said to be *equivalent* if they are in the same equivalence class.

A (*single-valued*) *function* from a set S to a set T is a binary relation between the sets S and T which is such that every element in S is related to exactly one element in T. For example, $\{(a,2),(b,1),(c,2)\}$ is a function from the set $\{a,b,c\}$ to the set $\{1,2\}$. The set S is called the *domain* of the function, and the set T is called the *range* of the function. Let f denote a function, and let $(a,2)$ be an ordered pair in the function. We shall write $f(a) = 2$ to mean that a is related to 2 by the function f. We say that 2 is the *value* (or *image*) of a under the function f, and also that f *maps a into* 2. A function is a *one-to-one function* if every element in the domain has a unique image. A function is an *onto function* if every element in the range is the image of at least one element in the domain. For example, Fig. 5-4*a* shows an arbitrary function, Fig. 5-4*b* shows a one-to-one function, and Fig. 5-4*c* shows an onto function.

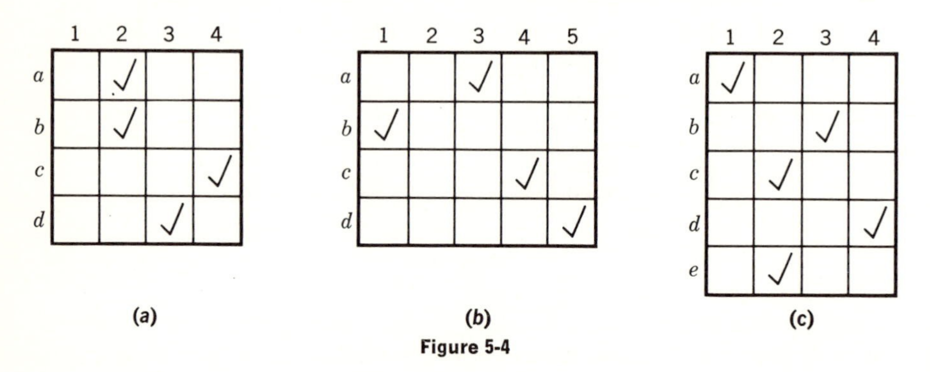

Figure 5-4

A *binary operation* on a set S is a function from the set $S \times S$ to a set T. For example, Fig. 5-5*a* shows a binary operation on the set

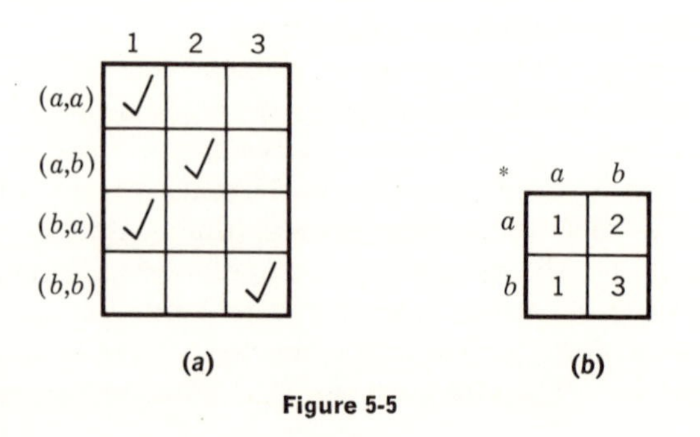

Figure 5-5

$S = \{a,b\}$ with $T = \{1,2,3\}$. An alternative way of representing a binary operation is shown in Fig. 5-5*b*. In that representation the value of an ordered pair under the binary operation is placed in the cell that is in the row corresponding to the first element of the ordered pair and is in the column corresponding to the second element of the ordered pair. Instead of the functional notation, we shall also let $*$ denote a binary operation and let $a * b$ denote the value of the ordered pair (a,b) under the binary operation. For example, in Fig. 5-5*a* (and Fig. 5-5*b*), we see that $a * a = 1$ and $a * b = 2$. A binary operation on a set S is said to be *closed* if it is a function from the set $S \times S$ to the set S.

A set S together with a binary operation $*$ on the set S is said to form a *group* if the following conditions are satisfied:

1. The binary operation $*$ is closed.
2. The binary operation $*$ is associative; that is, for any a, b, c in S,

   $$(a * b) * c = a * (b * c)$$

3. There is an element e in S which is such that $a * e = a$ for every a in S. This element is called an *identity element* of the group.
4. For any element a in S, there is another element in S, denoted by a^{-1} and called an *inverse* of a, which is such that $a * a^{-1} = e$.

As an example, Fig. 5-6 shows the binary operation for a group consisting of the five elements 0, 1, 2, 3, and 4. Notice that 0 is an identity

*	0	1	2	3	4
0	0	1	2	3	4
1	1	2	3	4	0
2	2	3	4	0	1
3	3	4	0	1	2
4	4	0	1	2	3

Figure 5-6

element, an inverse of the element 0 is 0 itself, an inverse of the element 1 is 4, and so on. Some of the basic properties of a group follow:

1. If b is an inverse of a, then a is an inverse of b. If b is an inverse of a, $a * b = e$. That gives $b * (a * b) = b * e = b$. Let b^{-1} denote an inverse of b. We have $(b * (a * b)) * b^{-1} = b * b^{-1}$. Since

   $$(b * (a * b)) * b^{-1} = b * (a * (b * b^{-1})) = b * (a * e) = b * a$$

   and $b * b^{-1} = e$, we have $b * a = e$. Therefore, a is an inverse of b.

2. For every a in S, $e * a = a$. According to property 1, $a^{-1} * a = e$. Thus, $a * (a^{-1} * a) = a * e$. However,

$$a * (a^{-1} * a) = (a * a^{-1}) * a = e * a$$

Therefore, $e * a = a * e = a$.
3. The identity element is unique. Suppose there are two elements e_1 and e_2 such that $a * e_1 = a$ and $a * e_2 = a$; that is, $a * e_1 = a * e_2$, or $a^{-1} * (a * e_1) = a^{-1} * (a * e_2)$, which simplifies to $e_1 = e_2$.
4. The inverse of any element is unique. Suppose there are two elements b and c such that $a * b = e$ and $a * c = e$; that is, $a * b = a * c$, or $a^{-1} * (a * b) = a^{-1} * (a * c)$, which simplifies to $b = c$.

5-3 EQUIVALENCE CLASSES UNDER A PERMUTATION GROUP

A one-to-one function from a set S to itself is called a permutation of the set S. We use the notation $\begin{pmatrix} abcd \\ bdca \end{pmatrix}$ for the permutation of the set $\{a,b,c,d\}$ that maps a into b, b into d, c into c, and d into a; that is, in the upper row the elements in the set are written down in an arbitrary order, and in the lower row the image of an element will be written below the element itself. Notice that the notion of a permutation of a set is the same as that discussed in Chap. 1, namely, an arrangement of a set of objects.

Let π_1 and π_2 be two permutations of a set S. The *composition* of π_1 and π_2, denoted by $\pi_1\pi_2$, is the successive permutations of the set S, first according to π_2 and then according to π_1. For example, let $\pi_1 = \begin{pmatrix} abcd \\ adbc \end{pmatrix}$ and $\pi_2 = \begin{pmatrix} abcd \\ bacd \end{pmatrix}$ be two permutations of the set $\{a,b,c,d\}$. Then $\pi_1\pi_2 = \begin{pmatrix} abcd \\ dabc \end{pmatrix}$. Notice that $\pi_1\pi_2$ maps a into d since π_2 maps a into b and π_1 maps b into d, and so on.

We see immediately that the composition of two permutations is also a permutation. Let π_1 and π_2 be two permutations of the set $S = \{a,b,c, \ldots ,x,y,z\}$. To show that $\pi_1\pi_2$ is also a permutation of the set S, we have only to show that no two elements in S are mapped into the same element by $\pi_1\pi_2$. Suppose that π_2 maps the element a into b and π_1 maps the element b into c. $\pi_1\pi_2$ will then map the element a into c. Let x be any element distinct from a. Since π_2 is a permutation of the set S, π_2 maps x into an element that is distinct from b, say y. Similarly, π_1 maps y into an element that is distinct from c, say z. We conclude that $\pi_1\pi_2$ always maps two distinct elements (for example,

a and x) into two distinct elements (for example, c and z) and is, therefore, a permutation of the set S.

Notice that the composition of permutations is noncommutative; that is, in general $\pi_1\pi_2 \neq \pi_2\pi_1$. This is illustrated by the fact that for $\pi_1 = \begin{pmatrix} abcd \\ adbc \end{pmatrix}$ and $\pi_2 = \begin{pmatrix} abcd \\ bacd \end{pmatrix}$, $\pi_1\pi_2 = \begin{pmatrix} abcd \\ dabc \end{pmatrix}$ and $\pi_2\pi_1 = \begin{pmatrix} abcd \\ bdac \end{pmatrix}$.

However, the composition of permutations is associative; that is, for any permutations π_1, π_2, and π_3 of a set, we have $(\pi_1\pi_2)\pi_3 = \pi_1(\pi_2\pi_3)$. This fact can be seen as follows: Suppose π_3 maps a into b, π_2 maps b into c, and π_1 maps c into d. Since $\pi_1\pi_2$ maps b into d, $(\pi_1\pi_2)\pi_3$ maps a into d. Similarly, since $\pi_2\pi_3$ maps a into c, $\pi_1(\pi_2\pi_3)$ maps a into d. For example, let

$$\pi_1 = \begin{pmatrix} abcd \\ adbc \end{pmatrix} \qquad \pi_2 = \begin{pmatrix} abcd \\ bacd \end{pmatrix} \qquad \pi_3 = \begin{pmatrix} abcd \\ bdac \end{pmatrix}$$

Then

$$(\pi_1\pi_2)\pi_3 = \left[\begin{pmatrix} abcd \\ adbc \end{pmatrix}\begin{pmatrix} abcd \\ bacd \end{pmatrix}\right]\begin{pmatrix} abcd \\ bdac \end{pmatrix} = \begin{pmatrix} abcd \\ dabc \end{pmatrix}\begin{pmatrix} abcd \\ bdac \end{pmatrix} = \begin{pmatrix} abcd \\ acdb \end{pmatrix}$$

and

$$\pi_1(\pi_2\pi_3) = \begin{pmatrix} abcd \\ adbc \end{pmatrix}\left[\begin{pmatrix} abcd \\ bacd \end{pmatrix}\begin{pmatrix} abcd \\ bdac \end{pmatrix}\right] = \begin{pmatrix} abcd \\ adbc \end{pmatrix}\begin{pmatrix} abcd \\ adbc \end{pmatrix} = \begin{pmatrix} abcd \\ acdb \end{pmatrix}$$

Let $G = \{\pi_1, \pi_2, \ldots\}$ be a set of permutations of a set S. Then G is said to be a *permutation group* of S if G and the binary operation of composition of permutations form a group. In other words, according to the definition of a group given in Sec. 5-2, the following conditions should be satisfied:

1. If π_1 and π_2 are in G, then $\pi_1\pi_2$ is also in G.
2. The binary operation, composition of permutations, is associative. However, this is known to be true.
3. The identity permutation that maps each element into itself, for example, $\begin{pmatrix} abcd \\ abcd \end{pmatrix}$, is in G. (Note that the identity permutation is the only permutation among all the permutations of a set that can be the identity element of the group.)
4. For every permutation π_1 in G, there is a permutation π_2, which is such that $\pi_1\pi_2$ is the identity permutation.

As an example, the reader can verify that

$$G = \left\{\begin{pmatrix} abc \\ abc \end{pmatrix}, \begin{pmatrix} abc \\ bca \end{pmatrix}, \begin{pmatrix} abc \\ cab \end{pmatrix}\right\}$$

is a permutation group of the set $\{a,b,c\}$.

Let G be a permutation group of a set $S = \{a,b, \ldots\}$. A binary relation on the set S, called the *binary relation induced by G*, is defined to be such that element a is related to element b if and only if there is a permutation in G that maps a into b. For example, let

$$G = \left\{ \begin{pmatrix} abcd \\ abcd \end{pmatrix}, \begin{pmatrix} abcd \\ bacd \end{pmatrix}, \begin{pmatrix} abcd \\ abdc \end{pmatrix}, \begin{pmatrix} abcd \\ badc \end{pmatrix} \right\}$$

The binary relation induced by G is shown in Fig. 5-7.

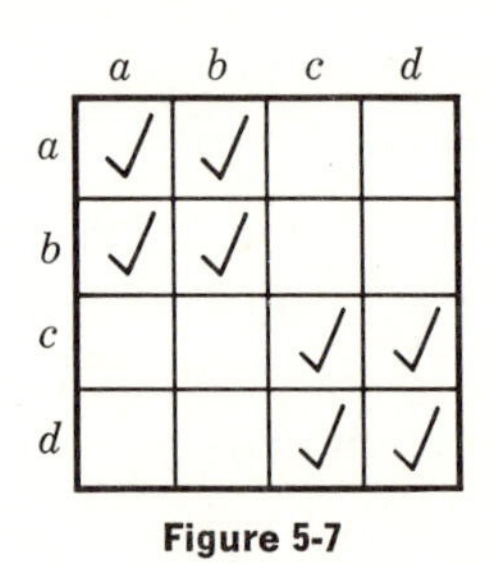

Figure 5-7

Theorem 5-1 *The binary relation on a set induced by a permutation group of the set is an equivalence relation.*

Proof Let G be a permutation group of the set $S = \{a,b, \ldots\}$.

1. Since the identity permutation is in G, every element in S is related to itself in the binary relation on S induced by G. Therefore, the reflexive law is satisfied.
2. If there is a permutation π_1 in G that maps a into b, the inverse of π_1, which is also in G, will map b into a. Therefore, the binary relation on S induced by G satisfies the symmetric law.
3. If there is a permutation π_1 mapping a into b and a permutation π_2 mapping b into c, the permutation $\pi_2\pi_1$, which is also in G, will map a into c. Therefore, the binary relation on S induced by G satisfies the transitive law. ■

After these preparations, we are now ready to prove a theorem due to Burnside. Given a set S and a permutation group G of S, we wish to find the number of equivalence classes into which S is divided by the equivalence relation on S induced by G. This problem can be solved most directly by finding the equivalence relation and then counting the number of equivalence classes. However, when the set S contains a large number of elements, such counting becomes prohibitively tedious. Burnside's theorem enables us to find the number of equivalence classes in an alternative way by counting the number of elements that are invariant under the permutations in the group. An element is said to be *invariant* under a permutation, or is called an *invariance*, if the permutation maps the element into itself.

Let us return to the example of 2×2 chessboards in Sec. 5-1 to see why we are interested in counting the number of equivalence classes into which a set is divided by the equivalence relation on the set induced by a permutation group. Observe in Fig. 5-1 that when the chessboards are rotated clockwise by 90°, C_1 remains as C_1, C_2 becomes C_3, C_3 becomes C_4, C_4 becomes C_5, C_5 becomes C_2, C_6 becomes C_9, C_7 becomes C_{10}, and so on. As a matter of fact, a 90° rotation amounts to a permutation of the chessboards since no two chessboards will become the same after the rotation. Let π_1 denote such a permutation of the chessboards. We have

$$\pi_1 = \begin{pmatrix} C_1C_2C_3C_4C_5C_6C_7C_8C_9C_{10}C_{11}C_{12}C_{13}C_{14}C_{15}C_{16} \\ C_1C_3C_4C_5C_2C_9C_{10}C_6C_{11}C_7C_8C_{15}C_{12}C_{13}C_{14}C_{16} \end{pmatrix}$$

Similarly, corresponding to a 180° clockwise rotation and a 270° clockwise rotation of the chessboards, there are the permutations π_2 and π_3. Thus,

$$\pi_2 = \begin{pmatrix} C_1C_2C_3C_4C_5C_6C_7C_8C_9C_{10}C_{11}C_{12}C_{13}C_{14}C_{15}C_{16} \\ C_1C_4C_5C_2C_3C_{11}C_7C_9C_8C_{10}C_6C_{14}C_{15}C_{12}C_{13}C_{16} \end{pmatrix}$$

$$\pi_3 = \begin{pmatrix} C_1C_2C_3C_4C_5C_6C_7C_8C_9C_{10}C_{11}C_{12}C_{13}C_{14}C_{15}C_{16} \\ C_1C_5C_2C_3C_4C_8C_{10}C_{11}C_6C_7C_9C_{13}C_{14}C_{15}C_{12}C_{16} \end{pmatrix}$$

Let π_4 be the identity permutation. Thus,

$$\pi_4 = \begin{pmatrix} C_1C_2C_3C_4C_5C_6C_7C_8C_9C_{10}C_{11}C_{12}C_{13}C_{14}C_{15}C_{16} \\ C_1C_2C_3C_4C_5C_6C_7C_8C_9C_{10}C_{11}C_{12}C_{13}C_{14}C_{15}C_{16} \end{pmatrix}$$

which corresponds to a 0° rotation. It can be shown that $G = \{\pi_1,\pi_2,\pi_3,\pi_4\}$ is a permutation group of the set of 2×2 chessboards. In the equivalence relation induced by G, we see that C_2 is related to C_3 (since π_1 maps C_2 into C_3), C_2 is related to C_4 (since π_2 maps C_2 into C_4), C_2 is related to C_5 (since π_3 maps C_2 into C_5), and C_2 is related to C_2 (since π_4 maps C_2 into C_2). Therefore, C_2, C_3, C_4, and C_5 are in the same equivalence class, which means that they become indistinguishable when rotations of the chessboards are allowed. It follows that the number of equivalence classes into which the chessboards are divided by the equivalence relation induced by G is the number of "distinct" chessboards. Here, two chessboards are distinct if one cannot be obtained from another through rotation. As shall be seen, there are many similar situations in which objects become equivalent under some classification which corresponds to the equivalence relation induced by a permutation group. In such cases, the enumeration theory developed in this chapter can be applied.

Theorem 5-2 (Burnside) *The number of equivalence classes into which a set S is divided by the equivalence relation induced by a permutation group G of S is given by*

$$\frac{1}{|G|}\sum_{\pi \in G} \psi(\pi)$$

where $\psi(\pi)$ is the number of elements that are invariant under the permutation π.

So that we can appreciate more the meaning of Theorem 5-2, let us illustrate its application before proceeding to the proof.

Example 5-1 Let $S = \{a,b,c,d\}$, and let G be the permutation group consisting of $\pi_1 = \begin{pmatrix} abcd \\ abcd \end{pmatrix}$, $\pi_2 = \begin{pmatrix} abcd \\ bacd \end{pmatrix}$, $\pi_3 = \begin{pmatrix} abcd \\ abdc \end{pmatrix}$, and $\pi_4 = \begin{pmatrix} abcd \\ badc \end{pmatrix}$. The equivalence relation on S induced by G is shown in Fig. 5-7. Clearly, S is divided into two equivalence classes, $\{a,b\}$ and $\{c,d\}$.

Since $\psi(\pi_1) = 4$, $\psi(\pi_2) = 2$, $\psi(\pi_3) = 2$, and $\psi(\pi_4) = 0$, according to Theorem 5-2, the number of equivalence classes can be computed as

$$\tfrac{1}{4}(4 + 2 + 2 + 0) = 2 \quad \blacksquare$$

Proof of Theorem 5-2 For any element s in S, let $\eta(s)$ denote the number of permutations under which s is invariant. Then

$$\sum_{\pi \in G} \psi(\pi) = \sum_{s \in S} \eta(s)$$

because both $\sum_{\pi \in G} \psi(\pi)$ and $\sum_{s \in S} \eta(s)$ count the total number of invariances under all the permutations in G. [One way to count the invariances is to go through the permutations one by one and count the number of invariances under each permutation. This gives $\sum_{\pi \in G} \psi(\pi)$ as the total count. Another way to count the invariances is to go through the elements one by one and count the number of permutations under which an element is invariant. That gives $\sum_{s \in S} \eta(s)$ as the total count.]

Let a and b be two elements in S that are in the same equivalence class. We want to show that there are exactly $\eta(a)$ permutations mapping a into b. Since a and b are in the same equivalence class, there is at least one such permutation which we shall denote by π_x. Let $\{\pi_1,\pi_2,\pi_3, \ldots\}$ be the set of the $\eta(a)$ permutations under which a is invariant. Then, the $\eta(a)$ permutations in the set $\{\pi_x\pi_1,\pi_x\pi_2,\pi_x\pi_3, \ldots\}$ are permutations that map a into b. First, we see that these permutations are all distinct because, if $\pi_x\pi_1 = \pi_x\pi_2$, we have

$$\pi_x^{-1}(\pi_x\pi_1) = \pi_x^{-1}(\pi_x\pi_2)$$

This gives $\pi_1 = \pi_2$, which is impossible. Secondly, we see that no other permutation in G maps a into b. Suppose that there is a permutation π_y that maps a into b. Then, $\pi_x^{-1}\pi_y$ is a permutation that maps a into a, because π_x^{-1} maps b into a. Since $\pi_x^{-1}\pi_y$ is a permutation in the set $\{\pi_1,\pi_2,\pi_3, \ldots\}$, $\pi_x(\pi_x^{-1}\pi_y) = \pi_y$ is a permutation in the set $\{\pi_x\pi_1,\pi_x\pi_2,\pi_x\pi_3, \ldots\}$. Therefore, we conclude that there are exactly $\eta(a)$ permutations in G that map a into b.

Let $a, b, c, \ldots, h$ be the elements in S that are in one equivalence class. All the permutations in G can be categorized as those that map a into a, those that map a into b, those that map a into c, $\ldots$, and those that map a into h. Since we have shown that there are exactly $\eta(a)$ permutations in each of these categories we have

$$\eta(a) = \frac{|G|}{\text{number of elements in the equivalence class containing } a}$$

Using a similar argument, we obtain

$$\eta(b) = \eta(c) = \cdots = \eta(h) = \frac{|G|}{\text{number of elements in the equivalence class containing } a}$$

and, therefore,

$$\eta(a) + \eta(b) + \eta(c) + \cdots + \eta(h) = |G|$$

It follows that, for any equivalence class of elements in S,

$$\sum_{\text{all } s \text{ in the equivalence class}} \eta(s) = |G|$$

and

$$\sum_{s\in S} \eta(s) = \begin{pmatrix}\text{number of equivalence classes} \\ \text{into which } S \text{ is divided}\end{pmatrix} \times |G|$$

Therefore, we have

$$\begin{matrix}\text{Number of equivalence classes} \\ \text{into which } S \text{ is divided}\end{matrix} = \frac{1}{|G|}\sum_{s\in S} \eta(s) = \frac{1}{|G|}\sum_{\pi\in G} \psi(\pi) \quad \blacksquare$$

Example 5-2 Find the number of distinct strings of length 2 that are made up of blue beads and yellow beads. The two ends of a string are not marked, and two strings are, therefore, indistinguishable if interchanging the ends of one will yield the other. Let b and y denote blue and yellow beads, respectively. Let bb, by, yb, and yy denote the four different strings of length 2 when equivalence between strings is not taken into consideration. The problem is to find the number of equivalence classes into which the set $S = \{bb,by,yb,yy\}$ is divided by the equivalence relation induced by the permutation group $G = \{\pi_1,\pi_2\}$, where

$$\pi_1 = \begin{pmatrix} bb & by & yb & yy \\ bb & by & yb & yy \end{pmatrix} \qquad \pi_2 = \begin{pmatrix} bb & by & yb & yy \\ bb & yb & by & yy \end{pmatrix}$$

The permutation π_1 merely indicates that every string is equivalent to itself, and the permutation π_2 indicates the equivalence between strings when the two ends of a string are interchanged. According to Burnside's theorem, the number of distinct strings is

$$\tfrac{1}{2}(4 + 2) = 3$$

Similarly, for the case of distinct strings of length 3 made up of blue beads and yellow beads, we have the set $S = \{bbb,bby,byb,ybb,byy,yby,yyb,yyy\}$ and the permutation group $G = \{\pi_1,\pi_2\}$, where π_1 is the identity permutation and π_2 is the permutation that maps a string into one that is obtained from the former by interchanging its ends; for example, bbb is mapped into bbb, bby is mapped into ybb, byb is mapped into byb, and so on. The number of elements that are invariant under π_1 is eight. The number of elements that are invariant under π_2 is four, since a string will be mapped into itself under π_2 if the beads at the two ends of a string are of the same color, and there are four such strings. Therefore, the number of distinct strings is equal to

$$\tfrac{1}{2}(8 + 4) = 6 \quad \blacksquare$$

Example 5-3 Find the number of distinct bracelets of five beads made up of yellow, blue, and white beads. Two bracelets are said to be indistinguishable if the rotation of one will yield another. However, to simplify the problem, we assume that the bracelets cannot be flipped over. Let S be the set of the 3^5 ($= 243$) distinct bracelets when rotational equivalence is not considered. Let $G = \{\pi_1,\pi_2,\pi_3,\pi_4,\pi_5\}$ be a permutation group, where π_1 is the identity permutation and π_2 is the permutation that maps a bracelet into one which is the former rotated clockwise by one bead position (for example, 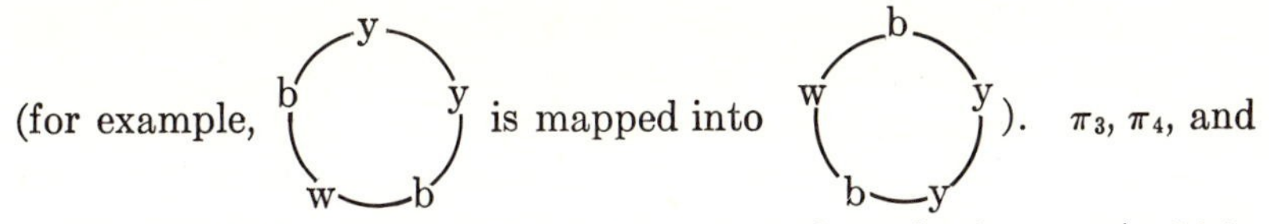

). π_3, π_4, and π_5 are, similarly, permutations that map a bracelet into one which is the former rotated clockwise by two, three, and four bead positions, respectively.

The number of elements that are invariant under π_1 is 243. The number of elements that are invariant under π_2 is three because only when all five beads in a bracelet are of the same color will its rotation by one bead position yield the same bracelet. Similarly, the number of elements that are invariant under each of π_3, π_4, and π_5 is also three. Therefore, the number of distinct bracelets is

$$\tfrac{1}{5}(243 + 3 + 3 + 3 + 3) = 51$$

It is interesting to observe that the problem of finding the number of ways to arrange n people around a circle, solved in Example 1-5, can also be solved using Burnside's theorem. Let S be the set of the $n!$ distinct ways to arrange n people around a circle when rotational equivalence is not considered. Let $G = \{\pi_1, \pi_2, \pi_3, \ldots, \pi_n\}$ be a permutation group where π_1 is the identity permutation, π_2 is the permutation that maps a circular arrangement into one which is the former rotated clockwise by one position, π_3 is the permutation that maps a circular arrangement into one which is the former rotated clockwise by two positions, . . . , and π_n is the permutation that maps a circular arrangement into one which is the former rotated clockwise by $n - 1$ positions. Since $\psi(\pi_1) = n!$ and $\psi(\pi_2) = \psi(\pi_3) = \cdots = \psi(\pi_{n-1}) = 0$, the number of distinct circular arrangements is

$$\frac{1}{n}(n! + 0 + 0 + \cdots + 0) = (n - 1)! \quad \blacksquare$$

Example 5-4 Suppose that we are to print all the five-digit numbers on slips of paper with one number on each slip. Clearly, there are 10^5 such slips. (For numbers smaller than 10,000, leading zeros are always filled in.) However, since the digits 0, 1, 6, 8, and 9 become 0, 1, 9, 8, and 6 when they are read upside down, there are pairs of numbers that can share the same slip if the slips will be read either right side up or upside down. For example, we can make up one slip for both the numbers 89166 and 99168. The question is then how many distinct slips will we have to make up for the 10^5 numbers. Here S is the set of the 10^5 numbers, $G = \{\pi_1, \pi_2\}$ is a permutation group of S, where π_1 is the identity permutation, and π_2 is the permutation that maps a number into itself if it is not readable as a number when turned upside down (for example, 13765 is mapped into 13765) and maps a number into the number obtained by reading the former upside down whenever it is possible (for example, 89166 is mapped into 99168). The number of invariances under π_1 is 10^5. The number of invariances under π_2 is $(10^5 - 5^5) + 3 \times 5^2$ because there are $10^5 - 5^5$ numbers that contain one or more of the digits 2, 3, 4, 5, and 7 and cannot be read upside down, and because there are 3×5^2 numbers that will read the same either right side up or upside down, for example, 16891 (the center digit of these numbers must be 0 or 1 or 8, the last digit must be the first digit turned upside down, and the fourth digit must be the second digit turned upside down). Therefore, the number of distinct slips to be made up is

$$\tfrac{1}{2}(10^5 + 10^5 - 5^5 + 3 \times 5^2) = 10^5 - \tfrac{1}{2} \times 5^5 + \tfrac{3}{2} \times 5^2 \quad \blacksquare$$

Theorem 5-2 can be generalized as Theorem 5-4 in the following.† Let Q be a group consisting of the elements $q_1, q_2, \ldots$ together with a binary operation $*$. Let $S = \{a, b, \ldots\}$. Suppose that every element q in Q is associated with a permutation π_q of the set S such that for any q_1 and q_2 in Q

$$\pi_{q_1 * q_2} = \pi_{q_1}\pi_{q_2}$$

That is, the permutation associated with the element $q_1 * q_2$ is equal to the composition of the permutations π_{q_1} and π_{q_2}, the permutations associated with the elements q_1 and q_2. This condition will be referred to as the *homomorphy condition.* Notice that different elements in Q need *not* be associated with distinct permutations.

Let us define a binary relation on the set S, called the binary relation induced by Q, such that elements a and b in S are related if and only if there is a permutation π_q, associated with an element q in Q, that maps a into b. We have the following theorems.

Theorem 5-3 *The binary relation induced by Q is an equivalence relation.*

Proof The proof is similar to that of Theorem 5-1.‡ $\blacksquare$

† Our reason for not stating Theorem 5-2 in a more general form is merely to simplify the issue.

‡ A reader who has no prior exposure to the theory of groups is encouraged to carry out the proofs of Theorems 5-3 and 5-4. Note that the homomorphy condition is essential.

Theorem 5-4 *The number of equivalence classes into which S is divided by the equivalence relation induced by Q is*

$$\frac{1}{|Q|}\sum_{q\in Q}\psi(\pi_q)$$

where $\psi(\pi_q)$ is the number of elements in S that are invariant under the permutation π_q, the permutation associated with the element q in Q.

Proof The proof is similar to that of Theorem 5-2. ■

It should be pointed out that the permutations associated with the elements in Q form a permutation group.† Moreover, the binary relation on the set S induced by this permutation group is the same as that induced by Q. (At this point, the reader might question the necessity of introducing Theorem 5-4, since Theorem 5-2 can be applied to the permutation group to count the number of equivalence classes. However, as will be seen later, it is more convenient in many occasions to consider the structure of the group Q than to consider the structure of the permutation group.)

5-4 EQUIVALENCE CLASSES OF FUNCTIONS

In applying Burnside's theorem to the counting of the number of equivalence classes into which a set is divided, one may find that the computation of the numbers of invariances under the permutations is still quite involved, especially when the set is large. Moreover, in addition to the number of equivalence classes, one may also wish to have further information about the properties of the equivalence classes. For example, in the problem of chessboards discussed in Sec. 5-1, one may wish to know the number of distinct chessboards consisting of two black cells and two white cells. (There are two such chessboards, C_6 and C_7.) Pólya's theory of counting, which we shall discuss in the remainder of this chapter, offers solutions to both of these problems.

In this section, the notion of equivalence classes of functions is introduced. Let f be a function from a set D, its domain, to a set R, its range. Since each element in D has a unique image in R, the function f corresponds to a way of distributing $|D|$ objects into $|R|$ cells. Therefore, the problem of enumerating the ways of distributing $|D|$ objects into $|R|$ cells is the same as that of enumerating the functions from D to R. For conciseness and clarity of notation, we shall conduct our discussion in terms of functions from one set to another set.

† The proof is left as an exercise.

Let D and R be two sets, and let G be a permutation group of the set D. We define a binary relation on the set of all the functions from D to R as follows: A function f_1 is related to another function f_2 if and only if there is a permutation π in G which is such that $f_1(d) = f_2[\pi(d)]$ for all d in D. This binary relation is an equivalence relation as shown below:

1. Because the identity permutation is in G, the reflexive law is satisfied.
2. If $f_1(d) = f_2[\pi(d)]$ for all d in D, then $f_2(d) = f_1[\pi^{-1}(d)]$ for all d in D. Since π^{-1} is a permutation in G, the symmetric law is satisfied.
3. If $f_1(d) = f_2[\pi_1(d)]$ and $f_2(d) = f_3[\pi_2(d)]$ for all d in D where π_1 and π_2 are permutations in G, then $f_1(d) = f_3[\pi_2\pi_1(d)]$ for all d in D. Since $\pi_2\pi_1$ is a permutation in G, the transitive law is satisfied.

It follows that the functions from D to R are divided into equivalence classes by the equivalence relation. These equivalence classes are also called *patterns*. The patterns correspond to the distinct ways of distributing $|D|$ objects into $|R|$ cells when equivalence between ways of distribution is introduced by the permutation group G.

As an example, let $D = \{a,b,c,d\}$ and $R = \{x,y\}$. Let G be the permutation group $\{\pi_1,\pi_2,\pi_3,\pi_4\}$, where $\pi_1 = \begin{pmatrix} abcd \\ bcda \end{pmatrix}$, $\pi_2 = \begin{pmatrix} abcd \\ cdab \end{pmatrix}$, $\pi_3 = \begin{pmatrix} abcd \\ dabc \end{pmatrix}$, and $\pi_4 = \begin{pmatrix} abcd \\ abcd \end{pmatrix}$. There are 16 functions, $f_1, f_2, \ldots, f_{16}$,

Table 5-1

	$f(a)$	$f(b)$	$f(c)$	$f(d)$
f_1	x	x	x	x
f_2	y	x	x	x
f_3	x	y	x	x
f_4	x	x	y	x
f_5	x	x	x	y
f_6	y	y	x	x
f_7	y	x	y	x
f_8	y	x	x	y
f_9	x	y	y	x
f_{10}	x	y	x	y
f_{11}	x	x	y	y
f_{12}	y	y	y	x
f_{13}	y	y	x	y
f_{14}	y	x	y	y
f_{15}	x	y	y	y
f_{16}	y	y	y	y

from D to R as shown in Table 5-1. For instance, since

$$\begin{aligned} f_3[\pi_1(a)] &= f_3(b) = y \quad &\text{and} \quad f_2(a) &= y \\ f_3[\pi_1(b)] &= f_3(c) = x \quad &\text{and} \quad f_2(b) &= x \\ f_3[\pi_1(c)] &= f_3(d) = x \quad &\text{and} \quad f_2(c) &= x \\ f_3[\pi_1(d)] &= f_3(a) = x \quad &\text{and} \quad f_2(d) &= x \end{aligned}$$

the functions f_2 and f_3 are equivalent. The reader can verify that the 16 functions are divided into six equivalence classes. They are $\{f_1\}$, $\{f_2,f_3,f_4,f_5\}$, $\{f_6,f_8,f_9,f_{11}\}$, $\{f_7,f_{10}\}$, $\{f_{12},f_{13},f_{14},f_{15}\}$, and $\{f_{16}\}$.

We recall now the problem of chessboards discussed in Sec. 5-1. Let the four cells in a 2×2 chessboard be labeled a, b, c, and d as shown in Fig. 5-8, and let the two colors, white and black, be denoted by x and y.

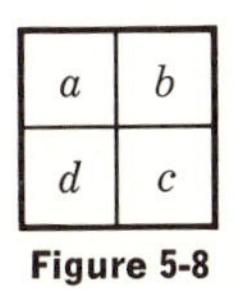

Figure 5-8

A function from the set $\{a,b,c,d\}$ to the set $\{x,y\}$ then corresponds to a chessboard, and the permutations in the group

$$G = \left\{ \begin{pmatrix} abcd \\ bcda \end{pmatrix}, \begin{pmatrix} abcd \\ cdab \end{pmatrix}, \begin{pmatrix} abcd \\ dabc \end{pmatrix}, \begin{pmatrix} abcd \\ abcd \end{pmatrix} \right\}$$

correspond to the rotations of the chessboards. For example, the permutation $\pi_1 = \begin{pmatrix} abcd \\ bcda \end{pmatrix}$ corresponds to the rotation of the chessboards in a clockwise direction by 90° since it specifies that two chessboards are equivalent if the cell a of one board and the cell b [which is equal to $\pi_1(a)$] of another board are of the same color, the cell b of one board and the cell c [which is equal to $\pi_1(b)$] of another board are of the same color, the cell c of one board and the cell d [which is equal to $\pi_1(c)$] of another board are of the same color, and the cell d of one board and the cell a [which is equal to $\pi_1(d)$] of another board are of the same color. The reader can verify that the 16 functions in Table 5-1 correspond to the 16 chessboards. As was shown above, they are divided into six equivalence classes by the equivalence relation induced by G.

5-5 WEIGHTS AND INVENTORIES OF FUNCTIONS

In addition to counting the number of equivalence classes of functions, we frequently wish to have information about the properties of the functions in the equivalence classes. For that purpose, we introduce the concept of the *weight of a function*. Let D and R be the domain and the range, respectively, of a set of $|R|^{|D|}$ functions. Suppose that a *weight* is assigned to each of the elements in R. The weights can be either numbers or symbols. Let r be an element in R, and let $w(r)$ denote the weight assigned to r. The *store enumerator* of the set R is defined to be the sum of the weights of the elements in R; that is,

$$\text{Store enumerator} = \sum_{r \in R} w(r)$$

The term "store enumerator" is actually very descriptive. Since the elements in the set R are the values that the elements in the set D can assume under functions from D to R, the store enumerator is a description of what is "in the store." For example, let $R = \{r_1,r_2,r_3\}$ and $w(r_1) = r_1$, $w(r_2) = r_2$, and $w(r_3) = r_3$. Then the store enumerator is $r_1 + r_2 + r_3$, which simply indicates that the value that an element in D can assume is either r_1 or r_2 or r_3. Suppose we let $w(r_1) = u$, $w(r_2) = v$, and $w(r_3) = u$. The store enumerator $2u + v$ means that there are two elements of type u and one element of type v in the set R from which the value for an element in D can be chosen. It should be pointed out that the notion of store enumerator is just a generalization of the notion of generating functions we have developed in Chap. 2. For the selection of one object from the three objects r_1, r_2, and r_3, according to our discussion in Chap. 2, the generating function is $r_1x + r_2x + r_3x$ where x is just the indicator which can be omitted when it is understood that exactly one of the three objects is selected. Furthermore, when objects r_1 and r_3 are of the same kind u and object r_2 is of another kind v, the generating function becomes $2u + v$.

For a function f from D to R, we define its *weight*, denoted by $W(f)$, as the product of the weights of the images of the elements in D under f; that is,

$$W(f) = \prod_{d \in D} w[f(d)]$$

The *inventory* of a set of functions is defined as the sum of their weights; that is,

$$\text{Inventory of a set of functions} = \sum_{\text{all } f \text{ in the set}} W(f)$$

As an example, let $D = \{d_1,d_2,d_3\}$, $R = \{r_1,r_2,r_3\}$, $w(r_1) = u$, $w(r_2) = v$, and $w(r_3) = u$. The weight of the function f_1 in Fig. 5-9 is $W(f_1) = uv^2$. Similarly, the inventory of the set of functions f_1, f_2, and f_3 in Fig. 5-9 is

$$W(f_1) + W(f_2) + W(f_3) = uv^2 + 2u^2v$$

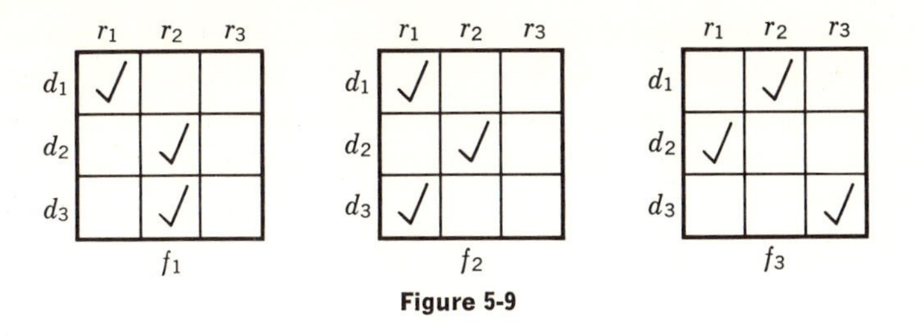

Figure 5-9

Again, recalling the discussion in Chap. 2, we see that the weight of a function is a representation of the way $|D|$ objects are distributed into $|R|$ cells as described by the function. Similarly, the inventory of a set of functions is a representation of the ways the objects are distributed.

Let G be a permutation group of D. As was shown in Sec. 5-4, the $|R|^{|D|}$ functions are divided into equivalence classes by the equivalence relation induced by G. Let f_1 and f_2 be two functions in the same equivalence class. Since there exists a permutation π in G such that

$$f_1(d) = f_2[\pi(d)]$$

for all d in D, we have

$$\prod_{d \in D} w[f_1(d)] = \prod_{d \in D} w[f_2(\pi(d))]$$

However,

$$\prod_{d \in D} w[f_2(\pi(d))] = \prod_{d \in D} w[f_2(d)]$$

because the two products contain the same factors, only in different orders. We conclude, therefore, that functions in the same equivalence class have the same weight. This weight is called the *weight of the pattern (equivalence class)*. Notice, however, that functions having the same weight might not be in the same equivalence class. Also, the *inventory of a set of patterns* is defined as the sum of the weights of the patterns in the set.

We present two examples illustrating the idea of weights of functions.

Example 5-5 Find all the possible ways of painting three distinct balls in solid colors when there are three kinds of paint available, an expensive kind of red paint, a cheap kind of red paint, and blue paint. Let D be the set of the three balls, and let R be the set of the three kinds of paint. Let r_1, r_2, and b be the weight assigned to the expensive red paint, cheap red paint, and blue paint, respectively. Since the store enumerator $r_1 + r_2 + b$ gives the ways in which one ball can be painted, $(r_1 + r_2 + b)^3$ gives all the possible ways in which the three balls can be painted. In other words, $(r_1 + r_2 + b)^3$ is the inventory of the set of all the functions from D to R. From the inventory

$$(r_1 + r_2 + b)^3 = r_1^3 + r_2^3 + b^3 + 3r_1^2r_2 + 3r_1r_2^2 + 3r_1^2b + 3r_2^2b + 3r_1b^2 + 3r_2b^2 + 6r_1r_2b$$

we have all the information about the different ways of painting the balls. For example, the term $3r_1r_2^2$ means that there are three ways of painting the three balls in which the expensive red paint is used for one ball and the cheap red paint is used for two balls.

Suppose we let the weights of both the expensive red paint and the cheap red paint be r and let the weight of the blue paint be b. The inventory of the set of all the functions from D to R is

$$(r + r + b)^3 = (2r + b)^3 = 8r^3 + 12r^2b + 6rb^2 + b^3$$

The store enumerator $2r + b$ indicates that there are two ways to paint a ball red and one way to paint a ball blue. In the inventory $(2r + b)^3$, the term $8r^3$ means that there are eight ways in which all three balls are painted red, and the term $12r^2b$ means that there are 12 ways in which two balls are painted red and one ball is painted blue, and so on. It should be emphasized that the two kinds of red paints are still two distinct kinds even though they are assigned the same weight. For example, painting all three balls with the expensive red paint is different from painting all three balls with the cheap red paint. They are counted as *two* ways of painting the balls in red. It is because we wish to look at the red paints as two kinds of paint having a common property that we assign to them the same weight. If the two kinds of red paint are indistinguishable (that means there is only one kind of red paint), the store enumerator should be $r + b$ instead. ∎

Example 5-6 Eight people are planning vacation trips. There are three cities they can visit. Three of these eight people are in one family, and two of them are in another family. If the people in the same family must go together, find the ways the eight people can plan their trips. Let $D = \{a,b,c,d,e,f,g,h\}$ be the set of the eight people, and suppose that a, b, and c are in one family, and d and e are in the other family. Let $R = \{c_1,c_2,c_3\}$ be the set of the three cities, and let α, β, and γ be the weights of c_1, c_2, and c_3. The symbolic representation of the different trips that a, b, and c can take is $\alpha^3 + \beta^3 + \gamma^3$ because they will either visit c_1 together, c_2 together, or c_3 together. Similarly, the symbolic representation of the different trips that d and e can take is $\alpha^2 + \beta^2 + \gamma^2$, and the symbolic representation of the different trips that each of f, g, and h can take is $\alpha + \beta + \gamma$. Therefore, the different ways in which the eight people can plan their trips are

$$(\alpha^3 + \beta^3 + \gamma^3)(\alpha^2 + \beta^2 + \gamma^2)(\alpha + \beta + \gamma)^3$$

As a matter of fact, this example illustrates a general result that is used in a later section. Let $\{D_1, D_2, \ldots, D_k\}$ be a partition on the set D where $D_1, D_2, \ldots, D_k$ are the disjoint subsets. The inventory of the set of all the functions from D to R which are such that the elements in the same subset will have the same value is

$$\prod_{i=1}^{k} \left[\sum_{r \in R} w(r)^{|D_i|} \right]$$

because $\sum_{r \in R} w(r)^{|D_i|}$ is the representation of the ways to distribute the objects in the subset D_i such that they will all be in the same cell. ■

5-6 PÓLYA'S FUNDAMENTAL THEOREM

Let D and R be two sets, and let G be a permutation group of D. Our problem is to find the inventory of the equivalence classes of the functions from D to R, which is also called the *pattern inventory*. As was pointed out in Secs. 5-4 and 5-5, the pattern inventory is a representation of all the distinct ways of distributing the objects in D into the cells in R.

Let us categorize the $|R|^{|D|}$ functions from D to R according to their weights. Let $F_1, F_2, \ldots, F_i, \ldots$ denote the sets of functions that have weights $W_1, W_2, \ldots, W_i, \ldots$, respectively. Associated with each permutation π in the group G, let us define a function $\pi^{(i)}$, mapping the set of functions F_i into itself, which is such that a function f_1 in F_i will be mapped into the function f_2, where $f_1(d) = f_2[\pi(d)]$ for all d in D. Notice that f_2 is, indeed, a function in the set F_i as both f_1 and f_2 have the same weight W_i.

Lemma 5-1 *The function $\pi^{(i)}$ is a permutation of the set of functions F_i.*

Proof We only have to prove that no two functions in F_i are mapped into the same function by $\pi^{(i)}$. Suppose there are two functions f_1 and f_3 both of which are mapped into f_2 under $\pi^{(i)}$; that is, $f_1(d) = f_2[\pi(d)]$ and $f_3(d) = f_2[\pi(d)]$ for all d in D. This means that $f_1(d) = f_3(d)$ for all d in D, and f_1 and f_3 are the same function. ■

Lemma 5-2 *For any π_1 and π_2 in G*

$$(\pi_1\pi_2)^{(i)} = \pi_1^{(i)}\pi_2^{(i)}$$

Proof Note that the condition to be proved is the homomorphy condition defined in Sec. 5-3.

Suppose that $\pi_2^{(i)}$ maps f_1 into f_2 and $\pi_1^{(i)}$ maps f_2 into f_3. That is, $f_1(d) = f_2[\pi_2(d)]$ and $f_2(d) = f_3[\pi_1(d)]$ for all d in D. It follows that $f_1(d) = f_3[\pi_1\pi_2(d)]$ for all d in D. Therefore, both $\pi_1^{(i)}\pi_2^{(i)}$ and $(\pi_1\pi_2)^{(i)}$ map f_1 into f_3. ∎

A *cycle* in a permutation is a subset of elements that are cyclically permuted. For example, in the permutation $\begin{pmatrix} abcdef \\ cedabf \end{pmatrix}$, $\{a,c,d\}$ forms a cycle since a is permuted into c, c is permuted into d, and d is permuted into a. Similarly, $\{b,e\}$ forms a cycle, and $\{f\}$ forms a cycle. The *length* of a cycle is the number of elements in the cycle. In the permutation $\begin{pmatrix} abcdef \\ cedabf \end{pmatrix}$, there is a cycle of length 3, a cycle of length 2, and a cycle of length 1.

Let π be a permutation that has b_1 cycles of length 1, b_2 cycles of length 2, . . . , b_k cycles of length k, and so on. We shall use x_1, x_2, . . . , x_k, . . . as formal variables and use the monomial $x_1^{b_1}x_2^{b_2} \cdots x_k^{b_k} \cdots$ to represent the number of cycles of various lengths in the permutation π. Such a representation is called the *cycle structure representation* of the permutation π. Given a permutation group G, we define the *cycle index* P_G of G as the sum of the cycle structure representations of the permutations in G divided by the number of permutations in G; that is,

$$P_G(x_1,x_2, \ldots ,x_k, \ldots) = \frac{1}{|G|} \sum_{\pi \in G} x_1^{b_1}x_2^{b_2} \cdots x_k^{b_k} \cdots$$

As an example, the cycle index of the permutation group consisting of the permutations $\begin{pmatrix} abcd \\ abcd \end{pmatrix}$, $\begin{pmatrix} abcd \\ bacd \end{pmatrix}$, $\begin{pmatrix} abcd \\ abdc \end{pmatrix}$, $\begin{pmatrix} abcd \\ badc \end{pmatrix}$ is

$$\tfrac{1}{4}(x_1^4 + x_1^2x_2 + x_1^2x_2 + x_2^2) = \tfrac{1}{4}(x_1^4 + 2x_1^2x_2 + x_2^2)$$

We are now ready to prove a fundamental theorem due to Pólya.

Theorem 5-5 (Pólya) *The inventory of the equivalence classes of functions from domain D to range R is*

$$P_G\left(\sum_{r \in R} w(r), \sum_{r \in R} [w(r)]^2, \ldots, \sum_{r \in R} [w(r)]^k, \ldots\right)$$

that is, the pattern inventory is obtained by substituting $\sum_{r \in R} w(r)$ *for* x_1, $\sum_{r \in R} [w(r)]^2$ *for* $x_2, \ldots, \sum_{r \in R} [w(r)]^k$ *for* $x_k, \ldots$ *in the expression of the cycle index* P_G *of the permutation group G.*

Proof Let m_i denote the number of equivalence classes of functions that have the weight W_i (in the set F_i). Clearly, the pattern inventory is equal to

$$\sum_i m_i W_i$$

According to Lemmas 5-1 and 5-2 and Theorem 5-4, we have

$$m_i = \frac{1}{|G|} \sum_{\pi \in G} \psi(\pi^{(i)})$$

Therefore,

$$\sum_i m_i W_i = \sum_i \left[\frac{1}{|G|} \sum_{\pi \in G} \psi(\pi^{(i)}) \right] W_i = \frac{1}{|G|} \sum_{\pi \in G} \left[\sum_i \psi(\pi^{(i)}) W_i \right]$$

The term $\sum_i \psi(\pi^{(i)}) W_i$ is the inventory of all the functions f which are such that $f(d) = f[\pi(d)]$ for all d in D. Notice that for a function f, $f(d) = f[\pi(d)]$ for all d in D if and only if the elements in D that are in one cycle in π have the same value under f. Therefore,

$$\sum_i \psi(\pi^{(i)}) W_i = \left[\sum_{r \in R} w(r) \right]^{b_1} \left[\sum_{r \in R} w(r)^2 \right]^{b_2} \cdots \left[\sum_{r \in R} w(r)^k \right]^{b_k} \cdots$$

where $b_1, b_2, \ldots, b_k, \ldots$ are the number of cycles of length $1, 2, \ldots, k, \ldots$ in π, respectively. (See Example 5-6.) It follows that

$$\sum_i m_i W_i = P_G \left(\sum_{r \in R} w(r), \sum_{r \in R} [w(r)]^2, \ldots, \sum_{r \in R} [w(r)]^k, \ldots \right) \quad ■$$

Corollary 5-5.1 *The number of equivalence classes of functions from D to R is*

$$P_G(|R|, |R|, \ldots, |R|, \ldots)$$

Proof If the weight 1 is assigned to each of the elements in R, the weight of any pattern is also equal to 1. Therefore, the pattern inventory gives the number of patterns. ■

Example 5-7 We now solve the second half of Example 5-2 using the result in Theorem 5-5. To find the number of distinct strings of three beads, let $D = \{1,2,3\}$ be the set of the three positions in a string, and let $R = \{b,y\}$ be the set of the two kinds of bead. Let

$w(b) = b$ and $w(y) = y$ be the weights of the elements in R. Let $G = \left\{\begin{pmatrix}123\\123\end{pmatrix}, \begin{pmatrix}123\\321\end{pmatrix}\right\}$. Clearly, the permutation $\begin{pmatrix}123\\123\end{pmatrix}$ corresponds to leaving a string as it is, and the permutation $\begin{pmatrix}123\\321\end{pmatrix}$ corresponds to interchanging the two ends of a string. Since the cycle index of the group G is

$$P_G(x_1,x_2) = \tfrac{1}{2}(x_1^3 + x_1x_2)$$

the pattern inventory is

$$\tfrac{1}{2}[(b + y)^3 + (b + y)(b^2 + y^2)] = b^3 + 2b^2y + 2by^2 + y^3$$

From the pattern inventory, we see that there is one string that is made up of three blue beads, two strings that are made up of two blue beads and one yellow bead, and so on. By assigning

$$w(b) = w(y) = 1$$

we find that the number of patterns is six. ■

Example 5-8 Find the number of ways of painting the four faces a, b, c, and d of the pyramid in Fig. 5-10 with two colors of paints, x and y.

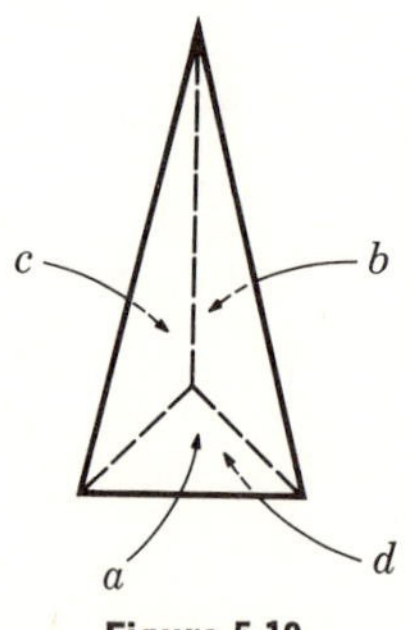

Figure 5-10

Let $D = \{a,b,c,d\}$ be the set of the four faces of the pyramid, and let $R = \{x,y\}$ be the set of the two colors with $w(x) = x$ and $w(y) = y$. The permutation group is $\left\{\begin{pmatrix}abcd\\abcd\end{pmatrix}, \begin{pmatrix}abcd\\bcad\end{pmatrix}, \begin{pmatrix}abcd\\cabd\end{pmatrix}\right\}$, where the permutation $\begin{pmatrix}abcd\\bcad\end{pmatrix}$ corresponds to the counterclockwise 120° rotation of the pyramid around the vertical axis, and the permutation $\begin{pmatrix}abcd\\cabd\end{pmatrix}$ corresponds to the counterclockwise 240° rotation of the pyramid around the vertical axis. Notice that in either rotation, face d remains in place. The cycle index of the group G is

$$\tfrac{1}{3}(x_1^4 + 2x_1x_3)$$

and the pattern inventory is

$$\tfrac{1}{3}[(x + y)^4 + 2(x + y)(x^3 + y^3)] = x^4 + y^4 + 2x^3y + 2x^2y^2 + 2xy^3$$

It follows that there are eight distinct ways of painting the four faces of the pyramid. ■

Example 5-9 Find the distinct ways of painting the eight vertices of a cube with two colors x and y. Let G be the permutation group corresponding to all possible rotations of the cube. There are 24 permutations in the group which can be divided into the following five categories:

1. The identity permutation. The cycle structure representation of this permutation is x_1^8.
2. Three permutations corresponding to 180° rotations around lines connecting the centers of opposite faces as shown in Fig. 5-11a.

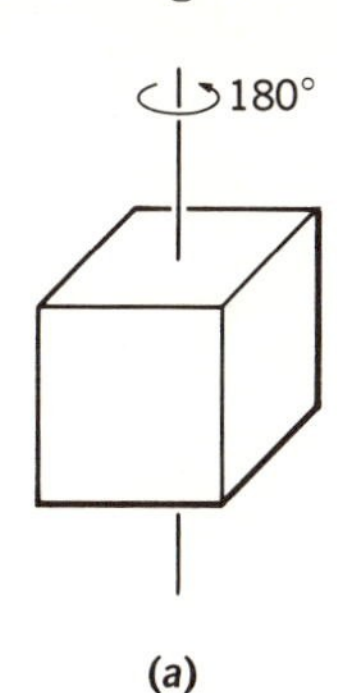

(a)

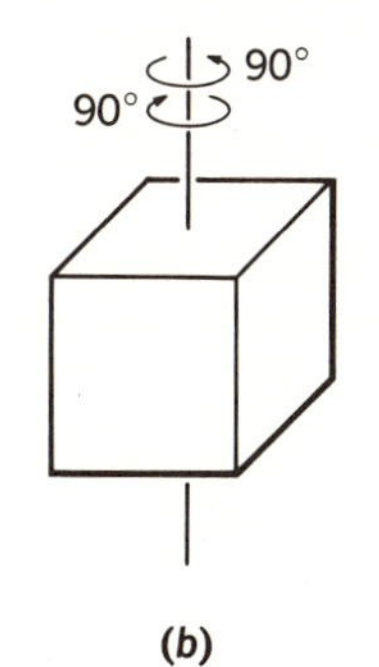

(b)

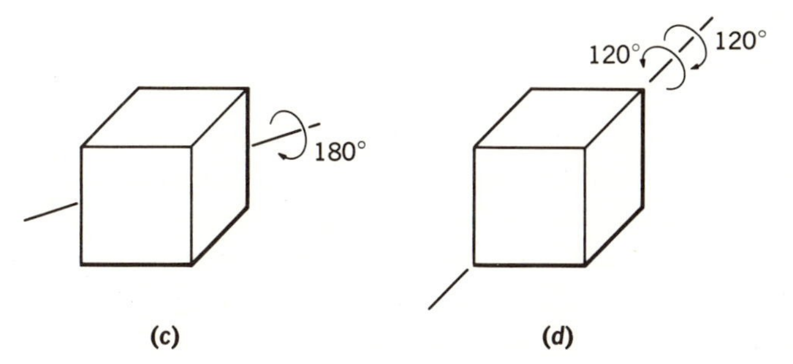

(c) (d)

Figure 5-11

The cycle structure representation of each of these permutations is x_2^4.

3. Six permutations corresponding to 90° rotations around lines connecting the centers of opposite faces as shown in Fig. 5-11*b*. The cycle structure representation of each of these permutations is x_4^2.
4. Six permutations corresponding to 180° rotations around lines connecting the midpoints of opposite edges as shown in Fig. 5-11*c*. The cycle structure representation of each of these permutations is x_2^4.
5. Eight permutations corresponding to 120° rotations around lines connecting opposite vertices as shown in Fig. 5-11*d*. The cycle structure representation of each of these permutations is $x_1^2x_3^2$.

Thus, the cycle index of the permutation group is

$$\tfrac{1}{24}(x_1^8 + 9x_2^4 + 6x_4^2 + 8x_1^2x_3^2)$$

and the pattern inventory is

$$\tfrac{1}{24}[(x + y)^8 + 9(x^2 + y^2)^4 + 6(x^4 + y^4)^2 + 8(x + y)^2(x^3 + y^3)^2]$$

By assigning $w(x) = w(y) = 1$, we compute the number of patterns as

$$\tfrac{1}{24}[2^8 + 9 \times 2^4 + 6 \times 2^2 + 8 \times 2^2 \times 2^2] = 23$$

which is the number of distinct ways of painting the eight vertices of a cube with two colors.† ■

Example 5-10 Consider the class of organic molecules of the form

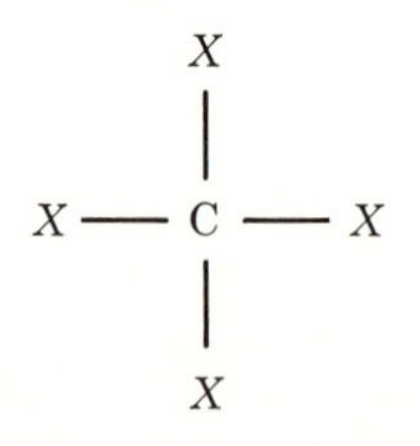

where C is a carbon atom, and each X denotes any one of the components CH_3 (methyl), C_2H_5 (ethyl), H (hydrogen), or Cl (chlorine). For example, the following is a typical molecule:

† This problem is closely related to the problem of counting the number of distinct Boolean functions. For further details on the subject, see Slepian [9] and Chap. 5 of Harrison [6].

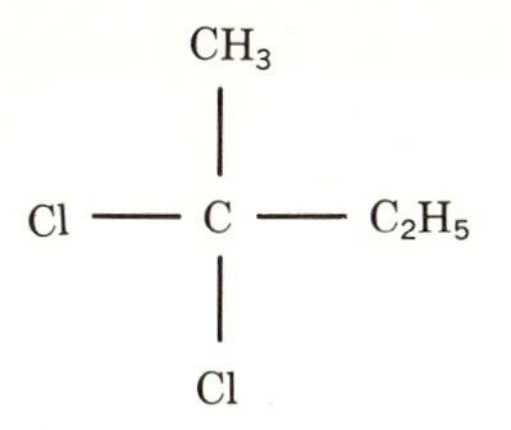

Each such molecule can be modeled as a regular tetrahedron with the carbon atom occupying the center position and the components labeled X at the corners. The problem of finding the number of different molecules of this form is the same as that of finding the number of equivalence classes of functions from the domain D containing the four corners of the tetrahedron to the range R containing the four components CH_3, C_2H_5, H, Cl, with the permutation group G consisting of the permutations corresponding to all the possible rotations of the tetrahedron.

To find the cycle index of the permutation group G, we notice that in G:

1. There is the identity permutation.
2. There are eight permutations corresponding to 120° rotations around lines connecting a vertex and the center of its opposite face as illustrated in Fig. 5-12*a*.

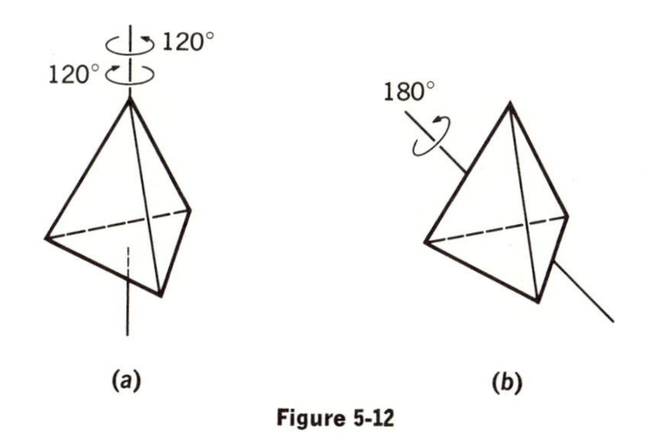

Figure 5-12

3. There are three permutations corresponding to 180° rotations around lines connecting the midpoints of opposite edges as illustrated in Fig. 5-12*b*.

It follows that

$$P_G(x_1,x_2,x_3) = \tfrac{1}{12}(x_1^4 + 8x_1x_3 + 3x_2^2)$$

Therefore, the number of different molecules is

$$P_G(4,4,4) = \tfrac{1}{12}(4^4 + 8 \times 4 \times 4 + 3 \times 4^2) = 36$$

Suppose we wish to find the number of molecules containing one or more hydrogen atoms. Let us assign the weight 1 to each of the components CH_3, C_2H_5, and Cl and the weight 0 to the component H. The pattern inventory is then

$$P_G(3,3,3) = \tfrac{1}{12}(3^4 + 8 \times 3 \times 3 + 3 \times 3^2) = 15$$

which is the number of molecules that *do not* contain the hydrogen atom. Therefore, there are

$$36 - 15 = 21$$

molecules containing the hydrogen atom.

If we assign the weight 1 to each of the components CH_3, C_2H_5, Cl and the weight h to the component H, the pattern inventory is

$$\begin{aligned} P_G(h + 3, h^2 + 3, h^3 + 3) &= \tfrac{1}{12}[(h + 3)^4 + 8(h + 3)(h^3 + 3) \\ &\qquad + 3(h^2 + 3)^2] \\ &= h^4 + 3h^3 + 6h^2 + 11h + 15 \end{aligned}$$

from which we find out immediately that there is one molecule containing four hydrogen atoms, three molecules containing three hydrogen atoms, six molecules containing two hydrogen atoms, 11 molecules containing one hydrogen atom, and 15 molecules containing no hydrogen atoms. ■

5-7 GENERALIZATION OF PÓLYA'S THEOREM

The notion of equivalence classes of functions is extended in this section. In addition to a permutation group G of the domain D, let there be a permutation group H of the range R. We define a binary relation on the functions from D to R as follows: A function f_1 is related to another function f_2 if and only if there is a permutation π in G and a permutation τ in H such that $\tau f_1(d) = f_2[\pi(d)]$ for all d in D.

We now show that such a binary relation is an equivalence relation.

1. Let both π and τ be the identity permutations in G and H. It follows that each function is related to itself and the reflexive law is satisfied.
2. Suppose that f_1 is related to f_2; that is, $\tau f_1(d) = f_2[\pi(d)]$ for all d in D. Since π^{-1} is a permutation of D, then $f_2[\pi(\pi^{-1}(d))] = \tau f_1[\pi^{-1}(d)]$ for all d in D; that is, $f_2(d) = \tau f_1[\pi^{-1}(d)]$, or $\tau^{-1}f_2(d) = f_1[\pi^{-1}(d)]$. Since π^{-1} is in G and τ^{-1} is in H, f_2 is related to f_1. Therefore, the symmetric law is satisfied.

3. Suppose that f_1 is related to f_2 and f_2 is related to f_3; that is, $\tau_1 f_1(d) = f_2[\pi_1(d)]$ and $\tau_2 f_2(d) = f_3[\pi_2(d)]$ for all d in D. Since π_1 is a permutation of D, $\tau_2 f_2(d) = f_3[\pi_2(d)]$ for all d in D is the same as

   $$\tau_2 f_2[\pi_1(d)] = f_3[\pi_2(\pi_1(d))]$$

   for all d in D. Thus, we have

   $$\tau_2\tau_1 f_1(d) = f_3[\pi_2\pi_1(d)]$$

   Since both $\pi_2\pi_1$ and $\tau_2\tau_1$ are in G and H, respectively, f_1 is related to f_3, and the transitive law is satisfied.

Similar to the case discussed in Sec. 5-4, such an equivalence relation divides the functions from D to R into equivalence classes. However, if we assign weights to the elements in R and compute the weights of the $|R|^{|D|}$ functions from D to R, we see that two functions in the same equivalence class may not have the same weight. This can be illustrated by a simple example. Let $D = \{a,b\}$ and $R = \{x,y\}$. Suppose that the permutation group G of the domain D contains the permutations $\pi_1 = \begin{pmatrix} ab \\ ab \end{pmatrix}$ and $\pi_2 = \begin{pmatrix} ab \\ ba \end{pmatrix}$, and suppose that the permutation group H of the range R contains the permutations $\tau_1 = \begin{pmatrix} xy \\ xy \end{pmatrix}$ and $\tau_2 = \begin{pmatrix} xy \\ yx \end{pmatrix}$. Clearly, the function f_1, with $f_1(a) = x$ and $f_1(b) = x$, and the function f_2, with $f_2(a) = y$ and $f_2(b) = y$, are equivalent because $\tau_2 f_1(d) = f_2[\pi_1(d)]$ for all d in D. However, for the assignment of weights $w(x) = x$ and $w(y) = y$, the weights of the functions f_1 and f_2 are x^2 and y^2, respectively. In order that we can still talk about the weight of a pattern and the pattern inventory, we shall have to impose the additional condition that weights should be assigned to the elements in the range R in such a way that functions in the same equivalence class will have the same weight. Here, we limit our discussion to the counting of the number of equivalence classes of functions. In this case, we assign the weight 1 to each element in R. Since the weight of any function is then equal to 1, the condition that the weights of the functions in the same equivalence class are the same is trivially satisfied. It follows that the pattern inventory will be the number of equivalence classes.

Theorem 5-6 *The number of equivalence classes of functions from D to R is given by*

$$\frac{1}{|G|}\frac{1}{|H|}\sum_{\pi\in G;\tau\in H} \psi[(\pi,\tau)']$$

where $\psi[(\pi,\tau)']$ *is the number of function* f*'s which are such that*

$$\tau f(d) = f[\pi(d)]$$

for all d *in* D.

Proof Let $G \times H$ be the set of $|G|\,|H|$ ordered pairs (π,τ), where π is a permutation in G and τ is a permutation in H. Let a binary operation $*$ on $G \times H$ be defined such that

$$(\pi_1,\tau_1) * (\pi_2,\tau_2) = (\pi_1\pi_2,\tau_1\tau_2).$$

It can be shown that $G \times H$ is a group under the binary operation $*$.

Associated with each ordered pair (π,τ) in $G \times H$, let us define a function $(\pi,\tau)'$ mapping the set of functions from D to R into the set itself, which is such that a function f_1 will be mapped into another function f_2, where $\tau f_1(d) = f_2[\pi(d)]$ for all d in D. Clearly, $(\pi,\tau)'$ is a permutation. Moreover, it can be shown that the homomorphy condition $(\pi_1\pi_2,\tau_1\tau_2)' = (\pi_1,\tau_1)'(\pi_2,\tau_2)'$ is satisfied.

According to Theorem 5-4, the number of equivalence classes into which the functions from D to R are divided by the equivalence relation induced by the group $G \times H$ is

$$\frac{1}{|G|\,|H|} \sum_{\pi \in G; \tau \in H} \psi[(\pi,\tau)']$$

where $\psi[(\pi,\tau)']$ is the number of invariances under the permutation $(\pi,\tau)'$. It follows that $\psi[(\pi,\tau)']$ is equal to the number of function f's which are such that $\tau f(d) = f[\pi(d)]$ for all d in D. ■

Lemma 5-3 *A function f from D to R is invariant under the permutation $(\pi,\tau)'$ if and only if f maps the elements of D that are in a cycle of length i in π into the elements of R that are in a cycle of length j in τ with j being a divisor of i. Moreover, within these two cycles, there must be a cyclic correspondence between the elements; that is, if $f(d) = r$, then $f[\pi(d)] = \tau(r)$, $f[\pi^2(d)] = \tau^2(r)$, . . . , $f[\pi^{i-1}(d)] = \tau^{i-1}(r)$.*

Proof That a function that satisfies these conditions is invariant under $(\pi,\tau)'$ is clear.

If f is invariant under $(\pi,\tau)'$, then $f(d) = r$ implies $f[\pi(d)] = \tau(r)$. It follows that $f[\pi(\pi(d))] = \tau f[\pi(d)]$, which is rewritten as $f[\pi^2(d)] = \tau^2(r)$. Similarly, $f[\pi^3(d)] = \tau^3(r)$, $f[\pi^4(d)] = \tau^4(r)$, . . . , $f[\pi^{i-1}(d)] = \tau^{i-1}(r)$, and $f[\pi^i(d)] = \tau^i(r)$. Since $\pi^i(d) = d$, it follows that $\tau^i(r) = r$ and i must be a multiple of j. ■

Theorem 5-7 *The number of equivalence classes of functions from D to R is the value of the expression*

$$P_G\left(\frac{\partial}{\partial z_1}, \frac{\partial}{\partial z_2}, \frac{\partial}{\partial z_3}, \ldots\right) \times P_H[e^{z_1+z_2+z_3+\cdots}, e^{2(z_2+z_4+z_6+\cdots)}, e^{3(z_3+z_6+z_9+\cdots)}, \ldots]$$

evaluated at $z_1 = z_2 = z_3 = \cdots = 0$.

Proof In view of Theorem 5-6, we have only to evaluate $\psi[(\pi,\tau)']$. According to Lemma 5-3, for a function that is invariant under the permutation $(\pi,\tau)'$, the elements in a cycle of length i in π must be mapped into elements in a cycle of length j in τ with j a divisor of i. Let b_i denote the number of cycles of length i in π, and let c_j denote the number of cycles of length j in τ. First of all, the elements in a cycle of length i in π can be mapped into the elements in any one of the c_j cycles of length j in τ. Second, for a cycle of length j in τ, there are j different ways in which a cyclic correspondence between the i elements in π and the j elements in τ can exist. Therefore, we have

$$\psi[(\pi,\tau)'] = \prod_i \Big(\sum_{\substack{j \\ j|i}} jc_j\Big)^{b_i}$$

$$= (c_1)^{b_1}(c_1 + 2c_2)^{b_2}(c_1 + 3c_3)^{b_3}(c_1 + 2c_2 + 4c_4)^{b_4}(c_1 + 5c_5)^{b_5} \cdots$$

Since

$$(c_1)^{b_1} = \left(\frac{\partial}{\partial z_1}\right)^{b_1} e^{c_1 z_1}\Big|_{z_1=0}$$

$$(c_1 + 2c_2)^{b_2} = \left(\frac{\partial}{\partial z_2}\right)^{b_2} e^{c_1 z_2} e^{2c_2 z_2}\Big|_{z_2=0}$$

$$(c_1 + 3c_3)^{b_3} = \left(\frac{\partial}{\partial z_3}\right)^{b_3} e^{c_1 z_3} e^{3c_3 z_3}\Big|_{z_3=0}$$

$$(c_1 + 2c_2 + 4c_4)^{b_4} = \left(\frac{\partial}{\partial z_4}\right)^{b_4} e^{c_1 z_4} e^{2c_2 z_4} e^{4c_4 z_4}\Big|_{z_4=0}$$

$$\cdots\cdots\cdots\cdots\cdots\cdots\cdots\cdots\cdots$$

we have

$$\psi[(\pi,\tau)'] = \left[\left(\frac{\partial}{\partial z_1}\right)^{b_1}\left(\frac{\partial}{\partial z_2}\right)^{b_2} \cdots \left(\frac{\partial}{\partial z_k}\right)^{b_k} \cdots\right] \times [e^{c_1(z_1+z_2+z_3+\cdots)} e^{2c_2(z_2+z_4+z_6+\cdots)} e^{3c_3(z_3+z_6+z_9+\cdots)} \cdots e^{mc_m(z_m+z_{2m}+z_{3m}+\cdots)} \cdots]\Big|_{z_1=z_2=z_3=\cdots=z_m=\cdots=0}$$

The theorem follows immediately. ■

Example 5-11 Let us have a final look at the example of the 2×2 chessboards in Sec. 5-1. Let $D = \{a,b,c,d\}$ be the set of the four cells and $R = \{x,y\}$ be the set of the two colors white and black. Let $G = \left\{\begin{pmatrix} abcd \\ abcd \end{pmatrix}, \begin{pmatrix} abcd \\ bcda \end{pmatrix}, \begin{pmatrix} abcd \\ cdab \end{pmatrix}, \begin{pmatrix} abcd \\ dabc \end{pmatrix}\right\}$, where the permutations correspond to the rotations of the chessboards. When we are interested only in the contrast patterns of the chessboards, we also have $H = \left\{\begin{pmatrix} xy \\ xy \end{pmatrix}, \begin{pmatrix} xy \\ yx \end{pmatrix}\right\}$, where the permutation $\begin{pmatrix} xy \\ yx \end{pmatrix}$ means the interchange of the two colors x and y. It follows from Theorem 5-7 that the number of distinct contrast patterns is

$$\frac{1}{8}\left(\frac{\partial^4}{\partial z_1^4} + \frac{\partial^2}{\partial z_2^2} + 2\frac{\partial}{\partial z_4}\right)\left[e^{2(z_1+z_2+z_3+z_4)} + e^{2(z_2+z_4)}\right]\Bigg|_{z_1=z_2=z_3=z_4=0}$$

$$= \tfrac{1}{8}(2^4 + 2^2 + 2^2 + 2 \times 2 + 2 \times 2)$$

$$= 4 \quad \blacksquare$$

Example 5-12 A certain number of messages are to be represented by n-digit quaternary sequences and transmitted through a communication channel. For each of the digits 0, 1, 2, and 3 received, a corresponding indicator light will be flashed so that the transmitted sequence can be recorded. Unfortunately, the indicator lights for the digits 2 and 3 were not labeled when the receiver was built, as illustrated in Fig. 5-13, and there is no way to tell which

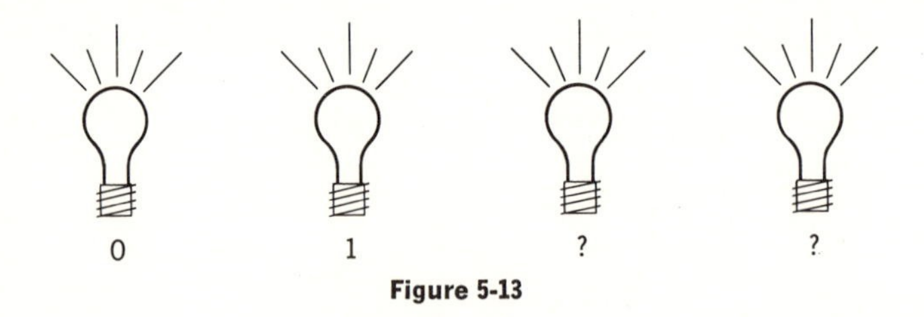

Figure 5-13

one of the two digits was transmitted. Therefore, we cannot expect to use all the 4^n n-digit sequences to represent 4^n distinct messages. For example, we cannot distinguish the two sequences 011023 and 011032 at the receiving end and one of them must be left unused. (Notice, however, that the two sequences 011022 and 011032 are distinguishable at the receiving end, since when the last two digits of the sequence 011022 are received, one of the two unlabeled lights will flash twice, whereas when the last two digits of the sequence 011032 are received, each of the two unlabeled lights will flash once.)

Let $D = \{a_1,a_2,a_3, \ldots ,a_n\}$ be the set of the n positions in the n-digit quaternary sequences. Let $R = \{0,1,2,3\}$ be the set of the four digits. Then $G = \left\{\begin{pmatrix} a_1a_2a_3 \cdots a_n \\ a_1a_2a_3 \cdots a_n \end{pmatrix}\right\}$ is the permutation group of D, and $H = \left\{\begin{pmatrix} 0123 \\ 0123 \end{pmatrix}, \begin{pmatrix} 0123 \\ 0132 \end{pmatrix}\right\}$ is the permutation group of R. The number of distinct messages one can transmit, which is equal to the number of distinct patterns from D to R, is

$$\frac{1}{2}\left(\frac{\partial^n}{\partial z_1^n}\right)(e^{4z_1} + e^{2z_1})\bigg|_{z_1=0} = \frac{1}{2}(4^n + 2^n)$$

Now suppose that at the transmitting end a sequence occasionally will be transmitted with the first two digits interchanged. Since there is no way to signal the receiver when this happens, how many distinct messages can be transmitted? In this case, the permutation group of D is $G = \left\{\begin{pmatrix} a_1a_2a_3 \cdots a_n \\ a_1a_2a_3 \cdots a_n \end{pmatrix}, \begin{pmatrix} a_1a_2a_3 \cdots a_n \\ a_2a_1a_3 \cdots a_n \end{pmatrix}\right\}$. The number of distinct messages that can be transmitted is then

$$\frac{1}{4}\left(\frac{\partial^n}{\partial z_1^n} + \frac{\partial^{n-2}}{\partial z_1^{n-2}}\frac{\partial}{\partial z_2}\right)[e^{4(z_1+z_2)} + e^{2(z_1+z_2)}e^{2z_2}]\bigg|_{z_1=z_2=0}$$

$$= \tfrac{1}{4}(4^n + 2^n + 4^{n-1} + 2^{n-2} \times 4)$$

$$= \tfrac{1}{4}(4^n + 4^{n-1} + 2^{n+1}) \quad \blacksquare$$

Example 5-13 In how many ways can five books, two of which are the same, be distributed to four children, if among them there is a set of identical twins? Let $D = \{a,b,c,d,e\}$ be the set of the five books with a and b being the two copies of the same book. Since two ways of distributing the books are equivalent if one becomes another when a and b are interchanged, we have $G = \left\{\begin{pmatrix} abcde \\ abcde \end{pmatrix}, \begin{pmatrix} abcde \\ bacde \end{pmatrix}\right\}$ as the permutation group of D. Let $R = \{u,v,x,y\}$ be the set of the four children with u and v being the twins. Since two ways of distributing the books are equivalent if the twins u and v interchange the books they receive, we have $H = \left\{\begin{pmatrix} uvxy \\ uvxy \end{pmatrix}, \begin{pmatrix} uvxy \\ vuxy \end{pmatrix}\right\}$ as the permutation group of R. The number of distinct patterns from D to R is

$$\frac{1}{4}\left(\frac{\partial^5}{\partial z_1^5} + \frac{\partial^3}{\partial z_1^3}\frac{\partial}{\partial z_2}\right)[e^{4(z_1+z_2)} + e^{2(z_1+z_2)}e^{2z_2}]\bigg|_{z_1=z_2=0}$$

$$= \tfrac{1}{4}(4^5 + 2^5 + 4^3 \times 4 + 2^3 \times 4)$$

$$= 336 \quad \blacksquare$$

Example 5-13 illustrates the relationship between the distinguishability of elements in the sets D and R and the permutation groups G and H. The permutation group G specifies that a and b are indistinguishable, and the permutation group H specifies that u and v are indistinguishable. However, notice that in our discussion in Chap. 1, objects are either distinct or totally indistinguishable. The introduction of the permutation groups of the sets D and R is a refinement of this notion. Objects become interchangeable under the permutations, whereas interchangeability does not always mean indistinguishability. For example, under the permutation group $G = \left\{\begin{pmatrix}abc\\abc\end{pmatrix}, \begin{pmatrix}abc\\bca\end{pmatrix}, \begin{pmatrix}abc\\cab\end{pmatrix}\right\}$, the elements a, b, and c do not become totally indistinguishable, but are only cyclically interchangeable. For instance, we cannot interchange the two elements a and b because the permutation $\begin{pmatrix}abc\\bac\end{pmatrix}$ is not in the group G. If a, b, and c are indeed totally indistinguishable, any interchange between them is certainly allowed. It follows that only when G is the group that contains all the possible permutations of the elements do the elements become totally indistinguishable. This point is illustrated in the following example.

Example 5-14 Let $D = \{a,b,c\}$, let $G = \left\{\begin{pmatrix}abc\\abc\end{pmatrix}, \begin{pmatrix}abc\\bca\end{pmatrix}, \begin{pmatrix}abc\\cab\end{pmatrix}, \begin{pmatrix}abc\\bac\end{pmatrix}, \begin{pmatrix}abc\\cba\end{pmatrix}, \begin{pmatrix}abc\\acb\end{pmatrix}\right\}$, and also let $R = \{x,y\}$ and $H = \left\{\begin{pmatrix}xy\\xy\end{pmatrix}, \begin{pmatrix}xy\\yx\end{pmatrix}\right\}$. The number of equivalence classes of functions from D to R is

$$\frac{1}{12}\left(\frac{\partial^3}{\partial z_1^3} + 2\frac{\partial}{\partial z_3} + 3\frac{\partial}{\partial z_1}\frac{\partial}{\partial z_2}\right)[e^{2(z_1+z_2+z_3)} + e^{2z_2}]\Big|_{z_1=z_2=z_3=0}$$

$$= \tfrac{1}{12}(2^3 + 2 \times 2 + 3 \times 2 \times 2) = 2$$

This is, of course, an expected result. The number of ways of distributing three indistinguishable objects into two indistinguishable cells is two. (The two ways are three in one cell, none in the other and two in one cell, one in the other.) ■

5-8 SUMMARY AND REFERENCES

In this chapter, we have studied Pólya's theory of counting, which is most useful in enumerating the equivalence classes into which a set of objects is divided by the equivalence relation induced by a permutation group. We see that the computation of the pattern inventory is simply another example of the application of the concept of generating functions studied in Chap. 2. As pointed out in Sec. 5-7, the notion of equivalence of objects is a refinement of the notion of indistinguishability of objects which was introduced at the start of our study of enumeration theory in Chap. 1. In this chapter, we hope that the reader not only sees how these basic concepts are tied together, but also appreciates the elegant and precise mathematical representation of many primitive intuitive notions.

Important applications of Pólya's theory of counting include the enumerations of graphs, trees, and Boolean functions. We shall see some of these in the exercises in this chapter and the next chapter. However, the interested reader is referred to the literature cited below for further details.

An excellent treatment of Pólya's theory of counting can be found in Chap. 5 of Beckenbach [1]. Pólya's fundamental theorem was presented in [7], and DeBruijn's generalization of the fundamental theorem was presented in [2]. Golomb's paper [3] contains a discussion of Burnside's theorem together with a number of examples. For applications of Pólya's theory of counting, see Pólya [7, 8], Harary [5], Slepian [9], Chap. 5 of Harrison [6], and Gilbert and Riordan [4].

1. Beckenbach, E. F. (ed.): "Applied Combinatorial Mathematics," John Wiley & Sons, Inc., New York, 1964.
2. DeBruijn, N. G.: Generalization of Pólya's Fundamental Theorem in Enumerative Combinatorial Analysis, *Ned. Akad. Wetenschap., Proc. Ser. A* **62,** *Indag. Math.,* **21**:59–79 (1956).
3. Golomb, S. W.: A Mathematical Theory of Discrete Classification, *Proc. Fourth London Symp. Inform. Theory,* Butterworth & Co. (Publishers), Ltd., London, 1961.
4. Gilbert, E. N., and J. Riordan: Symmetry Types of Periodic Sequences, *Illinois J. Math.,* **5**:657–665 (1961).
5. Harary, F.: The Number of Linear, Directed, Rooted and Connected Graphs, *Trans. Am. Math. Soc.,* **78**:445–463 (1955).
6. Harrison, M. A.: "Introduction to Switching and Automata Theory," McGraw-Hill Book Company, New York, 1965.
7. Pólya, G.: Kombinatorische Anzahlbestimmungen für Gruppen, Graphen und Chemische Verbindungen, *Acta Math.,* **68**:145–254 (1937).
8. Pólya, G.: Sur les Types des Propositions Composées, *J. Symbolic Logic,* **5**:98–103 (1940).
9. Slepian, D.: On the Number of Symmetry Types of Boolean Functions of n Variables, *Can. J. Math.,* **5**:185–193 (1953).

PROBLEMS

5-1. Let S be a set containing n elements.

(*a*) How many different binary relations on the set S are there?

(*b*) How many of them are reflexive?

(*c*) How many of them are symmetric?

(*d*) How many of them are neither reflexive nor symmetric?

(*e*) How many of them are equivalence relations? (This may be expressed in terms of Stirling numbers of the second kind.)

5-2. Let p_k denote the number of partitions of a set containing k elements, with $p_0 = 1$. Show that

$$p_{n+1} = \sum_{i=0}^{n} \binom{n}{i} p_i$$

5-3. The six faces of a cube are to be painted with six different colors, each face with a distinct color. In how many ways can this be done?

5-4. (*a*) The six faces of a cube are to be painted with one or more of six different colors. In how many ways can this be done?

(*b*) One of these six colors is red. Find the number of ways of painting the cube in which exactly three of the faces are painted red.

5-5. The six faces of a cube are to be painted with four different colors, A, B, C, and D. In how many ways can the cube be painted such that two of the faces are painted with color A, two of the faces are painted with color B, one of the faces is painted with color C, and one of the faces is painted with color D?

5-6. A rod divided into six segments is to be colored with one or more of n different colors. In how many ways can this be done?

5-7. Find the number of 2×4 red and white patterns made up with three red squares and five white squares.

5-8. (*a*) In how many distinct ways can the sectors of the circle in Fig. 5P-1 be painted with three colors?

(*b*) Repeat part (*a*) with the four radii extensions outside the circle removed.

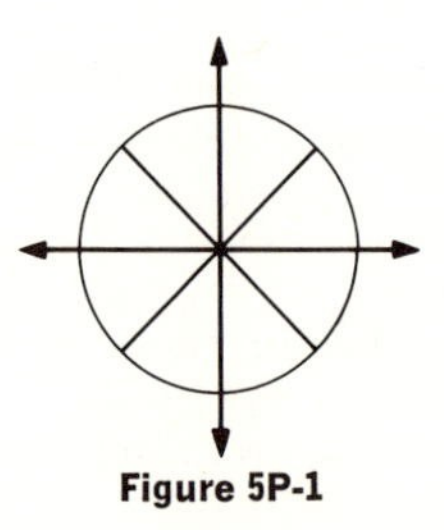

Figure 5P-1

5-9. Find the number of distinct ways of painting five of the eight vertices of a cube black and painting the remaining three white.

5-10. Show that $n^8 + 17n^4 + 6n^2$ is divisible by 24 for any positive integer n.

Hint: Consider the distinct ways of painting the vertices of a cube with n colors.

5-11. Let D be the set of the four faces of a regular pyramid, and let G be the group of all permutations of D that can be produced by the rotation of the pyramid.

(*a*) Find the cycle index of G.

(*b*) One or more of four colors, gold, red, white, and blue, are used to paint the faces of the pyramid. Find the number of distinct colorings.

5-12. Find the number of ways to distribute six balls, three red, two white, and one blue, into three distinct cells.

5-13. The sides of a square are colored using three colors. Two colorings are said to be equivalent if one can be obtained from the other by a rotation of the square and/or by a permutation of the colors. Find the number of equivalence classes of the coloring schemes.

5-14. (*a*) The cells of a 4×4 chessboard are to be painted with white and black paints. To generate all the black and white patterns, how many distinct drawings do we have to make?

(*b*) If each drawing can be used either as a pattern or as the negative of a pattern, how many distinct drawings do we have to make?

5-15. An electronic system consists of four different input units and four different output units. Connections between the various input units and output units are

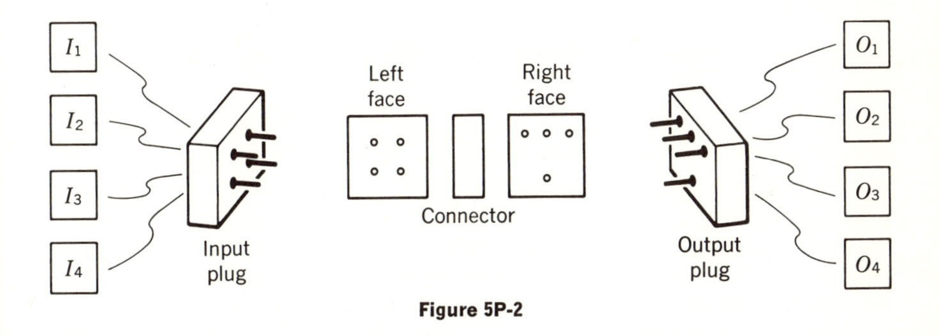

Figure 5P-2

made as shown in Fig. 5P-2. Connections between its left and right terminals are fixed internally within the connector.

(*a*) If each input unit must be connected to a distinct output unit, how many different connectors will be needed to permit all such system connections?

(*b*) If each input unit must be connected to an output unit, but any number of input units may share the same output unit, how many different connectors will be needed to permit all such system connections?

(*c*) Suppose the output plug is redesigned as shown in Fig. 5P-3 with a corresponding modification to the right face of the connector. Repeat part (*b*), with these modifications.

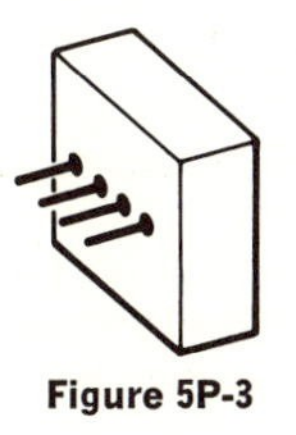

Figure 5P-3

5-16. From the 2^{2n} $2n$-digit binary sequences, subsets of k sequences are selected. Two subsets are said to be indistinguishable if they become identical subsets when the sequences in one subset are read from left to right and the sequences in the other are read from right to left. Find the number of distinct subsets of sequences.

5-17. To complement a binary digit means to change a 0 into a 1 and to change a 1 into a 0. Let D be the set of the 2^n n-digit binary sequences. Let G be a set of 2^n permutations of the set D, $\{\pi_{000\cdots0}, \pi_{000\cdots1}, \ldots, \pi_{111\cdots1}\}$, where the permutations are identified by the subscripts which are n-digit binary sequences. The permutation

$\pi_{a_1a_2\cdots a_n}$ permutes an n-digit binary sequence into another one by complementing its digits that correspond to the 1's in $a_1a_2 \cdots a_n$.

(*a*) Show that G is a permutation group.

(*b*) Show that the cycle index of G is

$$\tfrac{1}{2^n}[x_1^{2^n} + (2^n - 1)x_2^{2^{n-1}}]$$

(*c*) Let $R = \{0,1\}$. Find the number of equivalence classes of functions from D to R. (This is the number of distinct Boolean functions of n variables when complementation of the variables is permitted.)

(*d*) Find the number of equivalence classes of functions from D to R in which exactly k of the 2^n elements in D assume the value 1, for k being odd and for k being even.

5-18. In Prob. 5-17, find the number of equivalence classes of functions from D to R if there is also a permutation group $H = \left\{\begin{pmatrix}01\\01\end{pmatrix}, \begin{pmatrix}01\\10\end{pmatrix}\right\}$ acting on R.

5-19. Two Boolean functions are considered to be equivalent if one can be obtained from the other by permuting and/or complementing the independent variables. (The definition of a Boolean function is given in Prob. 1-17.) The set of all possible permutations and/or complementations of the independent variables can be represented by a permutation group G acting on the domain of the Boolean function.

(*a*) Find the number of equivalence classes of Boolean functions of two variables.

(*b*) Repeat part (*a*) for Boolean functions of three variables, given that the cycle index for G is

$$P_G = \tfrac{1}{48}(x_1^8 + 13x_2^4 + 8x_1^2x_3^2 + 8x_2x_6 + 6x_1^4x_2^2 + 12x_4^2)$$

(*c*) Repeat part (*a*) for Boolean functions of four variables, given that the cycle index for G is

$$P_G = \tfrac{1}{384}(x_1^{16} + 51x_2^8 + 48x_1^2x_2x_4^3 + 48x_8^2 + 12x_1^8x_2^4 + 84x_4^4 + 12x_1^4x_2^6 + 32x_1^4x_3^4 + 96x_2^2x_6^2)$$

(*d*) Repeat parts (*a*), (*b*), and (*c*), if two Boolean functions are considered to be equivalent if one can be obtained from the other by permuting and/or complementing the independent variables, and/or by complementing the value of the function.

5-20. Let S be a finite set of elements, and let G be a permutation group of S. Two subsets A and B of S are said to be equivalent if for some permutation π in G, the subset A is the set of all elements $\pi(b)$ obtained by letting b run through the elements in B.

(*a*) Show that all the subsets of S are partitioned into equivalence classes of subsets by the equivalence relation defined above.

(*b*) Show that the number of equivalence classes of subsets of S is equal to

$$P_G(2,2, \ldots ,2)$$

5-21. Let S be the set of all 2×2 matrices of 0's and 1's. Matrix multiplication is conducted in the field of integers modulo 2. Let T be a subset of S. A matrix t is in T if and only if t is nonsingular (the inverse of t exists).

(*a*) Show that the set T, together with the operation of matrix multiplication in the field of integers modulo 2, forms a group.

(*b*) Define a binary relation as follows. An element s_1 in S is said to be related to s_2 in S if there exists a matrix t in T such that $s_1 = t^{-1}s_2t$. Show that this relation is an equivalence relation.

(c) Find the number of equivalence classes in the set S under the equivalence relation defined above.

5-22. Let G be a permutation group of a set D. Two permutations π_1 and π_2 in G are said to be related by the self-conjugate relation if there exists a permutation π in G which is such that $\pi_1 = \pi^{-1}\pi_2\pi$.

(a) Show that the self-conjugate relation is an equivalence relation. (The equivalence classes of permutations are called self-conjugate classes.)

(b) Prove that if π_1 and π_2 are permutations in the same self-conjugate class, the number of invariances under π_1 is equal to that under π_2; that is, $\psi(\pi_1) = \psi(\pi_2)$.

(c) Show that the number of equivalence classes into which D is divided by the equivalence relation induced by G is equal to

$$\frac{1}{|G|} \sum \psi(\pi_K) \times |K|$$

where the summation is over all self-conjugate classes of permutations in G, π_K is any permutation in the self-conjugate class K, and $|K|$ is the number of permutations in K.

5-23. A group G is said to be a cyclic group if every element in the group is a power of some fixed element π in G. An element π_i is said to be of power j if j is the smallest positive integer such that $\pi_i = \pi^j$. If G contains n elements, it can be shown that these n elements can be expressed as $\pi, \pi^2, \pi^3, \ldots, \pi^{n-1}, \pi^n$, where π^n is the identity element in G.

(a) Let G be a cyclic permutation group of a set S. Show that if there are n permutations in G, the number of equivalence classes into which the set S is divided by the equivalence relation induced by G is

$$\frac{1}{n} \sum_{d|n} \psi(\pi_d)\phi\left(\frac{n}{d}\right)$$

where the summation is over all divisors d of n, π_d is the element in G of power d, and $\phi(n/d)$ is the number of positive integers which are less than or equal to n/d and are prime to n/d. [Recall that $\phi(x)$ is called the Euler function. See Prob. 4-11.]

(b) Find the number of distinct necklaces made up of eight beads of three distinct colors when cyclic rotations of the necklaces are allowed.

5-24. Consider a set D together with a permutation group G of D and a set R together with a permutation group H of R.

(a) Let π be a permutation in G, and let τ be a permutation in H. Let f be a function in the set of functions from D to R. Show that if

$$f(\pi(d)) = \tau f(d) \qquad \text{for all } d \in D$$

then

$$f(\pi^j(d)) = \tau^j f(d) \qquad \text{for all } d \in D;\ j = 1, 2, \ldots$$

(b) Let f be a one-to-one function. If $f(\pi(d)) = \tau f(d)$ for all d in D, prove that f maps any element in D that belongs to a cycle of length j in π into an element in R that belongs to a cycle of length j in τ.

(c) Let π in G be a permutation of the type $\{b_1, b_2, \ldots, b_{|D|}\}$ where b_j is the number of cycles of length j in π, and let τ in H be a permutation of the type $\{c_1, c_2, \ldots, c_{|R|}\}$ where c_j is the number of cycles of length j in τ. Show that the number of one-to-one functions f which are such that $f(\pi(d)) = \tau f(d)$ for all d in D is equal to

$$\prod_{j=1}^{|D|} j^{b_j} P(c_j, b_j) \qquad \text{where } P(c_j, b_j) = \begin{cases} \dfrac{c_j!}{(c_j - b_j)!} & c_j \geq b_j \\ 0 & c_j < b_j \end{cases}$$

Then show that

$$\prod_{j=1}^{|D|} j^{b_j} P(c_j, b_j) = \left[\left(\frac{\partial}{\partial z_1}\right)^{b_1} \left(\frac{\partial}{\partial z_2}\right)^{b_2} \cdots\right] [(1 + z_1)^{c_1} (1 + 2z_2)^{c_2} \cdots]$$

evaluated at $z_1 = z_2 = \cdots = 0$.

(*d*) Show that the number of equivalence classes of one-to-one functions from D to R, with equivalence induced by the permutation groups G and H, is equal to

$$P_G\left(\frac{\partial}{\partial z_1}, \frac{\partial}{\partial z_2}, \frac{\partial}{\partial z_3}, \ldots\right) P_H(1 + z_1, 1 + 2z_2, 1 + 3z_3, \ldots)$$

evaluated at $z_1 = z_2 = z_3 = \cdots = 0$.

(*e*) Show that if $|R| = |D|$, the number of equivalence classes of one-to-one functions from D to R is equal to

$$P_G\left(\frac{\partial}{\partial z_1}, \frac{\partial}{\partial z_2}, \frac{\partial}{\partial z_3}, \ldots\right) P_H(z_1, 2z_2, 3z_3, \ldots)$$

which in turn is equal to

$$P_H\left(\frac{\partial}{\partial z_1}, \frac{\partial}{\partial z_2}, \frac{\partial}{\partial z_3}, \ldots\right) P_G(z_1, 2z_2, 3z_3, \ldots)$$

5-25. Let D and R be two finite sets. Let f be a one-to-one function from D to R. The set of ordered pairs $(d, f(d))$ for all d in D is called a labeled set of R. Clearly, the labeled set is just an alternative representation of the one-to-one function f. Let H be a permutation group of R.

(*a*) Show that the number of equivalence classes of labeled sets of R is equal to

$$\left(\frac{d}{dz}\right)^{|D|} P_H(1 + z, 1, 1, \ldots) \bigg|_{z=0}$$

where equivalence between labeled sets is defined as equivalence between the corresponding one-to-one functions from D to R when the permutation group acting on D consists of only the identity permutation and the permutation group acting on R is H.

(*b*) If H also consists of the identity element only, find the number of equivalence classes of labeled sets of R.

(*c*) Let R be the set of the faces of a cube, and let H be the group of permutations of the faces obtained by the rotation of the cube. Let D be the set of integers 1, 2, 3, 4, 5, and 6. Find the number of equivalence classes of labeled sets of R. If the cube is a die, what is the physical significance of the labeled sets?

Chapter 6
Fundamental Concepts in the Theory of Graphs

6-1 INTRODUCTION

Let V be a set, and let E be a binary relation on V. As pointed out in Chap. 5, a binary relation on a set can be represented as a matrix. For example, the matrix representation for the relation $E = \{(a,b),(b,a),(b,d),(c,c),(d,a),(d,d)\}$ on the set $V = \{a,b,c,d\}$ is shown in Fig. 6-1a. An

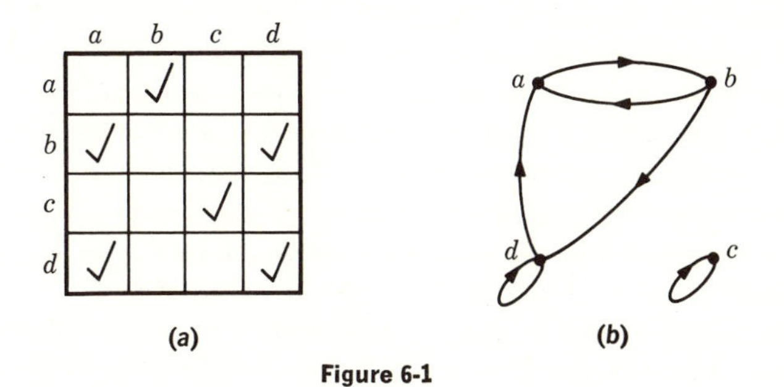

Figure 6-1

alternative way of representing the binary relation is shown in Fig. 6-1*b*: The elements in V are represented by the points marked a, b, c, and d. The ordered pair (a,b) in E is represented by an arrow from a to b, the ordered pair (b,d) in E is represented by an arrow from b to d, and so on. Such a representation of a set and a binary relation on it is called a *graph*. In other words, a graph G is defined abstractly to be an ordered pair (V,E), where V is a set and E is a binary relation on V. The elements in V are called the *vertices*, and the ordered pairs in E are called the *edges* of the graph. An edge is said to be *incident* with the vertices it joins. For example, the edge (a,b) is incident with the vertices a and b. Sometimes, when we wish to be more specific, we say that the edge (a,b) is incident *from* a and is incident *into* b. The vertex a is called the *initial vertex*, and the vertex b is called the *terminal vertex* of the edge (a,b). An edge that is incident from and into the same vertex, like (d,d), is called a *loop*. Two vertices are said to be *adjacent* if they are joined by an edge. Moreover, corresponding to an edge (a,b), the vertex a is said to be *adjacent to* the vertex b, and the vertex b is said to be *adjacent from* the vertex a. A vertex is said to be an *isolated* vertex if there is no edge incident with it.

For the graph $G = (V,E)$ shown in Fig. 6-2*a*, since E is a symmetric relation, there is always a pair of edges joining two vertices that are

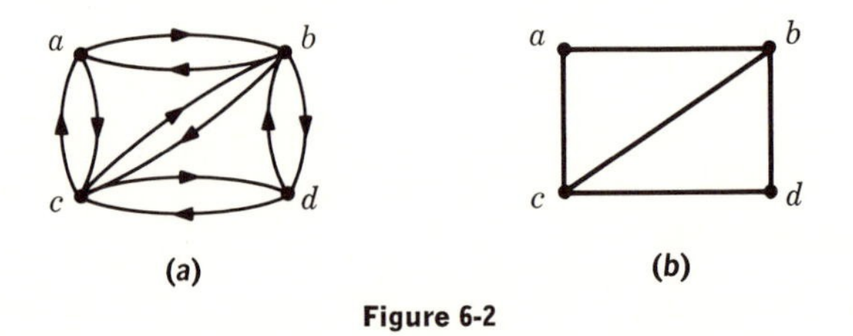

Figure 6-2

related. To represent the relation E on the set V, we can also draw just one edge between every two vertices that are related with the arrowheads omitted as in Fig. 6-2*b*. A graph is said to be a *directed* graph if directions are assigned to the edges. A graph is said to be *undirected* if directions are not assigned to the edges. Clearly, an undirected graph is a representation of a set and a symmetric binary relation on the set. (Note that a set and a symmetric relation on the set can be represented either as a directed or as an undirected graph. However, an undirected graph can represent only a set and a symmetric relation on it.) In an undirected graph, an edge joining the vertices a and b can be denoted either by (a,b) or by (b,a) as there is no need to make the distinction.

Since graphs can be used to represent a very general class of structures, the theory of graphs is an important area of study in combinatorial mathematics. As an example, consider the transmission of four messages a, b, c, and d through a communication channel where at the receiving end four corresponding messages a', b', c', and d' will be received. Because of noise interference in the communication channel, one message might be mistaken for another at the receiving end. The relation between transmitted messages and received messages can be represented by a directed graph as shown in Fig. 6-3. That either a' or b' will be

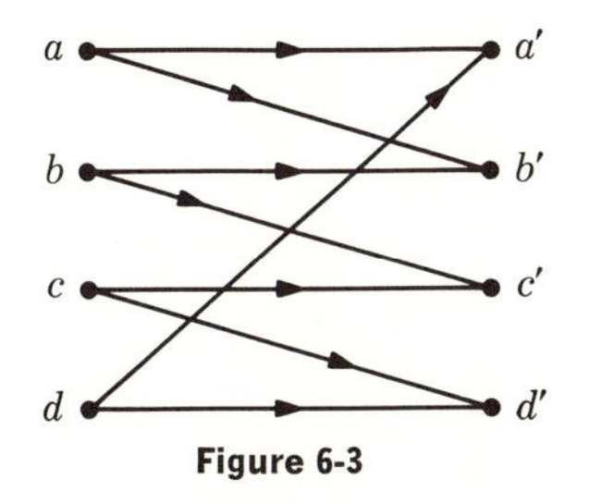

Figure 6-3

received when a is transmitted, that either b' or c' will be received when b is transmitted, that either c or d was transmitted when d' is received, and so on, are clearly depicted in the graph. As another example, for the chessboard shown in Fig. 6-4*a*, the adjacency relation between cells is

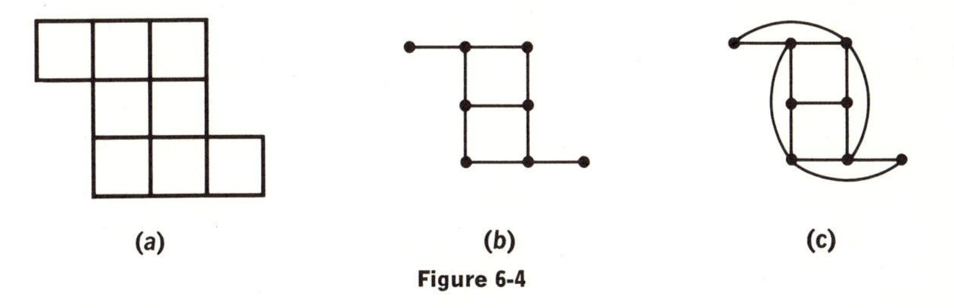

Figure 6-4

represented by the undirected graph in Fig. 6-4*b* where each cell is represented by a vertex. Similarly, the row-and-column dominance relation (cells dominated by a rook) is represented by the graph in Fig. 6-4*c*.

Two graphs are said to be *isomorphic* if there is a one-to-one correspondence between their vertices and between their edges such that incidences are preserved. In other words, if there is an edge between two vertices in one graph, there is a corresponding edge between the corresponding vertices in the other graph. For example, Fig. 6-5*a* shows a pair of isomorphic undirected graphs, and Fig. 6-5*b* shows a pair of isomorphic directed graphs. In these two figures, corresponding vertices in the two isomorphic graphs are labeled with the same letter, primed and

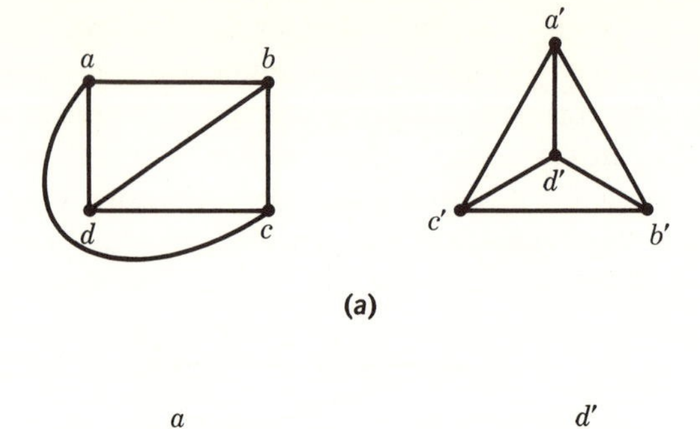

(a)

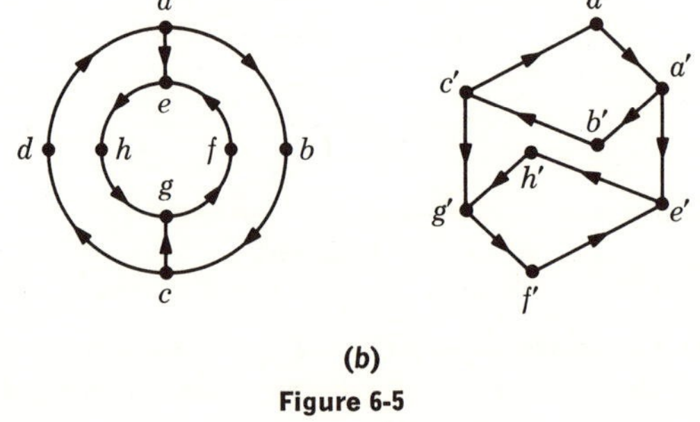

(b)

Figure 6-5

unprimed. The reader can convince himself that the graphs are isomorphic by checking the incidence relations.

Let $G = (V,E)$ be a graph. A graph $G' = (V',E')$ is said to be a *subgraph* of G if V' is a subset of V and E' is a subset of E. (Since G' *is* a graph, the vertices with which the edges in E' are incident must be in V'.) For example, Fig. 6-6*b* shows a subgraph of the graph in Fig. 6-6*a*.

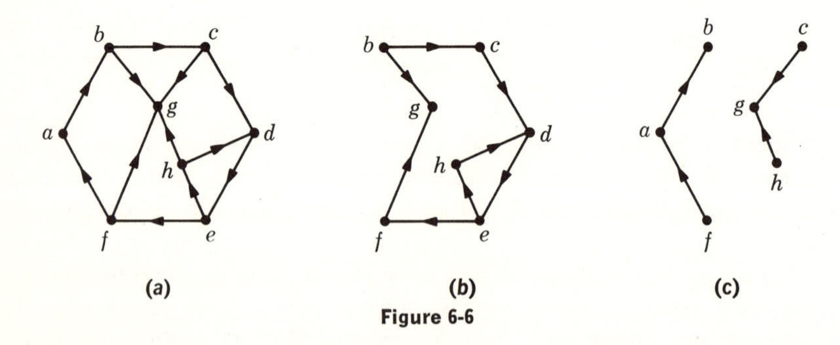

(a) (b) (c)

Figure 6-6

The *complement* of a subgraph $G' = (V',E')$ with respect to the graph G is another subgraph $G'' = (V'', E'')$ such that E'' is equal to $E - E'$ and V'' contains only the vertices with which the edges in E'' are incident. For example, Fig. 6-6*c* shows the complement of the subgraph in Fig. 6-6*b*.

A graph is a *finite* graph if it contains a finite number of edges (and thus, a finite number of nonisolated vertices). Although there are many interesting results in the theory of infinite graphs, we shall limit our discussion to finite graphs.

The definition of a graph can be extended by assigning a nonnegative integer, called the *multiplicity*, to every ordered pair of vertices. Instead of just one edge between two vertices, there may be several edges; the number of edges between two vertices in the graph is the multiplicity of this pair of vertices. Figure 6-7 shows an example of such graphs where,

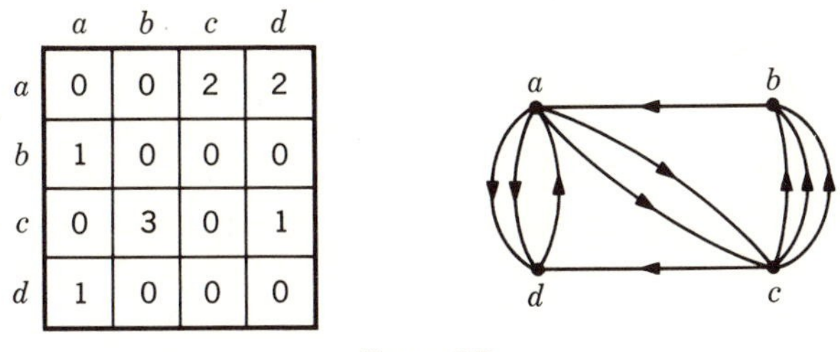

	a	b	c	d
a	0	0	2	2
b	1	0	0	0
c	0	3	0	1
d	1	0	0	0

Figure 6-7

in the matrix representation, the numbers in the cells indicate the multiplicities of the ordered pairs of vertices. A graph that contains pairs of vertices with multiplicities larger than 1 is sometimes called a *multigraph*. Also, the meanings of the terms *directed multigraph* and *undirected multigraph* are clear. A multigraph is said to be a *k-graph* if the multiplicities of the ordered pairs of vertices in the graph do not exceed k while there is at least one ordered pair of vertices with multiplicity k. In the following, we shall use the term *linear graph* when we refer to a graph the multiplicities of the ordered pairs of vertices of which are no larger than 1. The term "graph" will mean either a linear graph or a multigraph.

6-2 THE CONNECTEDNESS OF A GRAPH

In a directed graph, a *path* is a sequence of edges $(e_{i_1}, e_{i_2}, \ldots, e_{i_k})$† such that the terminal vertex of e_{i_j} coincides with the initial vertex of $e_{i_{(j+1)}}$ for $1 \leq j \leq k - 1$. A path is said to be *simple* if it does not use the same edge twice. A path is said to be *elementary* if it does not meet the same vertex twice. A *circuit* is a path in which the terminal vertex of e_{i_k} coincides with the initial vertex of e_{i_1}. Similarly, a circuit is said to be simple if it does not use the same edge twice, and a circuit is said to be elementary if it does not meet the same vertex twice. In Fig. 6-8,

† To simplify the notation, we identify the edges of a graph by letter names such as $e_1, e_2, \ldots$, as shown in Fig. 6-8.

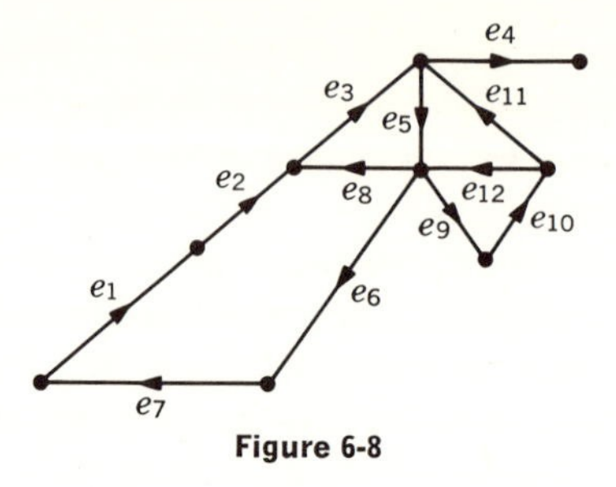

Figure 6-8

(e_1,e_2,e_3,e_4) is a path; $(e_1,e_2,e_3,e_5,e_8,e_3,e_4)$ is a path, but not a simple one; $(e_1,e_2,e_3,e_5,e_9,e_{10},e_{11},e_4)$ is a simple path, but not an elementary one. Also, $(e_1,e_2,e_3,e_5,e_9,e_{10},e_{12},e_6,e_7)$ is a simple circuit, but not an elementary one; $(e_1,e_2,e_3,e_5,e_6,e_7)$ is an elementary circuit. The *length* of a path (a circuit) is defined as the number of edges in the path (the circuit).

In an undirected graph, since we do not make the distinction between the initial vertex and the terminal vertex of an edge, a path (a circuit) is defined to be a sequence of edges $(e_{i_1},e_{i_2}, \ldots ,e_{i_k})$ to which directions can be assigned in such a way that the sequence becomes a directed path (a directed circuit). In an undirected graph, the notions of a simple path, an elementary path, a simple circuit, an elementary circuit, and the length of a path or a circuit are the same as those in a directed graph.

In a graph (directed or undirected), two vertices are said to be *connected* if there is a path between them. The following theorem can be used to determine the existence of a path between two vertices.

Theorem 6-1 *In a (directed or undirected) graph with n vertices, if two vertices are connected, there is a path of length less than or equal to $n - 1$ between them.*

Proof Suppose there is a path between the vertices v_1 and v_2. Let $(v_1, \ldots ,v_i, \ldots ,v_2)$ be the sequence of vertices that the path meets when it is traced from v_1 and v_2. If the length of the path is l, then there are $l + 1$ vertices in the sequence. For l larger than $n - 1$, there must be a vertex v_k that appears more than once in the sequence, that is, $(v_1, \ldots ,v_i, \ldots ,v_k, \ldots ,v_k, \ldots v_2)$. Deleting the edges in the path that leads v_k back to v_k, we have a path between v_1 and v_2 which is shorter than the original one.

This argument can be repeated until we have a path of length less than or equal to $n - 1$ between the vertices v_1 and v_2. ■

An undirected graph is said to be *connected* if every two vertices in the graph are connected and is said to be *unconnected* otherwise. A directed graph is said to be connected if the undirected graph derived from it by ignoring the directions of the edges is connected and is said to be unconnected otherwise. It follows that an unconnected graph consists of two or more components each of which is a connected graph. A directed graph is said to be *strongly connected* if for every two vertices a and b in the graph there is a path from a to b as well as a path from b to a. For example, Fig. 6-9a shows a connected graph which, however, is not strongly connected; Fig. 6-9b shows an unconnected graph.

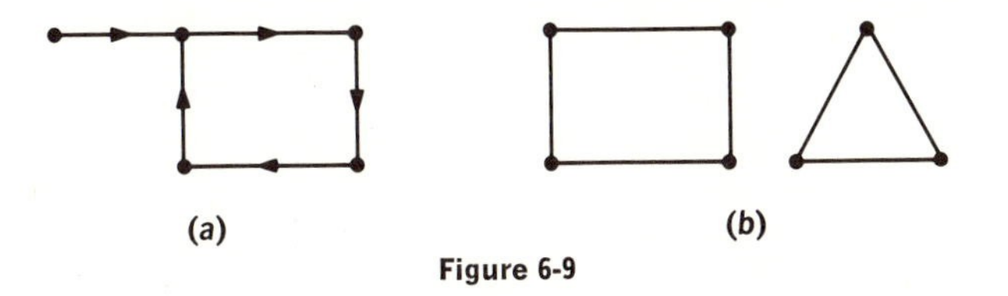

Figure 6-9

By splitting a vertex we mean dividing a vertex into two or more vertices. In a connected graph, a vertex is said to be an *articulation point* if the vertex can be split to yield an unconnected graph. For example, for the graph in Fig. 6-10a, the vertices b, c, f, h, and i are all articulation

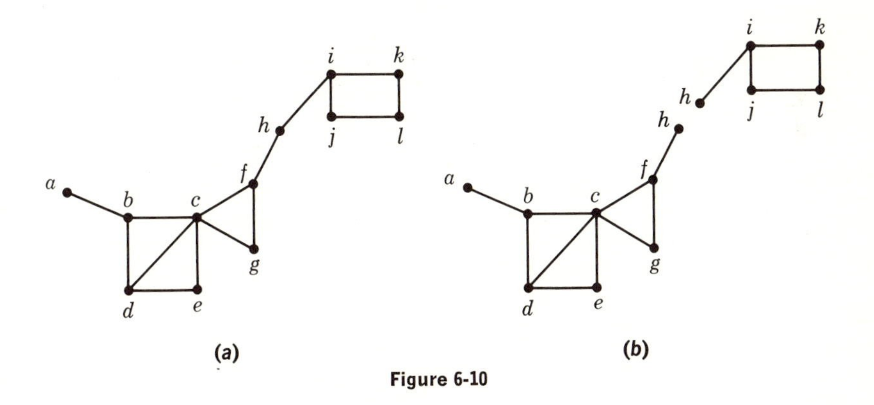

Figure 6-10

points. When an articulation point is split, the connected components in the resultant graph are called *pieces*. For example, Fig. 6-10b shows the two pieces of the graph in Fig. 6-10a when the articulation point h is split. A graph that contains no articulation point is said to be *biconnected.*

6-3 EULER PATH

There are some interesting problems concerning paths in a graph. An *Euler path* (*Euler circuit*) is a path (circuit) that traverses each edge in a graph exactly once. The problem of finding such a path in a graph was first investigated by Euler in 1736 when he proved that it is impossible to cross each of the seven bridges on the river Pregel in Königsberg, Germany, once and only once. A map of the Königsberg bridges is shown in Fig. 6-11*a*. As can be seen, the proof amounts to showing that the graph in Fig. 6-11*b*, where the edges correspond to the bridges, does not have an Euler path.

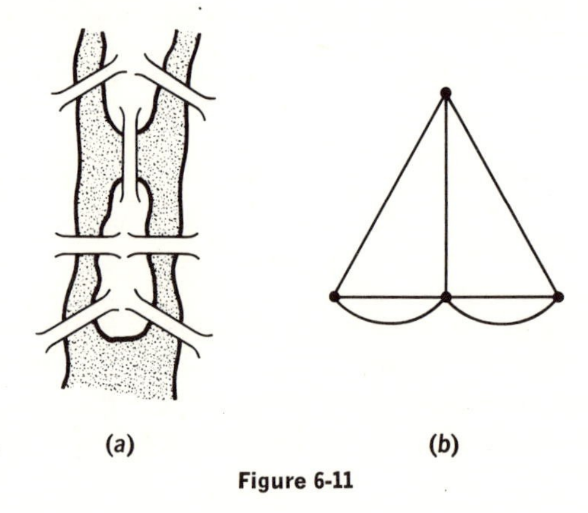

(*a*) (*b*)

Figure 6-11

The problem of drawing a figure in a continuous trace without repeating lines is also a problem of finding an Euler path in a graph. Clearly, a figure can be drawn in such a way if and only if it possesses an Euler path. For instance, both of the figures in Fig. 6-12 can be drawn in a continuous trace with each edge being traced exactly once.

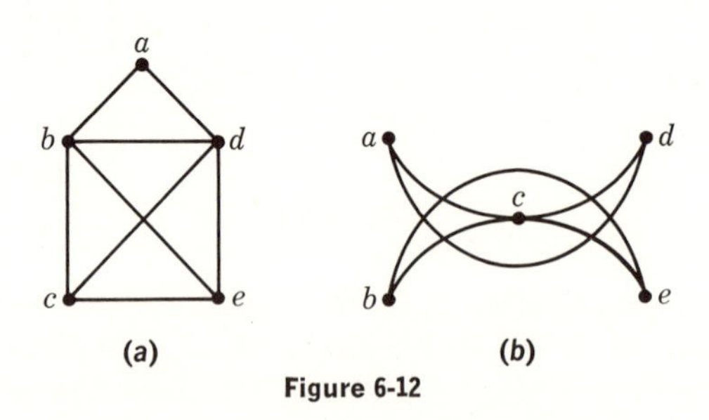

(*a*) (*b*)

Figure 6-12

Let us now study the problem of establishing the existence of Euler paths (circuits) in any arbitrary graph. To this end, we introduce the notion of the *degree* of a vertex. In a directed or an undirected graph, the degree of a vertex is the number of edges that are incident with it. In a directed graph, the *incoming* degree of a vertex is the number of edges that are incident into it, and the *outgoing* degree of a vertex is the number of edges that are incident from it. The following results concerning the degrees of the vertices in a graph are useful.

Theorem 6-2 *The sum of the degrees of the vertices in a graph (directed or undirected) is an even number, provided that a loop is considered to contribute a count of 2 to the degree of the vertex with which it is incident.*

Proof Since each edge contributes a count of 1 to the degree of each of the two vertices with which it is incident, the sum of the degrees of the vertices is equal to twice the number of edges in the graph. ■

Corollary 6-2.1 *In any graph, there is an even number of vertices that have odd degrees.*

The existence of Euler paths or Euler circuits in a graph is related to the degrees of the vertices as indicated by the following theorem and its corollaries.

Theorem 6-3 *An undirected graph possesses an Euler path if and only if it is connected and has no, or exactly two, vertices that are of odd degree.*

Proof Suppose that the graph possesses an Euler path. That the graph must be connected is obvious. When the Euler path is traced, we observe that every time the path meets a vertex, it goes through two edges which are incident with the vertex and have not been traced before. Except for the two terminal vertices of the path, the degree of any other vertex in the graph must be even. If the two terminal vertices of the Euler path are distinct, their degrees are odd. If the two terminal vertices coincide, their degrees are both even, and the Euler path becomes an Euler circuit. Thus, the necessity of the condition stated in the theorem is proved.

To prove the sufficiency of the condition, we construct an Euler path by starting at one of the two vertices that are of odd degree and going through the edges of the graph in such a way that no edge will be traced more than once. For a vertex of even degree, whenever the path "enters" the vertex through an edge, it can always "leave" the vertex through another edge that has not been traced before. Therefore, when the construction eventually comes to an end, we must have reached the other vertex the degree of which is odd. If all the edges in the graph were traced this way, clearly, we would have an Euler path. If not all of the edges in the graph were traced, we shall remove those edges that have been traced and obtain a subgraph formed by the remaining edges.

The degrees of the vertices of this subgraph are all even. Moreover, this subgraph must touch the path that we have traced at one or more vertices since the original graph is connected. Starting from one of these vertices, we can again construct a path that passes through the edges. Because the degrees of the vertices are all even, this path must return eventually to the vertex at which it starts. We can combine this path with the path we have constructed as one which starts and ends at the two vertices of odd degree. If necessary, the argument is repeated until we obtain a path that traverses all the edges in the graph. ∎

Corollary 6-3.1 *An undirected graph possesses an Euler circuit if and only if it is connected and its vertices are all of even degree.*

Corollary 6-3.2 *A directed graph possesses an Euler path if and only if it is connected and the incoming degree of every vertex is equal to its outgoing degree with the possible exception of two vertices. For these two vertices, the incoming degree of one is* 1 *larger than its outgoing degree, and the incoming degree of the other is* 1 *less than its outgoing degree.*

Corollary 6-3.3 *A directed graph possesses an Euler circuit if and only if it is connected and the incoming degree of every vertex is equal to its outgoing degree.*

We conclude from the above that the graph in Fig. 6-12*a* has an Euler path but not an Euler circuit, because the degree of both vertex c and vertex e is equal to 3. Also, the graph in Fig. 6-12*b* has an Euler circuit because the vertices are all of even degree. The directed graph in Fig. 6-13*a* does not have an Euler path, whereas the graph in Fig. 6-13*b* does.

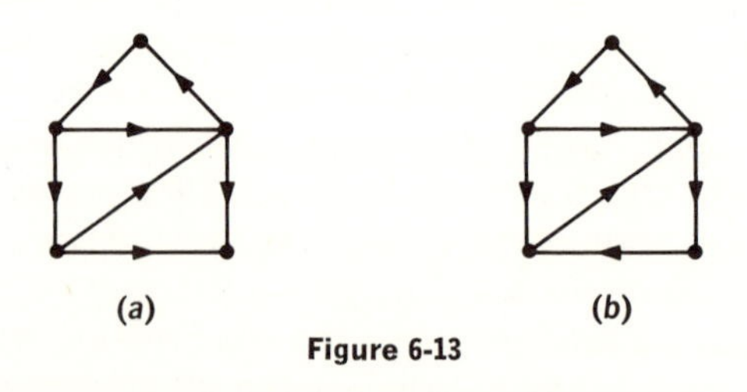

Figure 6-13

Example 6-1 The surface of a rotating drum is divided into 16 sectors as shown in Fig. 6-14*a*. The positional information of the drum is to be represented by the binary signals a, b, c, and d as shown in Fig. 6-14*b* where conducting (lined area) and nonconducting (white

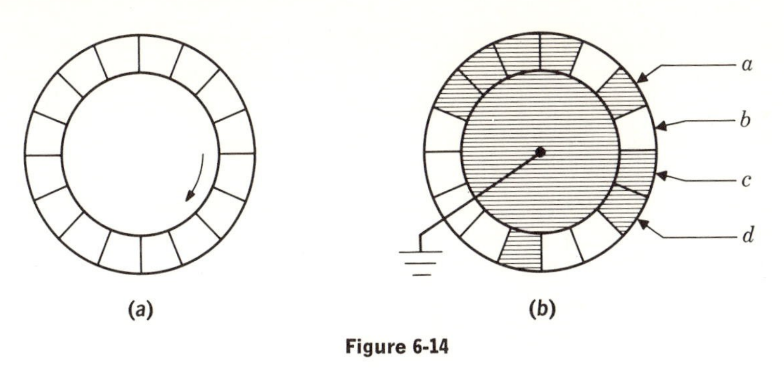

Figure 6-14

area) materials are used to make up the sectors. Depending on the position of the drum, the terminals a, b, c, and d will either be connected to the ground or be insulated from it. For example, when the position of the drum is that shown in Fig. 6-14b, terminals a, c, and d are connected to the ground, whereas terminal b is not. In order that the 16 different positions of the drum will be distinctly represented by the binary signals at the terminals, the sectors must be constructed in such a way that no two conducting and non-conducting patterns of four consecutive sectors are the same. The problem is to determine whether such an arrangement of conducting and nonconducting sectors exists, and if so, to determine how the sectors should be arranged. Letting the binary digit 0 denote a conducting sector and the binary digit 1 denote a nonconducting sector, we can rephrase the problem as follows: Arrange 16 binary digits in a circular array such that every sequence of four consecutive digits is distinct.

The answer to the question of the possibility of such an arrangement is affirmative and is actually quite obvious once the right point of view has been taken. We shall construct a directed graph with eight vertices which are labeled with the eight 3-digit binary numbers $\{000, 001, \ldots, 111\}$. From a vertex labeled $\alpha_1\alpha_2\alpha_3$, there is an edge to the vertex labeled $\alpha_2\alpha_3 0$ and an edge to the vertex labeled $\alpha_2\alpha_3 1$. The graph so constructed is shown in Fig. 6-15. Moreover, we shall label each edge of the graph with a four-digit binary number. In particular, the edge from the vertex $\alpha_1\alpha_2\alpha_3$ to the vertex $\alpha_2\alpha_3 0$ is labeled $\alpha_1\alpha_2\alpha_3 0$, and the edge from the vertex $\alpha_1\alpha_2\alpha_3$ to the vertex $\alpha_2\alpha_3 1$ is labeled $\alpha_1\alpha_2\alpha_3 1$. Since the vertices are labeled with the eight distinct three-digit binary numbers, the edges will be labeled with the 16 distinct four-digit binary numbers. In a path of the graph, the labels for any two consecutive edges must be of the form $\alpha_1\alpha_2\alpha_3\alpha_4$ and $\alpha_2\alpha_3\alpha_4\alpha_5$; namely, the three trailing digits of the label of the first

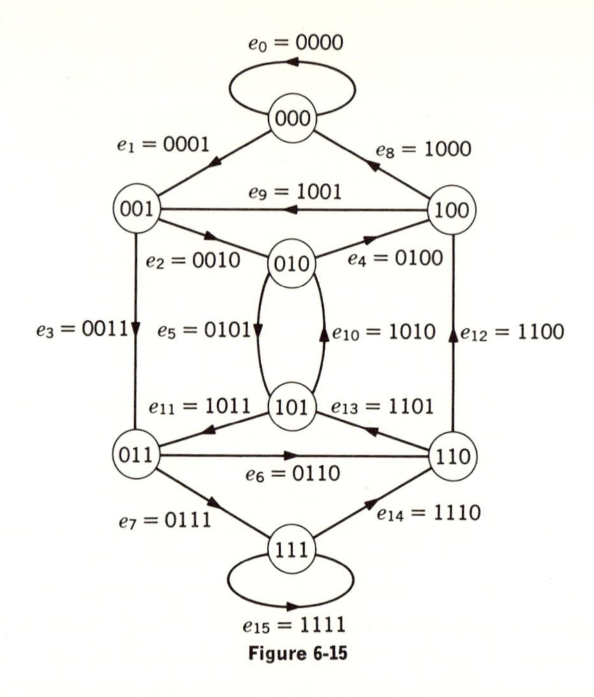

Figure 6-15

edge are identical to the three leading digits of the label of the second edge. Since the 16 edges in the graph are labeled with distinct binary numbers, it follows that corresponding to an Euler circuit of the graph, there is a circular arrangement of 16 binary digits that will give all the 16 four-digit binary numbers. For example, corresponding to the Euler circuit $(e_0,e_1,e_2,e_5,e_{10},e_4,e_9,e_3,e_6,e_{13},e_{11},e_7,e_{15},e_{14},e_{12},e_8)$, the sequence of 16 binary digits is 0000101001101111. (The circular arrangement is obtained by closing the two ends of the sequence.) According to Corollary 6-3.3, the existence of an Euler circuit in the graph is obvious, because every one of the vertices has an incoming degree equal to 2 and an outgoing degree equal to 2. Moreover, we can find an Euler circuit in the graph by following the construction procedure suggested in the proof of Theorem 6-3.

Using a similar argument, we can show that it is possible to arrange 2^n binary digits in a circular array such that the 2^n sequences of n consecutive digits in the arrangement are all distinct. To show this, we construct a directed graph with 2^{n-1} vertices which are labeled with the 2^{n-1} $(n-1)$-digit binary numbers. From the vertex $\alpha_1\alpha_2\alpha_3 \cdots \alpha_{n-1}$, there is an edge to the vertex $\alpha_2\alpha_3 \cdots \alpha_{n-1}0$ which is labeled $\alpha_1\alpha_2\alpha_3 \cdots \alpha_{n-1}0$, and an edge to the vertex $\alpha_2\alpha_3 \cdots \alpha_{n-1}1$ which is labeled $\alpha_1\alpha_2\alpha_3 \cdots \alpha_{n-1}1$. According to Corollary 6-3.3, the graph has an Euler circuit which corresponds to a circular arrangement of the 2^n binary digits. ■

6-4 HAMILTONIAN PATH

A *Hamiltonian path* (*Hamiltonian circuit*) is a path (circuit) that passes through each of the vertices in a graph exactly once. For example, Fig. 6-16a shows a graph and one of its Hamiltonian circuits (the edges

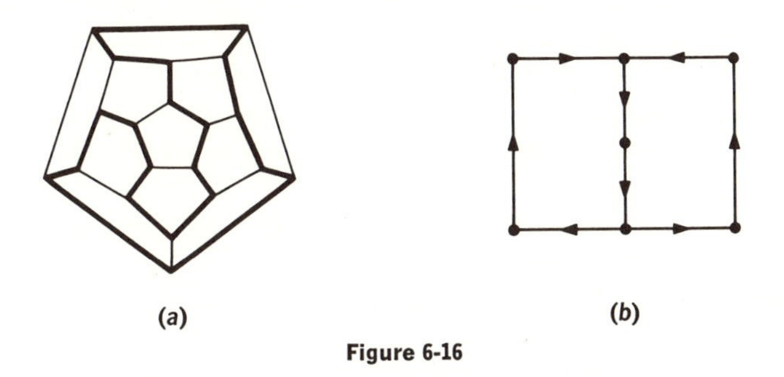

(*a*) (*b*)

Figure 6-16

in the circuit are drawn in heavy lines); Fig. 6-16b shows a directed graph that has a Hamiltonian path but not a Hamiltonian circuit.

Consider the vertices of a graph as cities and the edges of a graph as traveling routes between cities. (The graph is either directed or undirected, depending on whether there are restrictions on the directions in which the routes can be traveled.) A Hamiltonian circuit is then an itinerary by which one can visit each city exactly once and ultimately return to the point of departure. On the other hand, a Hamiltonian path is one by which one can visit each city once but not necessarily return to the point of departure. Another example is seating a group of people at a round table. If we let the vertices of an undirected graph denote the people and the edges represent the relation that two people are friends, a Hamiltonian circuit corresponds to a way of seating the people so that every one has two of his friends at his two sides.

Although the problem of Hamiltonian path closely resembles the problem of Euler path, there is no known criterion we can apply to determine the existence of a Hamiltonian path in a graph. The following two examples show some partial results in this direction.

Example 6-2 A directed graph is said to be *complete* if every two vertices of the graph are joined by at least an edge in one of the two directions. We want to show that a complete directed graph always contains a Hamiltonian path.

Let there be a path of length $p - 1$ in the graph which meets the sequence of vertices $(v_1,v_2,v_3, \ldots ,v_p)$. Let v_x be a vertex that is not included in this path. If there is an edge (v_x,v_1) in the graph, we can clearly augment the original path by the addition of the edge

(v_x,v_1) so that the vertex v_x will be included in the augmented path. If, on the other hand, there is no edge from v_x to v_1, then there must be an edge (v_1,v_x) in the graph. Suppose that (v_x,v_2) is also an edge in the graph. We can replace the edge (v_1,v_2) in the original path with the two edges (v_1,v_x) and (v_x,v_2) so that the vertex v_x will be included in the augmented path. On the other hand, if there is no edge from v_x to v_2, then there must be an edge (v_2,v_x) in the graph, and we can repeat the argument.

Eventually, if we find that it is not possible to include the vertex v_x in any augmented path by replacing an edge (v_k, v_{k+1}) in the original path with two edges (v_k,v_x) and (v_x, v_{k+1}) with $1 \leq k \leq p - 1$, we conclude that there must be an edge (v_p,v_x) in the graph. We can, therefore, augment the original path by adding to it the edge (v_p,v_x) so that the vertex v_x will be included in the augmented path.

We can repeat the argument until all the vertices in the graph are included in a path. ■

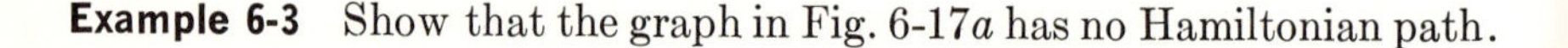

Example 6-3 Show that the graph in Fig. 6-17*a* has no Hamiltonian path.

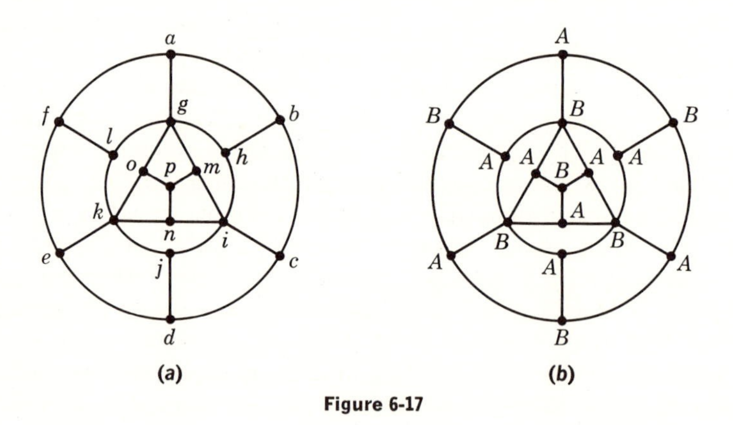

Figure 6-17

Method 1 We label the vertex *a* by *A* and label all the vertices that are adjacent to it by *B*. Continuing, we label all the vertices that are adjacent to an *A-vertex* by *B* and label all the vertices that are adjacent to a *B-vertex* by *A*, until all vertices are labeled. The labeled graph is shown in Fig. 6-17*b*. If there is a Hamiltonian path in the graph, then it must pass through the *A*-vertices and the *B*-vertices alternately. However, since there are nine *A*-vertices and seven *B*-vertices, the existence of a Hamiltonian path is impossible.

Method 2 Because a Hamiltonian path meets any vertex exactly once, among the edges that are incident with a vertex at most two of them can be included in a Hamiltonian path. For the graph in Fig. 6-17*a*, since the degree of the vertex g is 5, at least three of the five edges incident with g are not included in a Hamiltonian path. The same argument can be applied to the vertices k and i. Similarly, since the degrees of the vertices b, d, f, p are all equal to 3, at least one of the three edges incident with each of these vertices is not included in a Hamiltonian path. A simple computation reveals that the existence of a Hamiltonian path is impossible. Since the graph has 27 edges and $3 \times 3 + 1 \times 4 = 13$ of them cannot be included in a Hamiltonian path, only 14 of the edges can possibly be included. However, since the graph has 16 vertices, a Hamiltonian path must contain exactly 15 edges, and a Hamiltonian circuit must contain exactly 16 edges. Therefore, there is no Hamiltonian path or circuit in the graph. ■

A generalization of the problem of finding a Hamiltonian circuit is known as the traveling salesman's problem. The problem is usually stated as follows: A salesman is supposed to visit a certain number of cities during a trip. Given the distance between the cities, the problem is to find the shortest route that covers each of these cities once and returns to the point of departure; that is, we want to find a Hamiltonian circuit that has the minimal sum of distances. Although there is no efficient general algorithm for the solution to this problem, large-scale examples have been worked out by systematically searching for the minimal solution with the aid of digital computers.†

6-5 SUMMARY AND REFERENCES

Because of the generality in their mathematical structure, graphs can be used to model many problems that deal with the discrete arrangements of objects. For example, electrical networks, social-group structures, family trees, communication systems, transportation routes, production schedules, chemical-bond structures, and language structures can all be represented by graphs. Therefore, in addition to its richness and elegance in mathematical content, the theory of graphs also finds applications in many areas of study. It is beyond our scope of coverage to discuss the various applications in detail. We shall develop the results in general and abstract terms so that these results can be applied to the problems of the reader's special interest.

As a general reference for the theory of graphs, the book by Berge

† See, for example, Dantzig, Fulkerson, and Johnson [3].

[1] is excellent. Ore's book [5] is also recommended. Chapter 6 of Busacker and Saaty [2] covers a variety of interesting applications of the theory of graphs. Harary, Norman, and Cartwright [4] has an extensive coverage of the applications to sociology. Seshu and Reed [6] contains the most thorough treatment on the applications to electrical networks.

For the introductory material in this chapter, see Chaps. 1, 11, and 17 of Berge [1] and Chaps. 1, 2, and 3 of Ore [5].

1. Berge, C.: "The Theory of Graphs and Its Applications," John Wiley & Sons, Inc., New York, 1962.
2. Busacker, R. G., and T. L. Saaty: "Finite Graphs and Networks: An Introduction with Applications," McGraw-Hill Book Company, New York, 1965.
3. Dantzig, G. B., R. Fulkerson, and S. Johnson: Solution of a Large-scale Traveling Salesman Problem, *Operations Res.*, **2**:394–410 (1954).
4. Harary, F., R. Z. Norman, and D. Cartwright: "Structural Models: An Introduction to the Theory of Directed Graphs," John Wiley & Sons, Inc., New York, 1965.
5. Ore, O.: "Theory of Graphs," American Mathematical Society, Providence, R.I., 1962.
6. Seshu, S., and M. B. Reed: "Linear Graphs and Electrical Networks," Addison-Wesley Publishing Company, Inc., Reading, Mass., 1961.

PROBLEMS

6-1. Show that in a finite graph with exactly two vertices of odd degree there is a path connecting these two vertices.

6-2. An electronic circuit is built to recognize sequences of 0's and 1's. In particular it shall accept sequences of the form 010*10, where 0* means any number (including none) of 0's. For example, 0110, 01010, and 01000010 are all acceptable sequences. Construct a directed graph in which every vertex has two outgoing edges labeled 0 and 1, and in which there are two vertices v_i (the initial vertex) and v_f (the final vertex) such that every path from v_i to v_f is a sequence of the form 010*10. (For each sequence the circuit will start at v_i and trace the edges according to the 0's and 1's in the sequence. The circuit will accept the sequence if the path terminates at v_f.)

6-3. Repeat Prob. 2 for sequences of the form 01*(10)*10*, where (10)* means any number (including none) of 10 patterns, using each of the following modifications:

(*a*) For each vertex there is no restriction on the number of outgoing edges labeled 0 or 1. The sequence of labels on every path from v_i to v_f is a sequence of the form described above. Furthermore, for every sequence of the form described above there is a corresponding path from v_i to v_f.

(*b*) There may be several final vertices $v_{f1}, v_{f2}, \ldots$. The sequence of labels on every path from v_i to some final vertex v_{fj} is a sequence of the form described above. Furthermore, for every sequence of the form described above there is a corresponding path from v_i to some final vertex v_{fj}.

6-4. Let $V = \{1,2,3,4\}$. Let $D = \{\{1,2\},\{1,3\},\{1,4\},\{2,3\},\{2,4\},\{3,4\}\}$, that is, the set of all 2-subsets of the set V. Let G be the group consisting of all the permutations of the set V.

(*a*) Show that for each permutation π in G there is a corresponding permutation $\hat{\pi}$ on the set D such that $\hat{\pi}(\{a,b\}) = \{\pi(a),\pi(b)\}$.

(*b*) Show that the set of $\hat{\pi}$'s is a permutation group.

(*c*) Let $R = \{x,y\}$ with $w(x) = 1$ and $w(y) = y$. Find the pattern inventory of the equivalence classes of functions from D to R.

(*d*) In terms of undirected graphs with four vertices, what is the significance of the result in part (*c*)?

6-5. Repeat Prob. 4 for D being the set of all ordered pairs in $V \times V$ that contain two distinct elements.

6-6. Show that a linear graph G with no loops and n vertices is not bipartite if it has more than $[n^2/4]$ edges. (See Sec. 11-1 for a definition of bipartite graph.)

6-7. Show that a linear graph with no loops and n vertices is connected if it has more than $\frac{1}{2}(n-1)(n-2)$ edges.

6-8. An (undirected) complete graph is a graph in which there is an edge joining every two vertices.

(*a*) The edges of a complete graph with six vertices are to be painted either red or blue. Show that for any arbitrary way of painting the edges there is either a red triangle (all three edges are painted red) or a blue triangle in the graph.

(*b*) Show that among a group of six people there are either three who are mutual friends or three who are strangers to each other.

6-9. (*Krausz's Theorem*) The line graph $L(G)$ of a linear graph G is defined as follows:

1. There is a one-to-one correspondence between the vertices in $L(G)$ and the edges in G.
2. There is an edge joining two vertices in $L(G)$ when their corresponding edges in G are incident with a common vertex.

Prove that a linear graph H without loops is a line graph of some linear graph if and only if there is a partition of the edges of H which is such that each subset in the partition forms a complete graph and every vertex in H is incident with edges that are in at most two subsets in the partition.

6-10. Show that a vertex x is an articulation point of a connected graph if and only if there exist two vertices a and b such that every path joining a and b passes through x.

6-11. Show that a directed graph that contains an Euler circuit is strongly connected.

6-12. Prove that a connected graph possesses an Euler circuit if and only if it can be decomposed as a set of elementary circuits that have no edges in common.

6-13. Prove that if a connected graph has k vertices of odd degree ($k > 0$), there are $\frac{1}{2}k$ paths, no two of which have a common edge, that cover all the edges in the graph.

6-14. Find a circular arrangement of nine a's, nine b's, and nine c's such that each of the 27 words of length 3 from the alphabet $\{a,b,c\}$ appears exactly once.

6-15. Is it possible to move a knight on an 8×8 chessboard so that it completes every possible move exactly once? A move between two squares of the chessboard is completed when it is made in either direction.

6-16. Do there exist graphs in which an Euler path is also a Hamiltonian path? Characterize this class of graphs.

6-17. n teams play in a round-robin tournament. Show that they can be ordered according to their winning records so that each team immediately precedes a team it has beaten. (The ordering is by no means unique.)

6-18. Eleven students plan to have dinner together for several days. They will be seated at a round table, and the plan calls for each student to have different neighbors at every dinner. For how many days can this be done?

6-19. Let G be a linear graph with n vertices. Show that G has a Hamiltonian path if the sum of the degrees of any two vertices is equal to or larger than $n - 1$.

Hint: Show the existence of an elementary path of length $n - 1$ by induction on the length of the path.

6-20. Show that a group of people can be seated around a table such that every one will have two of his friends at his two sides if every one knows at least half of the people in the group.

Hint: Extend the condition in Prob. 19 to that for the existence of a Hamiltonian circuit in a graph.

Chapter 7
Trees, Circuits, and Cut-sets

7-1 TREES AND SPANNING TREES

In this chapter, we introduce the notions of tree, spanning tree, and cut-set in undirected graphs that contain no loops. (The inclusion of loops does not introduce any new concept but may only bring up exceptions in some definitions.) These notions can also be extended to directed graphs. However, we shall leave such extensions to the reader.

A *tree* is a connected graph that contains no circuit.† As an example, Fig. 7-1*a* shows a tree. A collection of disjoint trees is called, quite appropriately, a *forest*. A tree of a graph is a subgraph of the graph which is a tree. A *spanning tree* of a graph is a tree of the graph that contains all the vertices in the graph. For example, Fig. 7-1*c* shows a tree, and Fig. 7-1*d* shows a spanning tree of the graph in Fig. 7-1*b*. According to the definition of the complement of a subgraph in Chap. 6, the complement of a tree (spanning tree) consists of the edges that are not included in the tree (spanning tree). For example, Fig. 7-1*e* shows

† From now on, unless otherwise specified, the term circuit means elementary circuit.

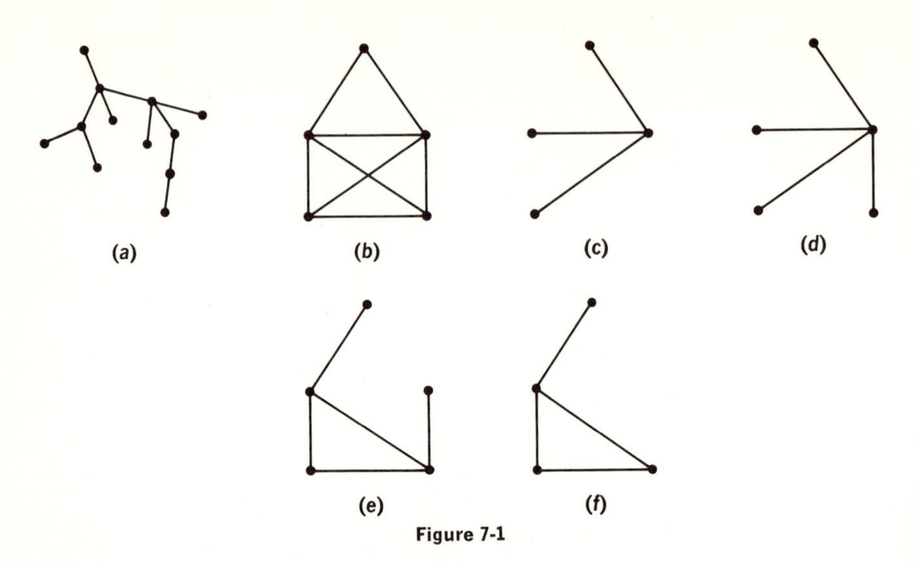

Figure 7-1

the complement of the tree in Fig. 7-1*c*, and Fig. 7-1*f* shows the complement of the spanning tree in Fig. 7-1*d*. A *branch* of a tree is an edge that is in the tree. A *chord*, or a *link* (relative to a tree), is an edge that is not in the tree.

We now examine some of the properties of trees.

Theorem 7-1 *A tree that has* $\mathbf{v}$ *vertices contains* $\mathbf{v} - 1$ *edges.*

Proof We prove the theorem by induction. As the basis of induction, we see that a tree having two vertices contains one edge. Suppose that the theorem is true for trees that have $\mathbf{v} - 1$ vertices. Let there be a tree that has $\mathbf{v}$ vertices. Notice that a tree must contain a vertex of degree 1 because if the degrees of the vertices are all equal to or larger than 2, there must exist a circuit in the tree, and if the degrees of the vertices are all equal to 0, the tree contains only isolated vertices. In either case, there is a contradiction to the definition of a tree. The removal of a vertex of degree 1 together with the edge incident with it results in a tree that has $\mathbf{v} - 1$ vertices which, according to the induction hypothesis, contains $\mathbf{v} - 2$ edges. Adding back the vertex and the edge we have removed, we conclude that a tree having $\mathbf{v}$ vertices contains $\mathbf{v} - 1$ edges. ■

Theorem 7-2 *Any two vertices in a tree are connected by a unique path.*

Proof Because a tree is a connected graph, there is at least one path connecting any two vertices. However, if there are two or more paths connecting a pair of vertices, there would be a circuit in the tree. We conclude, therefore, that there is one and only one path connecting any two vertices in a tree. ■

Theorem 7-3 *A graph is connected if and only if it contains a spanning tree.*

Proof. Clearly, if a graph contains a spanning tree, it must be connected.

A connected graph that does not contain any circuit is a tree by definition. For a connected graph that contains one or more circuits, we can remove an edge from one of the circuits and still have a connected subgraph. Such removal of edges from circuits can be repeated until we have a spanning tree. ■

From Theorems 7-1 and 7-3, we see that for a connected graph with $\mathbf{v}$ vertices and $\mathbf{e}$ edges, there are $\mathbf{v} - 1$ branches in any spanning tree. It follows that, relative to any spanning tree, there are $\mathbf{e} - \mathbf{v} + 1$ chords.

Because a spanning tree contains a unique path connecting any two vertices in the graph, the addition of a chord to the spanning tree yields a subgraph that contains exactly one circuit. Suppose that the chord (v_1,v_2) is added to a spanning tree. Because the spanning tree contains a path connecting the vertices v_1 and v_2, this path together with the edge (v_1,v_2) is a circuit in the graph. On the other hand, if the addition of the chord (v_1,v_2) yields two or more circuits, there must be two or more paths between the vertices v_1 and v_2 in the spanning tree, which is impossible according to Theorem 7-2. For a given spanning tree, the set of $\mathbf{e} - \mathbf{v} + 1$ circuits obtained by adding the $\mathbf{e} - \mathbf{v} + 1$ chords to the spanning tree one at a time is called the *fundamental system of circuits* relative to the spanning tree. A circuit in the fundamental system of circuits is called a *fundamental circuit.* Clearly, there is exactly one chord in a fundamental circuit; all the other edges are branches of the spanning tree. As an example, for the graph in Fig. 7-2*a* and the spanning tree in

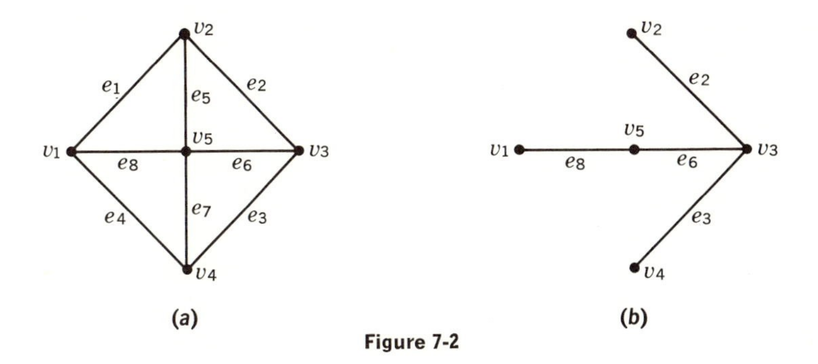

Figure 7-2

Fig. 7-2*b*, $\{e_1,e_2,e_6,e_8\}$, $\{e_5,e_2,e_6\}$, $\{e_4,e_8,e_6,e_3\}$, and $\{e_7,e_6,e_3\}$ are the fundamental circuits.

For an unconnected graph that has $\mathbf{k}$ components, there will be a forest of $\mathbf{k}$ spanning trees, one in each component, containing all the vertices in the graph. If the graph has $\mathbf{e}$ edges and $\mathbf{v}$ vertices, there will be $\mathbf{v} - \mathbf{k}$ branches in a forest of $\mathbf{k}$ spanning trees and, correspondingly, $\mathbf{e} - \mathbf{v} + \mathbf{k}$ chords. Similarly, for a given forest of spanning trees, we define the fundamental system of circuits as the set of $\mathbf{e} - \mathbf{v} + \mathbf{k}$ circuits obtained by adding the $\mathbf{e} - \mathbf{v} + \mathbf{k}$ chords to the spanning trees one at a time.

7-2 CUT-SETS

A *cut-set* is a (minimal) set of edges in a graph the removal of which will increase the number of connected components in the remaining subgraph, whereas the removal of any proper subset of which will not. It follows that in a connected graph, the removal of a cut-set will separate the graph into two parts. This suggests an alternative way of defining a cut-set. Let the vertices in a connected component of a graph be divided into two subsets such that every two vertices in one subset are connected by a path that meets only vertices in the subset. Then, the set of edges joining the vertices in the two subsets is a cut-set. As an example, for the graph in Fig. 7-2*a*, the set of edges $\{e_1,e_5,e_6,e_7,e_4\}$ is a cut-set, since its removal will leave an unconnected subgraph as shown in Fig. 7-3*a* while the removal

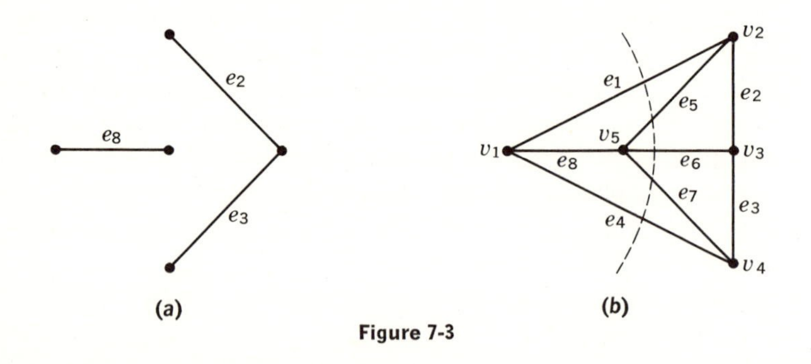

Figure 7-3

of any of its proper subsets will not. Also, this is the set of edges that join the vertices in the two subsets $\{v_1,v_5\}$ and $\{v_2,v_3,v_4\}$. Fig. 7-2*a* is redrawn as Fig. 7-3*b* to emphasize such a division of vertices.

Since the removal of any branch from a spanning tree breaks the spanning tree up into two trees (either or both of which may consist of a single vertex), we say that corresponding to a branch in a spanning tree there is a division of the vertices in the graph into two subsets, corresponding to the vertices in the two trees. It follows that for every branch in a spanning tree there is a corresponding cut-set. For example, for the graph in Fig. 7-2*a* the removal of the branch e_3 from the spanning tree in Fig. 7-2*b* divides the vertices into the two subsets $\{v_1,v_2,v_3,v_5\}$ and $\{v_4\}$. The corresponding cut-set is $\{e_3,e_7,e_4\}$. For a given spanning tree, the set of the $\mathbf{v} - 1$ cut-sets corresponding to the $\mathbf{v} - 1$ branches of the spanning tree is called the *fundamental system of cut-sets* relative to the spanning tree. A cut-set in the fundamental system of cut-sets is called a *fundamental cut-set.* Also, there is exactly one tree branch in a fundamental cut-set; all the other edges are chords. For the graph in Fig. 7-2*a* and the spanning tree in Fig. 7-2*b*, the fundamental cut-sets are $\{e_1,e_5,e_2\}$, $\{e_1,e_8,e_4\}$, $\{e_1,e_5,e_6,e_7,e_4\}$, and $\{e_4,e_7,e_3\}$.

We now discuss some of the properties of circuits and cut-sets. Unless otherwise stated, our discussion in the rest of this chapter will be limited to connected graphs, since its extension to unconnected graphs is straightforward.

Theorem 7-4 *A circuit and the complement of any spanning tree must have at least one edge in common.*

Proof If there is a circuit that has no common edge with the complement of a spanning tree, the circuit is contained in the spanning tree. However, this is impossible as a tree cannot contain a circuit. ∎

We have a similar theorem for the cut-sets.

Theorem 7-5 *A cut-set and any spanning tree must have at least one edge in common.*

Proof If there is a cut-set that has no common edge with a spanning tree, the removal of the cut-set will leave the spanning tree intact. However, according to Theorem 7-2, this means that the removal of the cut-set will not separate the graph into two components. This is in contradiction to the definition of a cut-set. ∎

Theorem 7-6 *Every circuit has an even number of edges in common with every cut-set.*

Proof Corresponding to a cut-set, there is a division of the vertices of the graph into two subsets which are the two sets of vertices in the two components of the graph when the edges in the cut-set are removed. Therefore, a path connecting a vertex in one subset and another vertex in the other subset must traverse an odd number of edges in the cut-set. Since a circuit is a path that starts and ends at the same vertex, it must have traversed the edges in the cut-set an even number of times, as illustrated in Fig. 7-4. (The edges in the circuit are drawn in heavy lines.) ■

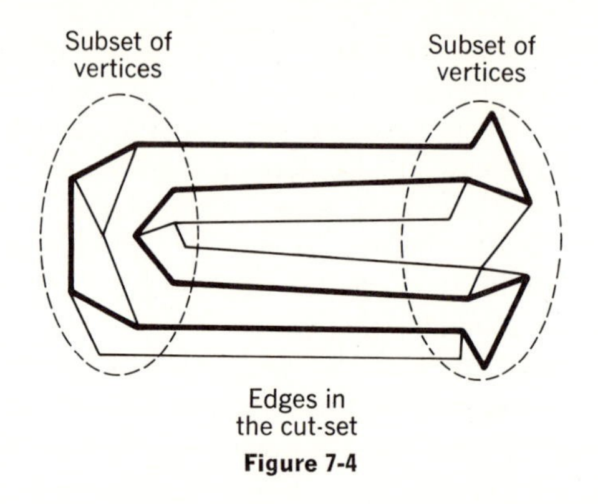

Figure 7-4

As was pointed out earlier, there is a fundamental circuit corresponding to each of the chords of a spanning tree, and there is a fundamental cut-set corresponding to each of the branches of a spanning tree. The following theorems point out a close relationship between the fundamental system of circuits and the fundamental system of cut-sets relative to a spanning tree.

Theorem 7-7 *For a given spanning tree, let* $L = \{e_1,e_2,e_3, \ldots ,e_k\}$ *be a fundamental circuit in which* e_1 *is a chord and* $e_2, e_3, \ldots , e_k$ *are branches of the spanning tree. Then,* e_1 *is contained in the fundamental cut-set corresponding to* e_i *for* $i = 2, 3, \ldots , k$. *Moreover,* e_1 *is not contained in any other fundamental cut-set.*

Proof Let C be the fundamental cut-set corresponding to the tree branch e_2. Since e_1 is the only chord in L and e_2 the only branch in C, C must also contain e_1 because L and C have an even number of edges in common. Similar argument can be applied to the fundamental cut-sets corresponding to the branches $e_3, e_4, \ldots , e_k$.

On the other hand, let C' be the fundamental cut-set corresponding to the tree branch e_{k+1}. C' cannot contain e_1, because otherwise, L and C' will have e_1 as the only common edge. ■

Theorem 7-8 *For a given spanning tree, let* $C = \{e_1,e_2,e_3, \ldots ,e_k\}$ *be a fundamental cut-set in which* e_1 *is a branch and* $e_2, e_3, \ldots , e_k$ *are chords of the spanning tree. Then,* e_1 *is contained in the fundamental circuit corresponding to* e_i *for* $i = 2, 3, \ldots , k$. *Moreover,* e_1 *is not contained in any other fundamental circuit.*

Proof Similar to that of Theorem 7-7. ■

•7-3 LINEAR VECTOR SPACES

This section introduces some fundamental concepts of algebra.

We have given the definition of a group in Chap. 5. An *abelian group* is a group in which the binary operation is also *commutative;* that is, $a * b = b * a$ for any two elements a and b in the group, where $*$ denotes the binary operation of the group. For example, Fig. 7-5*a* shows a

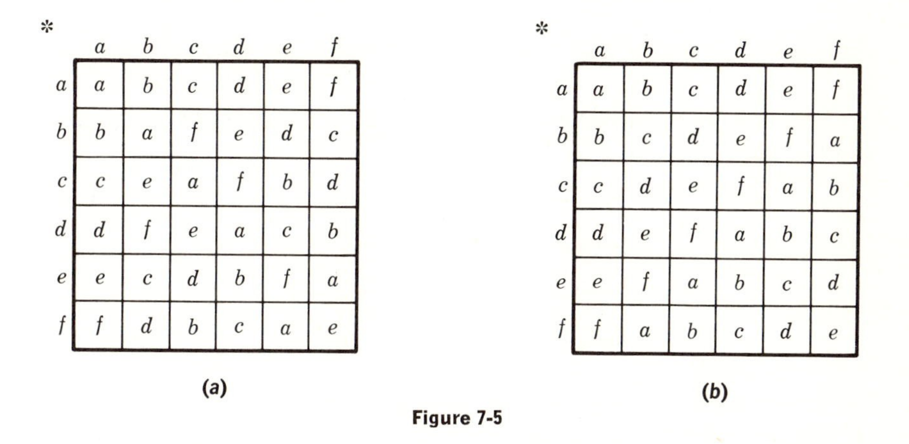

$*$	a	b	c	d	e	f
a	a	b	c	d	e	f
b	b	a	f	e	d	c
c	c	e	a	f	b	d
d	d	f	e	a	c	b
e	e	c	d	b	f	a
f	f	d	b	c	a	e

(*a*)

$*$	a	b	c	d	e	f
a	a	b	c	d	e	f
b	b	c	d	e	f	a
c	c	d	e	f	a	b
d	d	e	f	a	b	c
e	e	f	a	b	c	d
f	f	a	b	c	d	e

(*b*)

Figure 7-5

nonabelian group (for instance, $b * c \neq c * b$), and Fig. 7-5*b* shows an abelian group.

Let F be a set of elements $\{a,b,c, \ldots\}$ on which two binary operations, addition and multiplication,† are defined. The addition and multiplication operations will be denoted by $+$ and $\cdot$, respectively. If the following postulates hold, then F is said to be a *field.*

1. F is an abelian group under addition.
2. With the additive identity (the identity element in the group under addition) excluded, the set of remaining elements in F is an abelian group under multiplication.
3. Multiplication is distributive over addition; that is,

$$a \cdot (b + c) = a \cdot b + a \cdot c \qquad \text{for any } a, b, \text{ and } c \text{ in } F$$

For example, let $F = \{a,b,c,d\}$, and let addition and multiplication be defined as

† The reader should be reminded that these are just names we give to the two binary operations so that we can refer to them conveniently. That they are called addition and multiplication has no significance.

+	a	b	c	d
a	a	b	c	d
b	b	a	d	c
c	c	d	a	b
d	d	c	b	a

	a	b	c	d
a	a	a	a	a
b	a	b	c	d
c	a	c	d	b
d	a	d	b	c

The reader can check that F, under the binary operations + and ·, is a field.

As another example, let $F = \{0,1\}$, and let addition and multiplication be defined as

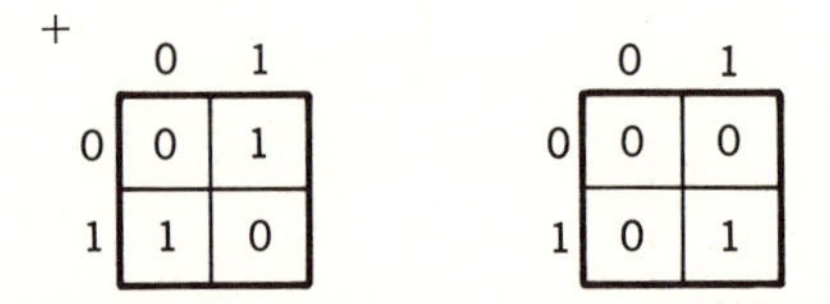

F is a field which is called the field of integers modulo 2.

Let S be a set of elements $\{\alpha,\beta,\gamma, \ldots\}$ on which a binary operation, addition, is defined. We shall denote this addition operation by $\boxplus$. Let F be a field consisting of the elements in the set $\{a,b,c, \ldots\}$ on which the operations + and · are defined. A multiplication operation, denoted by $\triangle$, between the elements in F and the elements in S which maps the elements in $F \times S$ into the elements in S is also defined. The set S is said to be a *vector space* over the field F if the following postulates hold:

1. S is an abelian group under $\boxplus$.
2. For any elements α and β in S and any elements a and b in F,

$$a \triangle (\alpha \boxplus \beta) = (a \triangle \alpha) \boxplus (a \triangle \beta)$$
$$(a + b) \triangle \alpha = (a \triangle \alpha) \boxplus (b \triangle \alpha)$$

3. For any element α in S and any elements a and b in F,

$$(a \cdot b) \triangle \alpha = a \triangle (b \triangle \alpha)$$

4. For any element α in S, $1 \triangle \alpha = \alpha$ where 1 is the multiplicative identity in F.

For a vector space S over a field F, the elements in S are called *vectors*, and the elements in F are called *scalars*. A vector α is said to be expressible as a linear combination of the vectors $\alpha_1, \alpha_2, \ldots, \alpha_t$ if there exists a set of scalars $a_1, a_2, \ldots, a_t$ such that

$$\alpha = (a_1 \triangle \alpha_1) \boxplus (a_2 \triangle \alpha_2) \boxplus \cdots \boxplus (a_t \triangle \alpha_t)$$

A set of vectors $\{\alpha_1, \alpha_2, \ldots, \alpha_t\}$ is said to be a *basis* of the vector space if any vector in S can be expressed as a linear combination of the vectors in the set and no vector in the set can be expressed as a linear combination of the other vectors in the set. It can be shown that all bases of the vector space contain the same number of vectors. (See Appendix 7-1.) The *dimension* of a vector space is defined as the number of vectors in a basis.

A subset S' of S is said to be a *subspace* of S if S' is also a vector space over the field F.

7-4 THE VECTOR SPACES ASSOCIATED WITH A GRAPH

This section is concerned with the properties of circuits and cut-sets and their relationship. We shall show that the set of subsets of edges in a graph is a vector space over the field of integers modulo 2 and that circuits and cut-sets are vectors in two subspaces of this vector space. Such an abstract point of view is a most useful one because of the insight it offers into our study of the several fundamental concepts in both this and the next chapter.

In the next two theorems, we prove that the set of circuits and edge-disjoint unions of circuits in a graph is an abelian group under the operation ring sum of sets $\oplus$. We use the term *edge-disjoint union of circuits* to mean the union of a set of circuits that have no common edges. Similarly, the term *edge-disjoint union of cut-sets* means the union of a set of cut-sets that have no common edges.

Theorem 7-9 *The ring sum of two circuits is a circuit or an edge-disjoint union of circuits.*

Proof Consider the subgraph containing the edges that are in the ring sum of two circuits. Since the degree of every vertex in a circuit is 2, the degree of any vertex in the subgraph is either 2 or 4. According to Corollary 6-3.1, the subgraph possesses an Euler circuit. Moreover, repeating the argument used in proving the sufficiency part of Theorem 6-3, we conclude that the subgraph is either a circuit or an edge-disjoint union of circuits. ∎

Corollary 7-9.1 *The ring sum of two circuits or edge-disjoint unions of circuits is a circuit or an edge-disjoint union of circuits.*

Theorem 7-10 *The set of all circuits (the empty set is considered a circuit) and edge-disjoint unions of circuits in a graph is an abelian group under the operation ring sum.*

Proof We shall check each of the postulates for an abelian group.

1. Since the ring sum of any set and the empty set is the set itself, the empty set is the identity element of the group.
2. Let A, B, and C be three sets. Since both the set $(A \oplus B) \oplus C$ and the set $A \oplus (B \oplus C)$ contain the elements that are in the set A alone, the set B alone, the set C alone, or in all of the three sets A, B, and C, it follows that ring sum is an associative operation.
3. According to the definition of the ring sum of two sets, it is clear that the operation is also commutative.
4. According to Theorem 7-9 and Corollary 7-9.1, the set of all circuits and edge-disjoint unions of circuits is closed under the operation ring sum.
5. Since the ring sum of a circuit (an edge-disjoint union of circuits) and itself is the identity element, the inverse of a circuit (an edge-disjoint union of circuits) is the circuit (the edge-disjoint union of circuits) itself. ■

Similarly, for cut-sets and edge-disjoint unions of cut-sets, we have the following theorems.

Theorem 7-11 *The ring sum of two cut-sets is a cut-set or an edge-disjoint union of cut-sets.*

Proof Let C_1 and C_2 be two cut-sets. Corresponding to C_1, there is a division of the vertices in the graph into two subsets which we denote by A_1 and B_1. Corresponding to C_2, there is a division of the vertices in the graph into two subsets which we denote by A_2 and B_2.

Note that the vertices in the graph are divided into two disjoint subsets $A_1 \oplus A_2$ and $B_1 \oplus A_2$. A vertex that is in $A_1 \oplus A_2$ either is in A_1 but not A_2 or is in A_2 but not A_1. If it is in A_1 but not A_2, it is not in $B_1 \oplus A_2$, as it is in neither B_1 nor A_2. If it is in A_2 but not A_1, it is not in $B_1 \oplus A_2$, as it is in both B_1 and A_2. A vertex that is not in $A_1 \oplus A_2$ either is in both A_1 and A_2 or is in neither A_1 nor A_2. If it is in both A_1 and A_2, it is in $B_1 \oplus A_2$, because it is not in B_1 but is in A_2. If it is in neither A_1 nor A_2, it is in $B_1 \oplus A_2$, because it is in B_1 but is not in A_2.

Let v_x be a vertex in $A_1 \oplus A_2$, and let v_y be a vertex in $B_1 \oplus A_2$. We claim that if there is an edge in the graph joining v_x and v_y, this edge must be included in the set $C_1 \oplus C_2$. There are four cases to be examined.

1. Vertex v_x is in A_1 but not in A_2, and vertex v_y is in B_1 but not in A_2. Since v_x is in A_1 and v_y is in B_1, the edge (v_x,v_y) is in C_1. Since both v_x and v_y are in B_2, the edge (v_x,v_y) is not in C_2. Therefore, the edge (v_x,v_y) is in $C_1 \oplus C_2$.
2. Vertex v_x is in A_1 but not in A_2, and vertex v_y is not in B_1 but in A_2. Since both v_x and v_y are in A_1, the edge (v_x,v_y) is not in C_1. Since v_x is in B_2 and v_y is in A_2, the edge (v_x,v_y) is in C_2. Therefore, the edge (v_x,v_y) is in $C_1 \oplus C_2$.
3. Vertex v_x is not in A_1 but in A_2, and vertex v_y is in B_1 but not in A_2. This case is similar to case 2.
4. Vertex v_x is not in A_1 but in A_2, and vertex v_y is not in B_1 but in A_2. This case is similar to case 1.

We have shown that any edge joining a vertex in $A_1 \oplus A_2$ and a vertex in $B_1 \oplus A_2$ is in the set $C_1 \oplus C_2$. It remains to be proved that any edge in the set $C_1 \oplus C_2$ joins a vertex in $A_1 \oplus A_2$ and a vertex in $B_1 \oplus A_2$. Suppose that there is an edge joining two vertices v_x and v_y that are both in $A_1 \oplus A_2$. If both v_x and v_y are in A_1 and are not in A_2, the edge (v_x,v_y) is in neither C_1 nor C_2. Therefore, the edge (v_x,v_y) is not in $C_1 \oplus C_2$. If v_x is in A_1 and not in A_2, while v_y is not in A_1 but in A_2, the edge (v_x,v_y) is in both C_1 and C_2. Therefore, the edge (v_x,v_y) is not in $C_1 \oplus C_2$. Using the same argument, we can show that an edge joining the two vertices v_x and v_y that are both in $B_1 \oplus A_2$ is not in $C_1 \oplus C_2$.

We conclude that $C_1 \oplus C_2$ is the set of edges joining the two subsets of vertices $A_1 \oplus A_2$ and $B_1 \oplus A_2$. Therefore, $C_1 \oplus C_2$ is either a cut-set or an edge-disjoint union of cut-sets. That $C_1 \oplus C_2$ is possibly an edge-disjoint union of cut-sets (instead of being just a cut-set) comes from the fact that the subgraph containing the vertices in $A_1 \oplus A_2$ and the subgraph containing the vertices in $B_1 \oplus A_2$ are possibly unconnected subgraphs after the edges in $C_1 \oplus C_2$ are removed. ■

Corollary 7-11.1 *The ring sum of two cut-sets or edge-disjoint unions of cut-sets is a cut-set or an edge-disjoint union of cut-sets.*

The proofs of the following two theorems are similar to that of Theorem 7-10 and will be left as exercises.

Theorem 7-12 *The set of all cut-sets and edge-disjoint unions of cut-sets is an abelian group under the operation ring sum.*

Theorem 7-13 *The set of all subsets of edges in a graph is an abelian group under the operation ring sum.*

We are now ready to state the main result of this section.

Theorem 7-14 *Let S be the set of all subsets of edges in a graph (including the empty set). Let F be the field of integers modulo 2. S is a vector space over F when the following definitions are made:*

1. *Addition between the elements in S is defined as the ring sum of sets.*
2. *Multiplication between the elements $\{0,1\}$ in F and the elements in S is defined such that 0 times any set is equal to the empty set and 1 times any set is equal to the set itself.*

Proof The algebraic structure defined this way satisfies the postulates of a vector space over a field. ∎

Corollary 7-14.1 *The set of all circuits and edge-disjoint unions of circuits in a graph is a subspace of S.*

Corollary 7-14.2 *The set of all cut-sets and edge-disjoint unions of cut-sets in a graph is a subspace of S.*

The subspace of circuits and edge-disjoint unions of circuits and the subspace of cut-sets and edge-disjoint unions of cut-sets are called the *circuit subspace* and the *cut-set subspace*, respectively.

7-5 THE BASES OF THE SUBSPACES

In the previous section we discussed the circuit subspace and the cut-set subspace associated with a graph. In this section we show that the bases of these subspaces are closely related to the spanning trees in the graph.

Theorem 7-15 *The fundamental system of circuits relative to a spanning tree is a basis of the circuit subspace.*

Proof First it is shown that any circuit can be expressed as a linear combination of the circuits in the fundamental system of circuits relative to a spanning tree. Let $L_1 = \{e_1,e_2,e_3, \ldots ,e_j,e_{j+1}, \ldots ,e_k\}$

be a circuit where $e_1, e_2, e_3, \ldots, e_j$ are chords and $e_{j+1}, \ldots, e_k$ are branches of a spanning tree. In the fundamental system of circuits relative to the spanning tree, there is a circuit containing the chord e_1, a circuit containing the chord $e_2, \ldots$, and a circuit containing the chord e_j. Let L_2 denote the ring sum of these fundamental circuits. We claim that both L_1 and L_2 contain the same set of edges. If this is not the case, then the set $L_1 \oplus L_2$ is nonempty. According to Theorem 7-9, L_2 is a circuit or an edge-disjoint union of circuits and so is $L_1 \oplus L_2$. However, since L_2 contains the chords $e_1, e_2, \ldots, e_j$ but no other chords of the spanning tree, $L_1 \oplus L_2$ contains only the branches of the spanning tree. That a tree contains no circuits gives us a contradiction.

Now it is shown that a circuit in a fundamental system of circuits cannot be expressed as the ring sum of the other circuits in the fundamental system. Because each circuit in a fundamental system of circuits contains a chord that no other circuit in the fundamental system contains, it is impossible for a circuit to be expressed as a linear combination of the other circuits in the fundamental system. ■

Corollary 7-15.1 *The dimension of the circuit subspace is* $\mathbf{e} - \mathbf{v} + 1$.

The proof of the following theorem on the cut-set subspace is similar to that of Theorem 7-15 and is left as an exercise.

Theorem 7-16 *The fundamental system of cut-sets relative to a spanning tree is a basis of the cut-set subspace.*

Corollary 7-16.1 *The dimension of the cut-set subspace is* $\mathbf{v} - 1$.

We also have the following theorems characterizing the vectors in the circuit subspace and the cut-set subspace.

Theorem 7-17 *A set of edges is a vector in the circuit subspace if and only if the set has an even number of common edges with every vector in the cut-set subspace.*

Proof According to Theorem 7-6, every circuit has an even number of common edges with every cut-set. It follows that every circuit or edge-disjoint union of circuits has an even number of common edges with every cut-set or edge-disjoint union of cut-sets.

Let $L_1 = \{e_1, e_2, \ldots, e_j, e_{j+1}, \ldots, e_k\}$ be a set of edges that has an even number of common edges with every vector in the cut-set subspace. Without loss of generality, assume that $e_1, e_2, \ldots, e_j$ are chords and $e_{j+1}, \ldots, e_k$ are branches of an arbitrarily chosen spanning tree. Let L_2 denote the ring sum of the fundamental circuits corresponding to the chords $e_1, e_2, \ldots, e_j$. Since L_2 is a vector in the circuit subspace, L_2 has an even number of common edges with every vector in the cut-set subspace. It follows that $L_1 \oplus L_2$ also has an even number of common edges with every vector in the cut-set subspace. If $L_1 \oplus L_2$ is a nonempty set of edges, it contains only branches of the spanning tree. Let C be the fundamental cut-set corresponding to one of the branches in $L_1 \oplus L_2$. Then, $L_1 \oplus L_2$ and C will have exactly one common edge, which is impossible. Therefore, we conclude that L_1 is equal to L_2 and is a vector in the circuit subspace. ■

Theorem 7-18 *A set of edges is a vector in the cut-set subspace if and only if the set has an even number of common edges with every vector in the circuit subspace.*

Proof Similar to that of Theorem 7-17. ■

7-6 MATRIX REPRESENTATION

This section is concerned with the matrix representation of the bases of the circuit subspace and the cut-set subspace. Such representations are useful when studying the properties of a graph from an algebraic point of view.

Undoubtedly, most readers are familiar with the notation of representing a vector in an n-dimensional vector space as an ordered n-tuple. For instance, the notation (8,7,4) represents a vector in the three-dimensional Euclidean space. The same notation can be used for the vectors in the vector space of the subsets of edges in a graph.† For a graph that has **e** edges, we number the edges from 1 through **e** in an arbitrary way. A vector in the vector space of the subsets of edges shall have **e** components which are 0's and 1's (the two elements in the field of integers modulo 2). Specifically, the ith component of the vector is a 0 if the ith edge of the graph is not included in the subset; the ith component is a 1 if the ith edge is included in the subset. For example, the subset of edges $\{e_1,e_6,e_8,e_{10}\}$ in the graph shown in Fig. 7-6a is represented as the ordered 11-tuple (1,0,0,0,0,1,0,1,0,1,0). In the ordered **e**-tuple notation, the ring sum of two subsets is obtained by componentwise addition carried out in the field of integers modulo 2. For example, corresponding to

† Those readers who are familiar with the subject of vector spaces are reminded that every n-dimensional vector space S over a field F is isomorphic to the space of all the ordered n-tuples over the field F.

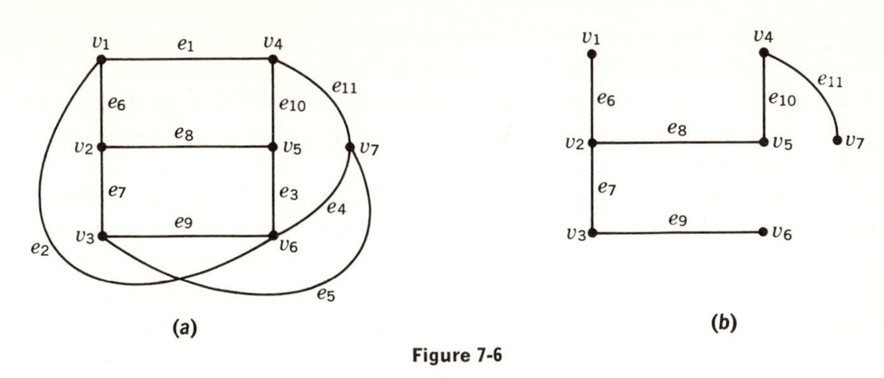

Figure 7-6

$\{e_1,e_6,e_8,e_{10}\} \oplus \{e_1,e_2,e_3,e_{10}\} = \{e_2,e_3,e_6,e_8\}$, we have

$$(1,0,0,0,0,1,0,1,0,1,0) \oplus (1,1,1,0,0,0,0,0,0,1,0) = (0,1,1,0,0,1,0,1,0,0,0)$$

Also, when a vector is multiplied by a scalar, the scalar multiplies each of the components, where, again, multiplication is carried out in the field of integers modulo 2. For example,

$$0 \triangle (1,0,0,0,0,1,0,1,0,1,0) = (0,0,0,0,0,0,0,0,0,0,0)$$
$$1 \triangle (1,0,0,0,0,1,0,1,0,1,0) = (1,0,0,0,0,1,0,1,0,1,0)$$

Clearly, the ordered **e**-tuple notation for vectors is consistent with the subset notation.

For a graph, a *circuit matrix* is an $(\mathbf{e} - \mathbf{v} + 1) \times \mathbf{e}$ matrix that has the circuits in a basis of the circuit subspace in their ordered **e**-tuple representation as rows. Similarly, a *cut-set matrix* is a $(\mathbf{v} - 1) \times \mathbf{e}$ matrix that has the cut-sets in a basis of the cut-set subspace in their ordered **e**-tuple representation as rows. For example, for the graph in Fig. 7-6*a* and the fundamental systems of circuits and cut-sets corresponding to the spanning tree in Fig. 7-6*b*, a circuit matrix M_L and a cut-set matrix M_C are

$$M_L = \begin{bmatrix} 1 & 0 & 0 & 0 & 0 & 1 & 0 & 1 & 0 & 1 & 0 \\ 0 & 1 & 0 & 0 & 0 & 1 & 1 & 0 & 1 & 0 & 0 \\ 0 & 0 & 1 & 0 & 0 & 0 & 1 & 1 & 1 & 0 & 0 \\ 0 & 0 & 0 & 1 & 0 & 0 & 1 & 1 & 1 & 1 & 1 \\ 0 & 0 & 0 & 0 & 1 & 0 & 1 & 1 & 0 & 1 & 1 \end{bmatrix}$$

$$M_C = \begin{bmatrix} 1 & 1 & 0 & 0 & 0 & 1 & 0 & 0 & 0 & 0 & 0 \\ 0 & 1 & 1 & 1 & 1 & 0 & 1 & 0 & 0 & 0 & 0 \\ 1 & 0 & 1 & 1 & 1 & 0 & 0 & 1 & 0 & 0 & 0 \\ 0 & 1 & 1 & 1 & 0 & 0 & 0 & 0 & 1 & 0 & 0 \\ 1 & 0 & 0 & 1 & 1 & 0 & 0 & 0 & 0 & 1 & 0 \\ 0 & 0 & 0 & 1 & 1 & 0 & 0 & 0 & 0 & 0 & 1 \end{bmatrix}$$

According to Theorem 7-6, the product of a circuit matrix and the transpose of a cut-set matrix is a matrix that contains 0's as all of its entries, as does the product of a cut-set matrix and the transpose of a circuit matrix. (Addition and multiplication are carried out in the field of integers modulo 2.)†

The *incidence matrix* of a graph is a $\mathbf{v} \times \mathbf{e}$ matrix in which the rows correspond to the vertices and the columns correspond to the edges. The (i,j)th entry in the matrix is a 1 if the ith vertex is incident with the jth edge and is a 0 otherwise. For example, the incidence matrix of the graph in Fig. 7-6a is

	e_1	e_2	e_3	e_4	e_5	e_6	e_7	e_8	e_9	e_{10}	e_{11}
v_1	1	1	0	0	0	1	0	0	0	0	0
v_2	0	0	0	0	0	1	1	1	0	0	0
v_3	0	0	0	0	1	0	1	0	1	0	0
v_4	1	0	0	0	0	0	0	0	0	1	1
v_5	0	0	1	0	0	0	0	1	0	1	0
v_6	0	1	1	1	0	0	0	0	1	0	0
v_7	0	0	0	1	1	0	0	0	0	0	1

The incidence matrix is obviously an alternative representation of a graph. Given the incidence matrix, one can always reconstruct the graph from it, and vice versa. Some of the properties of the incidence matrix are given in the following theorems.

Theorem 7-19 *If the incidence matrices of two graphs differ only by a permutation of rows and columns, the two graphs are isomorphic.*

Proof This theorem comes directly from the definition of two graphs being isomorphic. ■

Theorem 7-20 *Any set of* $\mathbf{v} - 1$ *rows of the incidence matrix is a basis of the cut-set subspace of the graph.*

Proof Since the dimension of the cut-set subspace is known to be $\mathbf{v} - 1$, it is sufficient to prove that any cut-set can be expressed as a linear combination of any $\mathbf{v} - 1$ of the rows. Because each column of the matrix contains exactly two 1's (the graph contains no loops), any row of the matrix can be expressed as the sum of all the other rows (componentwise addition carried out in the field of integers modulo 2). Therefore, we need only prove that any cut-set is

† For those readers who are familiar with the topic of linear algebra, it should be pointed out that the circuit subspace is the null space of the cut-set subspace, and the cut-set subspace is the null space of the circuit subspace. Such a point of view is very useful when we introduce the notion of duality in Chap. 8. However, our development in the main body shall not rely on this particular point of view.

expressible as a linear combination of the **v** rows of the incidence matrix. Let there be a cut-set that partitions the vertices into two subsets A and B. Adding the rows in the incidence matrix corresponding to the vertices in the subset A, we obtain the ordered **e**-tuple representation for the cut-set because a 1 in the sum of the rows corresponds to an edge joining two vertices, viz., one in A and another in B. ∎

7-7 SUMMARY AND REFERENCES

Far from being two unrelated notions, the circuits and the cut-sets in a graph have many similar properties. Moreover, the notion of spanning trees ties these notions together, as we have seen in this chapter. Our effort in establishing the vector spaces associated with a graph is a worthwhile one. First of all, the similarity between the properties of circuits and the properties of cut-sets becomes a direct and natural consequence when the point of view of vector spaces is taken. Secondly, we shall see that the known results in vector spaces can be applied to derive many of the properties of circuits and cut-sets. Finally, as will be seen in Chap. 8, such a point of view is most useful in our discussion of the concept of duality.

For further discussion on trees and related topics, see Chap. 16 of Berge [1] and Chap. 4 of Ore [5]. For the topic of circuits and cut-sets and their associated vector spaces, see Chaps. 2 and 4 of Seshu and Reed [6] and also the paper by Gould [3]. Applications to electrical network analysis can also be found in Chap. 1 of Guillemin [4]. For matrix representations of a graph, see Chaps. 14 and 15 of Berge [1], Chaps. 4 and 5 of Seshu and Reed [6], and Chap. 5 of Busacker and Saaty [2].

1. Berge, C.: "The Theory of Graphs and Its Applications," John Wiley & Sons, Inc., New York, 1962.
2. Busacker, R. G., and T. L. Saaty: "Finite Graphs and Networks: An Introduction with Applications," McGraw-Hill Book Company, New York, 1965.
3. Gould, R. L.: Graphs and Vector Spaces, *J. Math. Phys.*, **38**:193–214 (1958).
4. Guillemin, E. A.: "Introductory Circuit Theory," John Wiley & Sons, Inc., New York, 1953.
5. Ore, O.: "Theory of Graphs," American Mathematical Society, Providence, R.I., 1962.
6. Seshu, S., and M. B. Reed: "Linear Graphs and Electrical Networks," Addison-Wesley Publishing Company, Inc., Reading, Mass., 1961.

APPENDIX 7-1 THE NUMBER OF VECTORS IN THE BASES OF A VECTOR SPACE

Let α_0 denote the identity element in the group S under $\boxplus$. For an element α in S, let $-\alpha$ denote the inverse of α. Let 0 denote the additive identity, and let 1 denote the multiplicative identity in the field F. For an element a in F, let $-a$ denote the additive inverse of a, and let $1/a$ denote the multiplicative inverse of a (when a is not the additive identity).

Theorem 7A-1 *Let* α *be any vector and* a *be any scalar. Then*

1. $0 \triangle \alpha = \alpha_0$
2. $a \triangle \alpha_0 = \alpha_0$
3. $(-\alpha) = (-1) \triangle \alpha$
4. $-(a \triangle \alpha) = (-a) \triangle \alpha$

Proof 1.

$$\begin{aligned} 0 \triangle \alpha &= (0 \triangle \alpha) \boxplus \alpha \boxplus (-\alpha) \\ &= (0 \triangle \alpha) \boxplus (1 \triangle \alpha) \boxplus (-\alpha) \\ &= ((0+1) \triangle \alpha) \boxplus (-\alpha) \\ &= \alpha \boxplus (-\alpha) \\ &= \alpha_0 \end{aligned}$$

2. $a \triangle \alpha_0 = a \triangle (0 \triangle \alpha) = (a \cdot 0) \triangle \alpha = 0 \triangle \alpha = \alpha_0$†

3.

$$\begin{aligned} (-\alpha) &= (-\alpha) \boxplus \alpha_0 \\ &= (-\alpha) \boxplus (0 \triangle \alpha) \\ &= (-\alpha) \boxplus ((1 + (-1)) \triangle \alpha) \\ &= (-\alpha) \boxplus \alpha \boxplus ((-1) \triangle \alpha) \\ &= (-1) \triangle \alpha \end{aligned}$$

4.

$$\begin{aligned} -(a \triangle \alpha) &= (-(a \triangle \alpha)) \boxplus \alpha_0 \\ &= (-(a \triangle \alpha)) \boxplus (0 \triangle \alpha) \\ &= (-(a \triangle \alpha)) \boxplus ((a + (-a)) \triangle \alpha) \\ &= (-(a \triangle \alpha)) \boxplus (a \triangle \alpha) \boxplus ((-a) \triangle \alpha) \\ &= (-a) \triangle \alpha \quad \blacksquare \end{aligned}$$

Theorem 7A-2 *All the bases of a vector space contain the same number of vectors.*

Proof Let $\{\alpha_1,\alpha_2, \ldots ,\alpha_t\}$ and $\{\alpha'_1,\alpha'_2, \ldots ,\alpha'_r\}$ be two bases of a vector space. The vector α'_1 can be expressed as a linear combination of the vectors $\alpha_1, \alpha_2, \ldots, \alpha_t$. Thus,

$$\alpha'_1 = (a_1 \triangle \alpha_1) \boxplus (a_2 \triangle \alpha_2) \boxplus \cdots \boxplus (a_t \triangle \alpha_t)$$

† Since $a \cdot 0 = a \cdot (0 + 0) = a \cdot 0 + a \cdot 0$, $a \cdot 0$ must equal the additive identity 0 of the field F.

Since not all of $a_1, a_2, \ldots, a_t$ are 0, let us assume that a_t is not 0. It follows that we can express α_t as a linear combination of the vectors $\alpha_1', \alpha_1, \alpha_2, \ldots, \alpha_{t-1}$. Thus,

$$\alpha_t = \left(\frac{1}{a_t} \triangle \alpha_1'\right) \boxplus \left(\frac{-a_1}{a_t} \triangle \alpha_1\right) \boxplus \left(\frac{-a_2}{a_t} \triangle \alpha_2\right) \boxplus \cdots \boxplus \left(\frac{-a_{t-1}}{a_t} \triangle \alpha_{t-1}\right)$$

Therefore, any vector in the vector space can be expressed as a linear combination of the vectors in the set $\{\alpha_1',\alpha_1,\alpha_2, \ldots ,\alpha_{t-1}\}$. For the set $\{\alpha_1',\alpha_1,\alpha_2, \ldots ,\alpha_{t-1}\}$ we repeat the same argument to replace one of $\alpha_1, \alpha_2, \ldots, \alpha_{t-1}$ by α_2'. (α_1' will not be replaced because α_2' cannot be expressed in terms of α_1' alone.) It follows that any vector in the vector space can be expressed as a linear combination of the vectors in the set $\{\alpha_2',\alpha_1',\alpha_1,\alpha_2, \ldots ,\alpha_{t-2}\}$. Again, by the same argument, we can repeatedly replace an α by an α'. We conclude that t must be equal to or larger than r because if $t < r$, the replacement procedure will yield a proper subset of the set $\{\alpha_1',\alpha_2', \ldots ,\alpha_r'\}$, and any vector in the vector space can be expressed as a linear combination of the vectors in the subset. However, this is impossible because $\{\alpha_1',\alpha_2', \ldots ,\alpha_r'\}$ is a basis.

An analogous argument shows that t must be equal to or less than r. Therefore, we conclude that $t = r$. ∎

PROBLEMS

7-1. Prove that the complement of a spanning tree does not contain a cut-set and that the complement of a cut-set does not contain a spanning tree.

7-2. Let L be a circuit in a graph G. Let a and b be any two edges in L. Prove that there exists a cut-set C such that $L \cap C = \{a,b\}$.

7-3. Let T_1 and T_2 be two spanning trees of a connected graph G. Let a be an edge that is in T_1 but not T_2. Prove that there is an edge b in T_2 but not T_1 such that both $(T_1 - \{a\}) \cup \{b\}$ and $(T_2 - \{b\}) \cup \{a\}$ are spanning trees of G.

7-4. (*a*) Let L_1 and L_2 be two circuits in a graph G. Let a be an edge that is in both L_1 and L_2, and let b be an edge that is in L_1 but not L_2. Prove that there exists a circuit L_3 which is such that $L_3 \subseteq (L_1 \cup L_2) - \{a\}$ and $b \in L_3$.

(*b*) Repeat part (*a*) when the term "circuit" is replaced by the term "cut-set."

7-5. Let D be a set of edges in a graph G having the following two properties:

1. There is no cut-set C in G for which $D \cap C$ consists of a single edge.
2. For any two edges a and b in D, there is a cut-set C in G for which $D \cap C = \{a,b\}$.

Prove that D is a circuit or an edge-disjoint union of circuits.

7-6. Are there graphs for which the cut-set subspace is the same as the vector space of all the subsets of edges? If so, characterize this class of graphs.

7-7. Let G be a graph that has $\mathbf{v}$ vertices and $\mathbf{v} - 1$ edges. Prove that G is a tree if and only if any $\mathbf{v} - 1$ rows of the incidence matrix of G form a nonsingular matrix. (Addition and multiplication are conducted in the field of integers modulo 2.)

7-8. Show that there exists no graph having a circuit matrix equal to M_A. Show also that there exists no graph having a cut-set matrix equal to M_B.

$$M_A = \begin{bmatrix} 1 & 1 & 1 & 0 & 0 & 0 & 0 \\ 1 & 0 & 0 & 1 & 1 & 0 & 0 \\ 0 & 1 & 0 & 1 & 0 & 1 & 0 \\ 1 & 1 & 0 & 1 & 0 & 0 & 1 \end{bmatrix} \qquad M_B = \begin{bmatrix} 1 & 0 & 1 & 0 & 1 & 0 & 1 \\ 0 & 1 & 1 & 0 & 0 & 1 & 1 \\ 0 & 0 & 0 & 1 & 1 & 1 & 1 \end{bmatrix}$$

7-9. The adjacency matrix P of a graph with $\mathbf{v}$ vertices is a $\mathbf{v} \times \mathbf{v}$ matrix, the (i,j)th entry of which is 1 if there is an edge joining the ith and the jth vertices, and is 0 other-

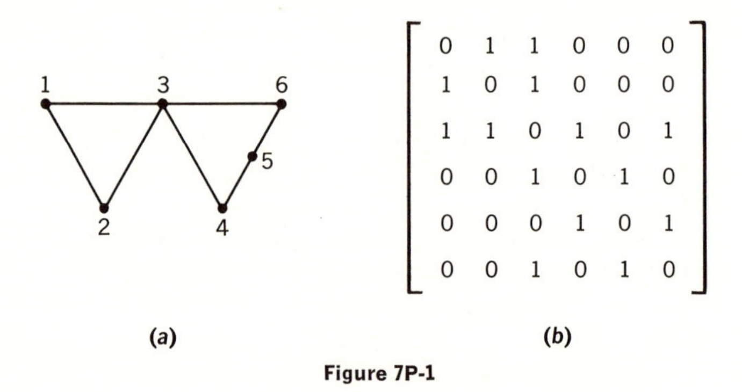

Figure 7P-1

wise. For example, the adjacency matrix of the graph in Fig. 7P-1*a* is shown in Fig. 7P-1*b*.

(*a*) Define the addition and multiplication operations for the elements 0 and 1 such that when P^k is computed, a 1 in the (i,j)th entry of P^k means that there is a path of length k between the ith and the jth vertices. Check your result by computing P^2 for the matrix in Fig. 7P-1*b*.

(*b*) Instead of the 0's in the diagonal, we shall fill the diagonal of P with 1's. For the addition and multiplication operations defined in part (*a*), what are the significances of a 0 entry and of a 1 entry in P^k?

7-10. Let G be a biconnected graph, and let a and b be two distinct vertices in G. Show that there exists a basis of the cut-set subspace which is such that every cut-set in this basis places a and b in two different connected components.

7-11. (*Whitney's Theorem*) Two graphs G and G' are said to be edge isomorphic when there is a one-to-one correspondence between the sets of edges in G and G' such that e_1 and e_2 in G are adjacent (incident with a common vertex) if and only if the corresponding edges e_1' and e_2' in G' are also adjacent. Note that the fact that G and G' are edge isomorphic does not necessarily imply that they are isomorphic. The two graphs shown in Fig. 7P-2 are edge isomorphic, but not isomorphic. However, if G and G' are isomorphic, then clearly they are edge isomorphic.

Figure 7P-2

Whitney's theorem implies that any graph which is edge isomorphic to a linear connected graph G, other than the two graphs shown above, is isomorphic to G.

(*a*) Show that the theorem holds for all nonisomorphic linear connected graphs having four or fewer vertices, other than the two shown above.

(*b*) Prove that for any linear connected graph G having five or more vertices, a graph which is edge isomorphic to G is isomorphic to G and the edge isomorphism is that specified by the isomorphism. Use induction on the number of edges in G, with trees containing five vertices as a basis.

7-12. In directed graphs, an oriented circuit is defined as a set of edges that is a circuit in the undirected graph obtained from the directed graph by ignoring the direction of the edges. Similarly, an oriented cut-set is a set of edges that is a cut-set in the undirected graph obtained in the same manner. The orientation of an oriented circuit is specified by a cyclic ordering of the vertices it encounters, that of an oriented cut-set by an ordering of the two subsets of vertices it separates. An oriented circuit vector or an oriented cut-set vector in a directed graph with $\mathbf{e}$ edges is represented by an ordered $\mathbf{e}$-tuple. The jth component of an oriented circuit vector $\mathbf{L}$ is either 1, -1, or 0 accordingly as the jth edge is in the oriented circuit and its direction agrees with the circuit orientation, the jth edge is in the oriented circuit and its direction is opposite to the circuit orientation, or the jth edge is not in the oriented circuit. Similarly for an oriented cut-set vector $\mathbf{C}$. These vectors are considered as vectors in an $\mathbf{e}$-dimensional vector space over the field of real numbers.

(*a*) Let $\mathbf{L}$ and $\mathbf{C}$ be an oriented circuit vector and an oriented cut-set vector, respectively, in a directed graph. Show that $\mathbf{L} \cdot \mathbf{C} = 0$.

(*b*) Prove that any oriented circuit vector $\mathbf{L}$ can be expressed as a linear combination of fundamental oriented circuit vectors.

(*c*) Prove that any oriented cut-set vector $\mathbf{C}$ can be expressed as a linear combination of fundamental oriented cut-set vectors.

The orientation of a fundamental oriented circuit is taken to agree with that of its defining chord. Similarly for a fundamental oriented cut-set and its defining branch.

Chapter 8
Planar and Dual Graphs

8-1 INTRODUCTION

A graph is said to be *planar* if it can be mapped on a plane in such a way that two edges meet one another only at the vertex, or vertices, with which the edges are incident (that is to say, no two edges cross one another). Figure 8-1*a* shows a planar graph. Notice that the graph in Fig. 8-1*b* is also planar because it can be redrawn as that in Fig. 8-1*c*. Figure 8-1*d* shows a nonplanar graph. As a matter of fact, the graph in Fig. 8-1*d* corresponds to the well-known problem of determining whether it is possible to connect three houses a, b, and c to three utilities d, e, and f in such a way that no two connecting pipelines meet one another except at their initial or terminal points. Experience shows that this cannot be done. In other words, the graph representing all of the connections among the houses and the utilities is indeed a nonplanar graph. (The reader may feel a little bit uncomfortable when we say that a graph is nonplanar simply because, after a certain number of attempts, we find

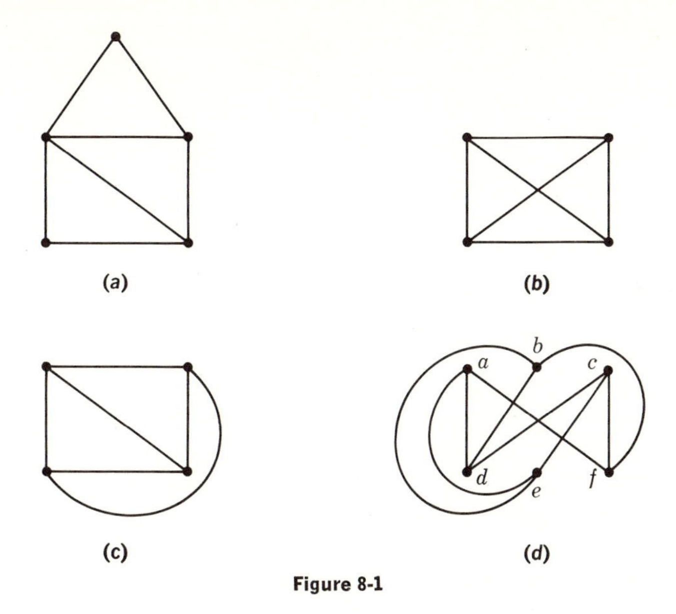

Figure 8-1

that we cannot map the graph on a plane. This issue will be settled rigorously in Secs. 8-2 and 8-3.)

We now show some of the properties of planar graphs.

Theorem 8-1 *A graph is planar if and only if it can be mapped onto the surface of a sphere such that no two edges meet one another except at the vertex, or vertices, with which the edges are incident.*

Proof Suppose that a graph is mapped onto the surface of a sphere. Let us place the sphere on a plane and call the point of contact the south pole and the diametrically opposing point the north pole. We want to place the sphere on the plane in such a way that the north pole is neither a vertex nor a point on an edge of the graph. Since the vertices are geometric points and the edges are geometric lines which do not occupy any area, such a placement is always possible. For any point P on the surface of the sphere, we join the north pole N and P by a straight line and extend the line to meet the plane at P' as shown in Fig. 8-2. The point P' is called the stereographic projection of the point P. Since there is a one-to-one correspondence between points on the surface of the sphere and their stereographic projections on the plane, a graph that can be mapped onto the surface of a sphere such that no two edges cross one another will be projected on the plane as a planar graph.

The converse is proved similarly. ■

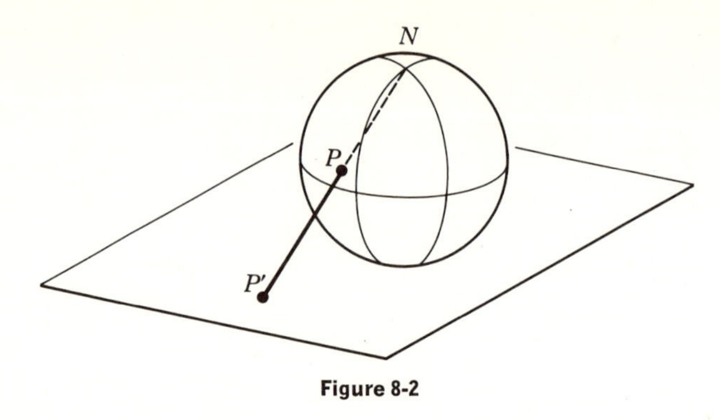

Figure 8-2

We shall state without proof a theorem due to Fary.†

Theorem 8-2 (Fary) *A linear planar graph that has no loops can be drawn on a plane with straight line segments as edges.*

As an example, the graph in Fig. 8-3*a* can be redrawn as that in Fig. 8-3*b* in which all the edges are straight-line segments. Besides being

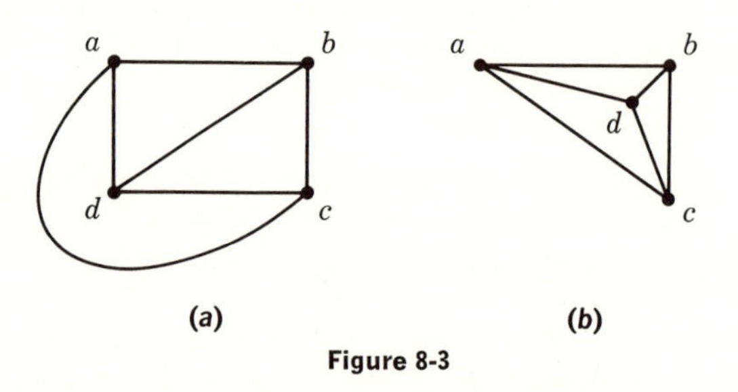

Figure 8-3

an interesting result, Fary's theorem also has some useful implications. Consider a planar graph as the wiring diagram of an electrical circuit; the theorem asserts that all the connections can be made with rigid straight wire. When graphs are to be constructed in some mechanical way (e.g., using a mechanical plotter or a digital computer), it is sufficient for the mechanical device to construct straight-line segments if only planar graphs are to be constructed.

8-2 EULER'S FORMULA

A *region* of a planar graph is an area of the plane that is bounded by edges and contains neither edges nor vertices. For example, the cross-hatched areas in Fig. 8-4*a* and Fig. 8-4*b* are regions of a planar graph.

† For a proof of the theorem, see Fary [3].

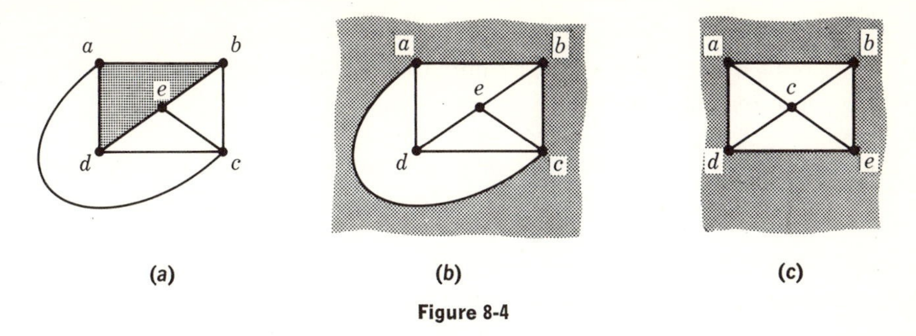

Figure 8-4

(We do not define "regions" for nonplanar graphs.) A finite region is a region the area of which is finite, and an infinite region is a region the area of which is infinite. Clearly, a planar graph has exactly one infinite region, the region external to all the edges. The crosshatched region in Fig. 8-4a is a finite region and the crosshatched region in Fig. 8-4b is an infinite region. Notice that a planar graph can be mapped onto a plane in such a way that any chosen region will become the infinite region. In particular, if the graph is mapped onto the surface of a sphere and the sphere is placed on the plane in such a way that the north pole is inside of the chosen region, then the stereographic projection of the graph from the surface of the sphere onto the plane will have the chosen region as the infinite region. For example, the graph in Fig. 8-4a is redrawn in Fig. 8-4c such that the region bounded by the edges (a,b), (b,e), (e,d), and (d,a) becomes the infinite region.

The *contour* of a region is the set of edges which bound the region. Two regions are said to be *adjacent* if their contours have at least one edge in common. Note that two regions are not adjacent if they meet only at a vertex. Since a planar graph can be mapped on a plane in many different ways, the following question arises: Does a planar graph have a fixed number of regions, independent of the way the graph is mapped on the plane? The answer to this question is yes, and the following theorem is a precise statement of this fact.

Theorem 8-3 *The set of contours of the finite regions of a planar graph is a basis of the circuit subspace of the graph.*

Proof A circuit of a planar graph encircles one or more finite regions. The set of edges in the circuit is, therefore, equal to the ring sum of the contours of the encircled regions. This proves that any circuit or edge-disjoint union of circuits can be expressed as a linear combination of the contours of the finite regions.

On the other hand, since the ring sum of the contours of two or more finite regions is a circuit (or an edge-disjoint union of circuits) that encircles the areas of these regions, no contour of a finite region can be expressed as a linear combination of the contours of the other finite regions. ■

Corollary 8-3.1 *A connected planar graph with* $\mathbf{e}$ *edges and* $\mathbf{v}$ *vertices has* $\mathbf{e} - \mathbf{v} + 1$ *finite regions.*

Corollary 8-3.2 *The relation* $\mathbf{v} - \mathbf{e} + \mathbf{r} = 2$ *is satisfied in any connected planar graph where* $\mathbf{r}$ *is the number of regions in the graph.*

Proof Since there is one infinite region in addition to the finite regions, we have $\mathbf{r} - 1 = \mathbf{e} - \mathbf{v} + 1$. ■

The relation $\mathbf{v} - \mathbf{e} + \mathbf{r} = 2$ is known as the Euler formula and is quite useful in showing whether or not a given graph is planar and in deducing some of the properties of planar graphs.

Example 8-1 In any linear planar graph that has no loops and has two or more edges, show that

$$\tfrac{3}{2}\mathbf{r} \leq \mathbf{e} \leq 3\mathbf{v} - 6$$

Let us count the number of edges in the boundary of a region and then compute the total count for all the regions. Because each region is bounded by at least three edges (the graph is a linear one), the total count is larger than or equal to $3\mathbf{r}$. On the other hand, in a planar graph an edge is in the boundaries of at most two regions; the total count is less than or equal to $2\mathbf{e}$. Thus, we have $2\mathbf{e} \geq 3\mathbf{r}$, or $\mathbf{e} \geq \tfrac{3}{2}\mathbf{r}$.† According to Euler's formula, we have

$$\mathbf{v} - \mathbf{e} + \tfrac{2}{3}\mathbf{e} \geq 2$$

or

$$3\mathbf{v} - 6 \geq \mathbf{e} \quad ■$$

Example 8-2 Show that the graph in Fig. 8-5*a* is not a planar graph. (The graph is called the star graph.)

Suppose that the graph is planar. Since $\mathbf{v} = 5$ and $\mathbf{e} = 10$, according to the result $\mathbf{e} \leq 3\mathbf{v} - 6$ proved in Example 8-1 for any linear graph, we have $10 \leq 3 \times 5 - 6 = 9$, which is impossible. ■

Example 8-3 Show that the graph in Fig. 8-5*b*, which is a repetition of Fig. 8-1*d*, is not a planar graph. (The graph is called the utility

† Because the graph has two or more edges, this inequality is still valid in the case where the graph has no finite region ($\mathbf{r} = 1$).

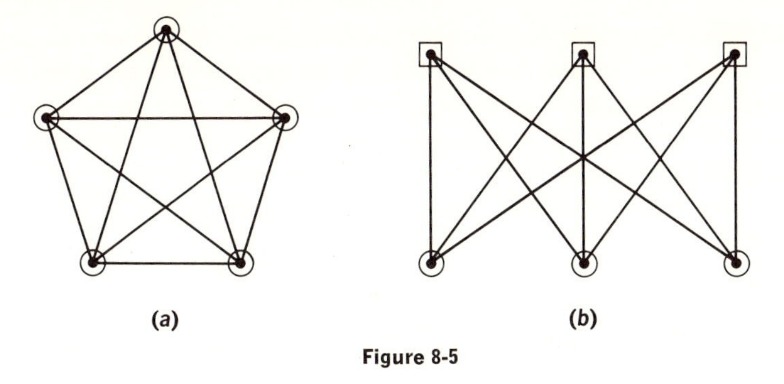

Figure 8-5

graph where the houses are marked with squares and the utilities are marked with circles.)

Suppose that the graph is planar. With $\mathbf{v} = 6$ and $\mathbf{e} = 9$, Euler's formula gives $\mathbf{r} = 2 - 6 + 9 = 5$. Since a region in this graph is bounded by four or more edges, using the same argument as that in Example 8-1, we have the inequality $2\mathbf{e} \geq 4\mathbf{r}$. However, this leads to a contradiction, because $2\mathbf{e} = 18$ and $4\mathbf{r} = 20$. ■

Example 8-4 Show that in a linear planar graph with no loops there is a vertex with degree equal to or less than 5.

Suppose that the degree of every vertex is larger than or equal to 6. The sum of the degrees of the vertices is larger than or equal to $6\mathbf{v}$. On the other hand, since each edge connects exactly two vertices, the sum of the degrees is equal to $2\mathbf{e}$. This gives $2\mathbf{e} \geq 6\mathbf{v}$, or $\mathbf{e} \geq 3\mathbf{v}$. Together with the inequality $\mathbf{e} \leq 3\mathbf{v} - 6$ obtained in Example 8-1, we have a contradiction. ■

8-3 KURATOWSKI'S THEOREM

We saw in the previous section that Euler's formula can sometimes be applied to assert that a given graph is nonplanar. However, such applications of the formula are rare because the argument becomes quite involved and tricky for graphs containing even a moderate number of vertices and edges. Moreover, except by actually mapping a graph on the plane, we have seen no way of asserting that a given graph is planar. In this section, we shall prove a theorem due to Kuratowski which enables us to determine the planarity of a graph unequivocally.

The planarity of a graph is clearly not affected if an edge is divided into two edges by the insertion of a new vertex of degree 2, as illustrated in Fig. 8-6*a*, or if two edges that are incident with a vertex of degree 2 are combined as a single edge by the removal of that vertex, as illustrated in Fig. 8-6*b*. This suggests the following definition: Two graphs G_1 and G_2

are said to be *isomorphic to within vertices of degree* 2 if they are isomorphic or if they can be transformed into isomorphic graphs by repeated insertions and/or removals of vertices of degree 2, as illustrated in Fig. 8-6*a* and Fig. 8-6*b*. For example, the two graphs in Fig. 8-6*c* are isomorphic to within vertices of degree 2.

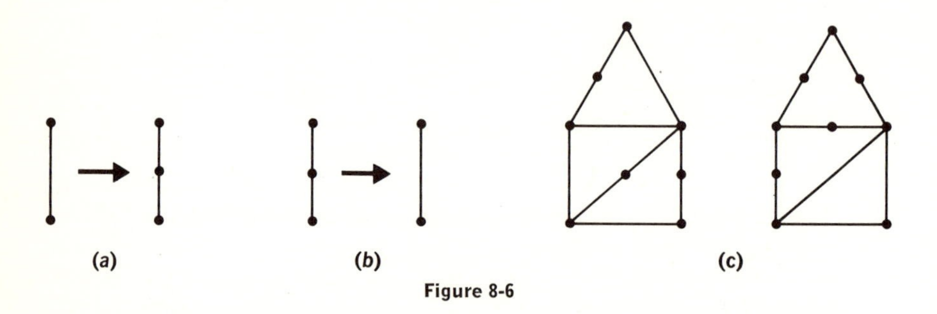

Figure 8-6

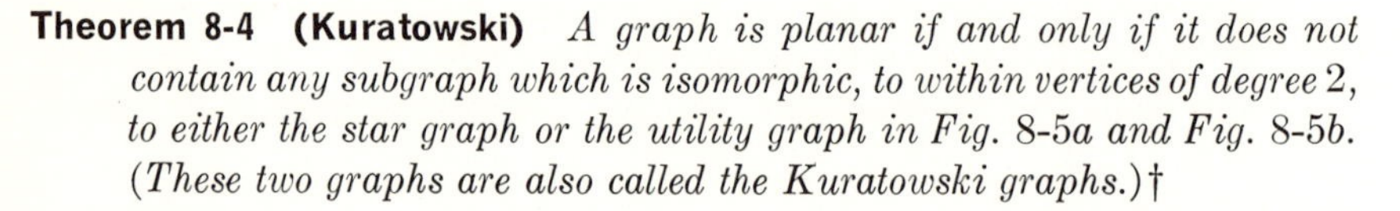

Theorem 8-4 (Kuratowski) *A graph is planar if and only if it does not contain any subgraph which is isomorphic, to within vertices of degree* 2, *to either the star graph or the utility graph in Fig.* 8-5*a and Fig.* 8-5*b*. (*These two graphs are also called the Kuratowski graphs.*)†

Proof We have already seen in Examples 8-2 and 8-3 that the Kuratowski graphs are nonplanar graphs. Therefore, a graph containing a subgraph that is isomorphic, to within vertices of degree 2, to either one of the Kuratowski graphs is a nonplanar graph.

Conversely, we shall show that a nonplanar graph must contain a subgraph that is isomorphic, to within vertices of degree 2, to one of the Kuratowski graphs. The proof is carried out by induction on the number of edges in a graph. As the basis of induction, it is clear that for graphs with one, two, or three edges, the statement that a nonplanar graph must contain a subgraph that is isomorphic, to within vertices of degree 2, to one of the Kuratowski graphs is true. As the induction hypothesis, let us assume that the statement is true for graphs with $m - 1$ or fewer edges. Suppose that there is a nonplanar graph G with m edges that does not contain a subgraph that is isomorphic, to within vertices of degree 2, to either one of the Kuratowski graphs. We now show that such a supposition is contradictory (that is, we examine all cases and show that either G is planar or that G contains a subgraph that is isomorphic, to within vertices of degree 2, to a Kuratowski graph).

† The proof of this theorem is quite long; the reader may want to skip it in the first reading.

First, note that there is a contradiction if G is not a connected graph. If G is not connected, all of its components will have fewer than m edges. Since G does not contain a subgraph that is isomorphic, to within vertices of degree 2, to either one of the Kuratowski graphs, none of its components does. According to the induction hypothesis, the components are all planar graphs. Therefore, G is also a planar graph, and we have a contradiction to the supposition.

Moreover, we can show that there is a contradiction if G is not biconnected. If G contains an articulation point, we can divide G into pieces relative to the articulation point. Since each of the pieces contains fewer than m edges, according to the induction hypothesis, the pieces are all planar graphs. As was shown in Sec. 8-2, by means of a stereographic projection, we can map each piece of the graph in such a way that the articulation point is on the boundary of the infinite region. These planar pieces can then be rejoined at the articulation point to yield a planar graph. Again, this is a contradiction to the supposition. (The construction is illustrated by an example in Fig. 8-7. Figure 8-7a shows a graph

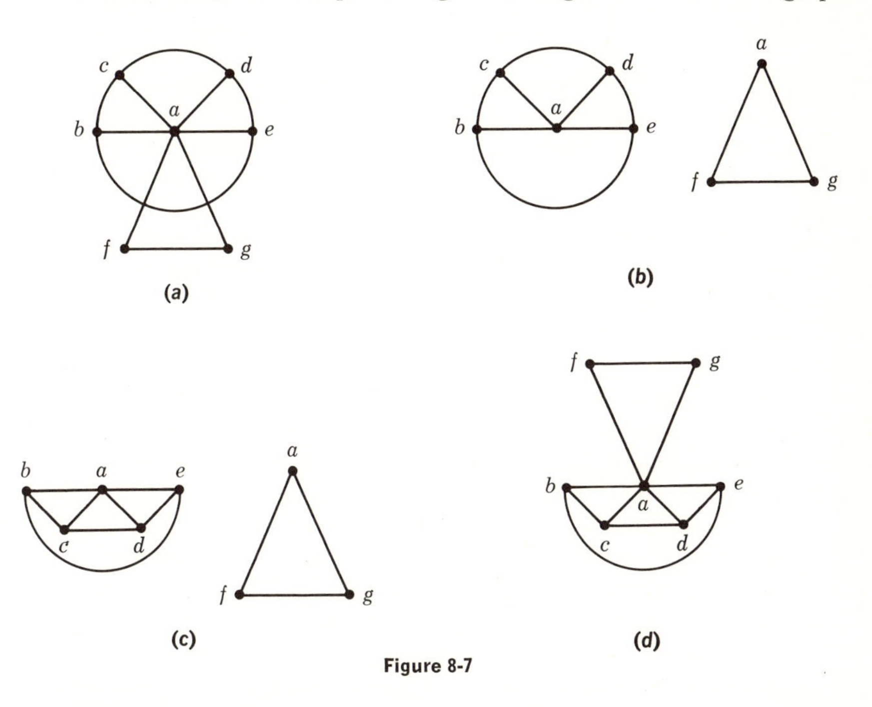

Figure 8-7

that has an articulation point a. Figure 8-7b shows the two pieces relative to the articulation point a. Figure 8-7c shows how the pieces can be redrawn with the articulation point a on the boundary of the infinite region in each piece. Figure 8-7*d* shows the two pieces in Fig. 8-7*c* rejoined at the articulation point a.)

What remains now are the more involved steps of showing a contradiction in the supposition when G is a biconnected graph. Let (a,b) be an edge in G. Let G' denote the subgraph of G with the edge (a,b) removed. We want to show that there is an elementary circuit in G' containing the vertices a and b. Suppose that there is no such elementary circuit. Since G is biconnected, G' must be connected. That there is no elementary circuit containing a and b means that every two paths between a and b in G' have a common vertex as illustrated in Fig. 8-8*a* (where x is the

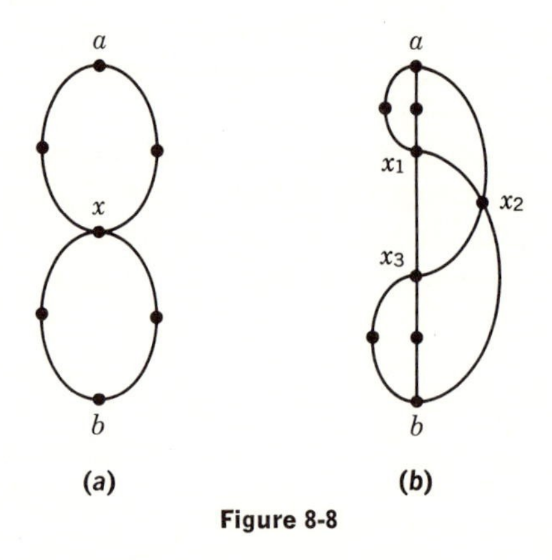

Figure 8-8

common vertex).† It follows that all the paths between a and b in G' must have a common vertex, because if the common vertices between every two paths do not coincide, as illustrated in Fig. 8-8*b* where x_1, x_2, and x_3 are the common vertices, then there is an elementary circuit containing a and b. This common vertex, denoted by x, clearly is an articulation point in G' (see Prob. 6-10). We can divide G' into two pieces, relative to x, such that a is in one piece G_1' and b is in another piece G_2'. Let G_1'' denote the graph obtained from G_1' by adding an edge (a,x) to G_1'. Let G_2'' denote the graph obtained from G_2' by adding an edge (b,x) to G_2'. We claim that both G_1'' and G_2'' do not contain a subgraph that is isomorphic, to within vertices of degree 2, to one of the Kuratowski graphs. If either one of them does, then the graph G will also contain a subgraph that is isomorphic, to within vertices of degree 2, to one of the Kuratowski graphs, and there is a contradiction to the suppo-

† The following argument is still applicable if there is only one path between a and b in G'.

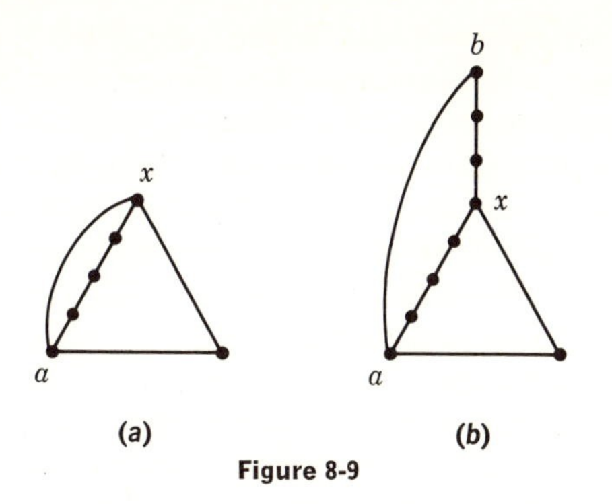

(a) (b)
Figure 8-9

sition. (The argument is illustrated in Fig. 8-9. Figure 8-9*a* shows G_1''. Figure 8-9*b* shows a portion of G; note that there is a path connecting a and b through x in G'. Clearly, if G_1'' contains a subgraph that is isomorphic, to within vertices of degree 2, to one of the Kuratowski graphs, G will also contain a subgraph that is isomorphic, to within vertices of degree 2, to that graph.) According to the induction hypothesis, both G_1'' and G_2'' are planar graphs. (G' contains $m - 1$ edges, and G_1' and G_2' contain fewer than $m - 1$ edges.) We can map G_1'' and G_2'', by means of stereographic projection, such that the edges (a,x) and (b,x) are in the boundaries of the infinite regions of G_1'' and G_2''. G_1'' and G_2'' can then be joined at the vertex x to yield a planar graph as illustrated in Fig. 8-10*a*.

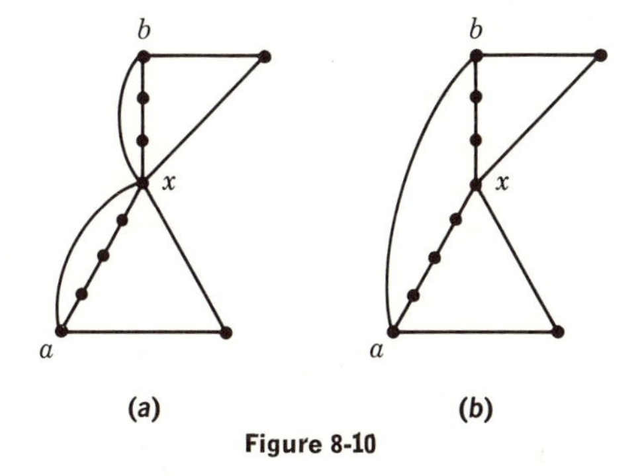

(a) (b)
Figure 8-10

It follows that G is also a planar graph as illustrated in Fig. 8-10*b*. However, this is in contradiction to our supposition.

Consider now the graph G' which is planar according to the induction hypothesis. Among all the elementary circuits that contain the vertices a and b, let S denote the one that includes the greatest number of regions in its interior. A connected subgraph that has all its edges in the interior of S will be called an *internal piece,* and a connected subgraph that has all its edges in the exterior of S will be called an *external piece.* To have a convenient notation for referring to the vertices along the circuit S, let us assign an arbitrary direction (say, counterclockwise) to S. Let p and q be any two vertices along S. We shall let $S[p,q]$ denote the sequence of vertices along the path from p to q in the counterclockwise direction with p and q included, and let $S]p,q[$ denote the sequence of vertices along the path from p to q in the counterclockwise direction with p and q excluded.

We see that no external piece contains more than one vertex in $S[a,b]$ or $S[b,a]$. If this is not the case, then one can find another elementary circuit S' containing a and b that includes more regions in its interior. (The situation is illustrated in Fig. 8-11a where the cir-

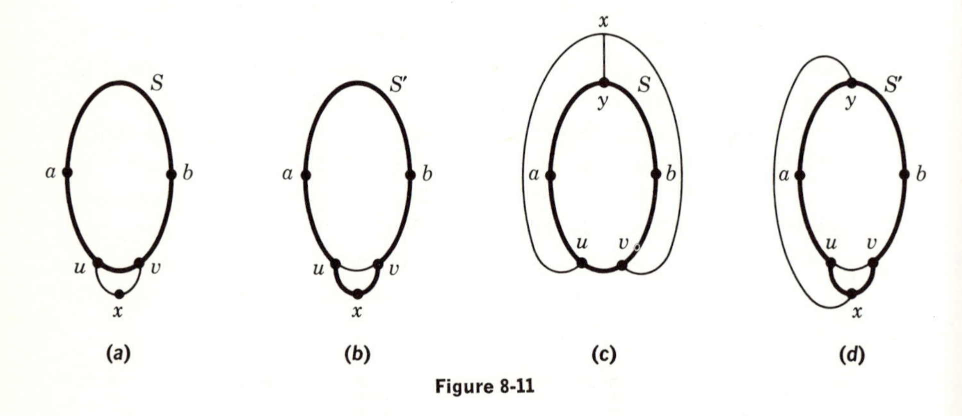

Figure 8-11

cuit S is indicated by the heavy line and there is an external piece containing the vertices u and v along $S[a,b]$. Figure 8-11b shows the circuit S' in a heavy line which includes more regions in its interior than does S. The situation illustrated in Fig. 8-11c is similar to that in Fig. 8-11a after the graph is redrawn as that in Fig. 8-11d where again S' is drawn in a heavy line.)

We note that there exist an internal piece and an external piece that meet both $S]a,b[$ and $S]b,a[$. If this is not the case, then one can add the edge (a,b) back to G' and have G as a planar graph. Moreover, we shall show that there exists an internal piece I and an external piece E which meet both $S]a,b[$ and $S]b,a[$ and are such that the points of contact c and d of E with S and two of the points of contact e and f of I with S are on S in alternating order: c, e, d, f. (Note that E can have only two points of contact along S, and I can have two or more points of contact along S. Moreover, to save any possible confusion later on, let us point out that e and f may coincide with a or b. In that case, I must have another point(s) of contact along $S]a,b[$ or $S]b,a[$.) Let us assume that the converse is true. Let I_1 be an internal piece, and let e_1 and f_1 be its points of contact with S as shown in Fig. 8-12a. Since it is assumed that there is

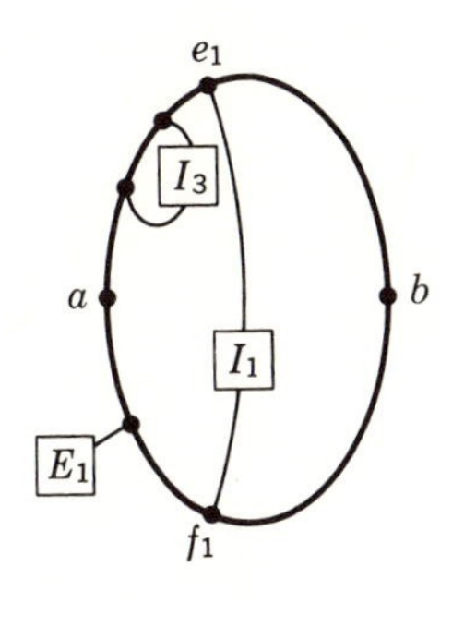

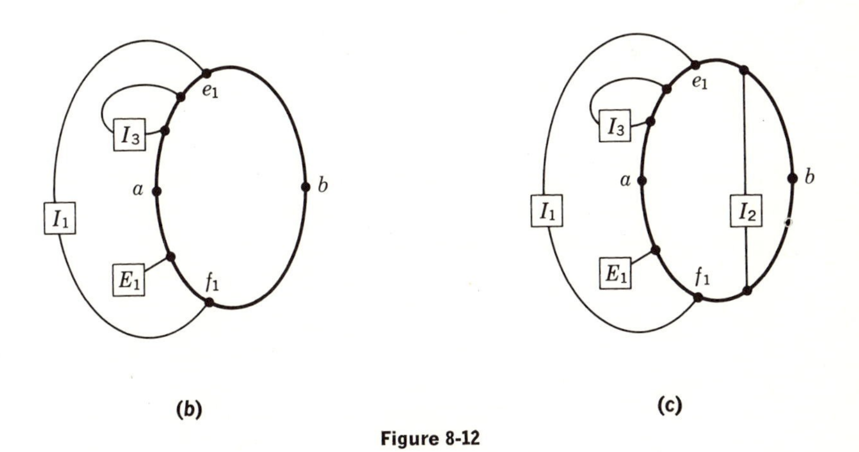

Figure 8-12

no external piece that meets both $S]e_1,f_1[$ and $S]f_1,e_1[$, every internal piece that meets S only along $S[e_1,f_1]$ can be transferred from the interior of S to the exterior of S as illustrated in Fig. 8-12b. Let I_2 be an internal piece that was not transferred, as illustrated in Fig. 8-12c. However, the argument we have used for the internal piece I_1 can be repeated for I_2. Moreover, this same argument can be applied to all the internal pieces. This means that no internal piece will be left in the interior of S, and the edge (a,b) can be added to G' so that we shall have G as a planar graph. This, of course, is a contradiction to the supposition.

We shall examine several cases in which the points of contact, e and f, of the internal piece I are at different segments of the circuit S. In each of these cases, we contradict the supposition by showing the existence of a subgraph of G which is isomorphic, to within vertices of degree 2, to one of the Kuratowski graphs.

Case 1 Vertex e is in $S]a,b[$, and vertex f is in $S]b,a[$. Regardless of the other points of contact, we have the configuration illustrated in Fig. 8-13. When the edge (a,b) is added to G', we see that G

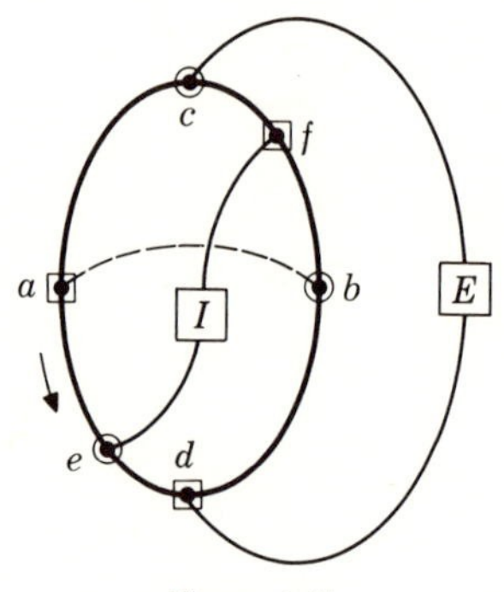

Figure 8-13

contains a subgraph which is isomorphic, to within vertices of degree 2, to the utility graph. (The vertices are marked with squares and circles as in Fig. 8-5*b*.)

Case 2 Both vertex e and vertex f are in $S]a,b[$. In this case, the internal piece I must have another point of contact along $S]b,a[$. If the point of contact is along $S]c,a[$, the case is reduced to case 1, as illustrated in Fig. 8-14*a*. If the point of contact is along $S]b,c[$, the case is again reduced to case 1, as illustrated in Fig. 8-14*b*. If the point of contact is c, we have the configuration illustrated in Fig. 8-14*c*. Again, we see that G contains a subgraph that is isomorphic, to within vertices of degree 2, to the utility graph.

Case 3 Vertex e coincides with vertex a, and vertex f is in $S]a,b[$. The internal piece I must have another point of contact along $S]b,a[$. If the point of contact is along $S]b,c[$, we have the configuration shown in Fig. 8-15*a*. If the point of contact is c or is along $S]c,a[$, we have the configuration shown in Fig. 8-15*b*. In either case,

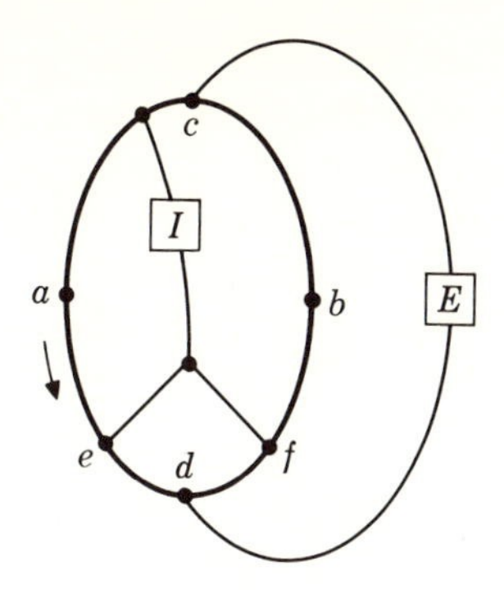

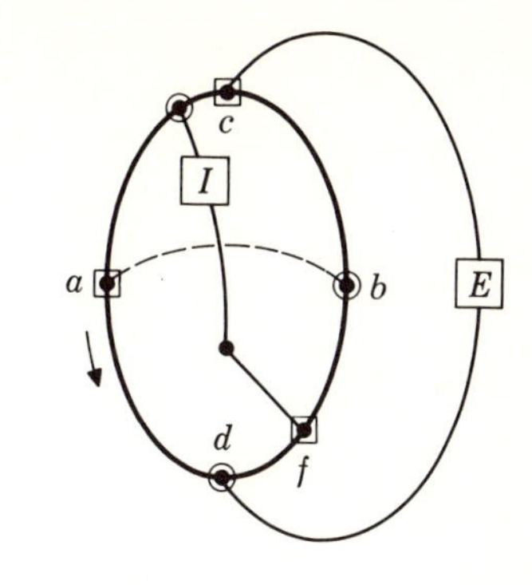

(a)

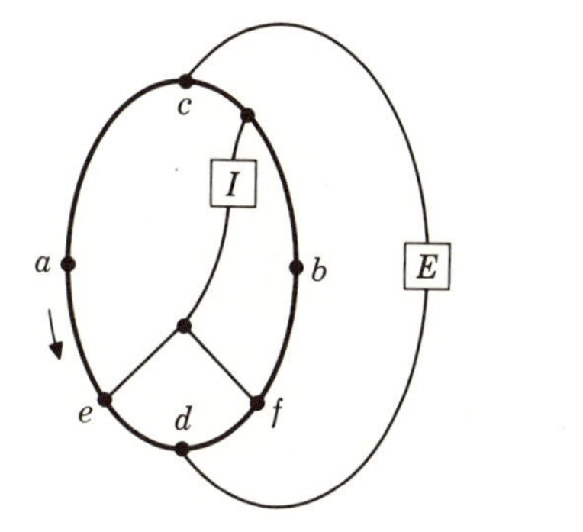

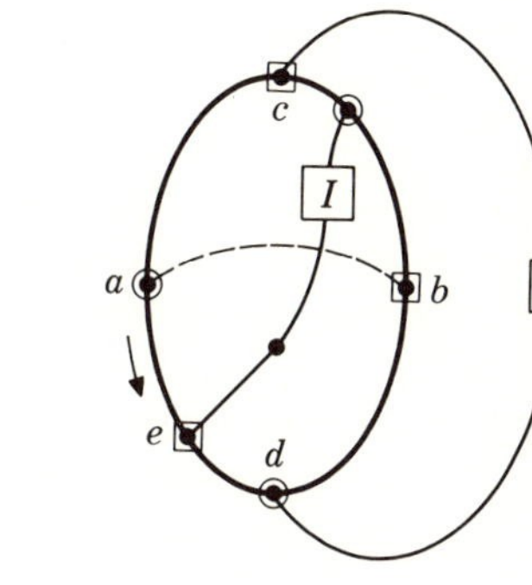

(b)

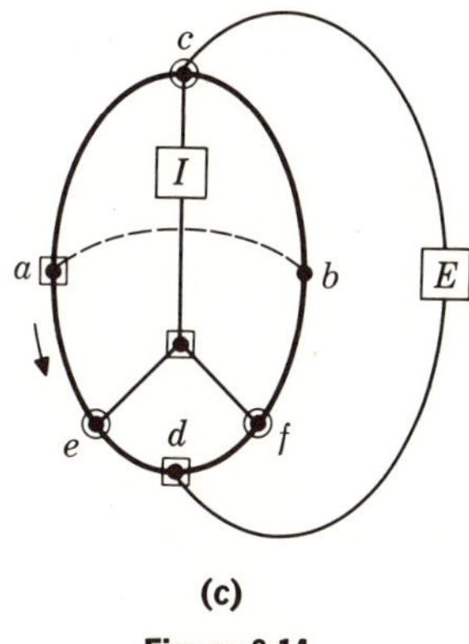

(c)

Figure 8-14

we see that G contains a subgraph that is isomorphic, to within vertices of degree 2, to the utility graph.

Case 4 Vertex e coincides with vertex a, and vertex f coincides with vertex b. The internal piece I must have two other points of contact, h and g, such that h is in $S]b,a[$ and g is in $S]a,b[$. If neither

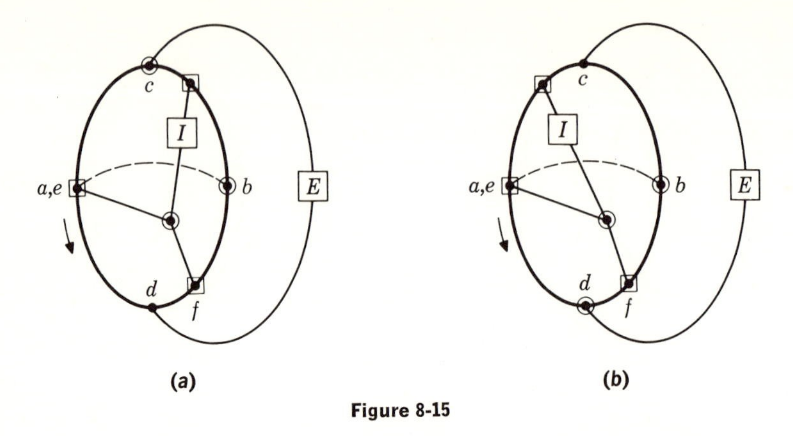

Figure 8-15

h nor g coincides with c or d, we have the configuration in which h and g are in $S]c,d[$ and $S]d,c[$, respectively, as illustrated in Fig. 8-16a, or the configuration in which g and h are both in $S]d,c[$ (or $S]c,d[$) as illustrated in Fig. 8-16b. The configuration in Fig. 8-16a is the same as that in case 1. The configuration in Fig. 8-16b is the same as that in case 3. If g coincides with d as illustrated in Fig. 8-16c, the configuration is the same as that in case 3. Therefore, we have only to consider the case where h and g coincide with c and d. There are two possibilities. If the path connecting a and b and the path connecting c and d have two (or more) vertices in common, we have the configuration illustrated in Fig. 8-16d. It follows that G will contain a subgraph that is isomorphic, to within vertices of degree 2, to the utility graph. If the path connecting a and b and the path connecting c and d have only one vertex in common, we have the configuration illustrated in Fig. 8-16e. It follows that G will contain a subgraph that is isomorphic, to within vertices of degree 2, to the star graph (the vertices are marked with circles as in Fig. 8-5a).

We have thus shown the supposition is contradictory and proved the theorem. ■

Example 8-5 An (undirected) complete graph is defined as a graph in which there is an edge between every two vertices. Let G be a graph with n vertices. The complementary graph of G, denoted by $\bar{G}$, is the complement of G with respect to the complete graph with n vertices. We want to show that for any graph G with seven or less vertices, either G or its complementary graph $\bar{G}$ is a planar graph.

If G has six vertices, one of the graphs G and $\bar{G}$ must have

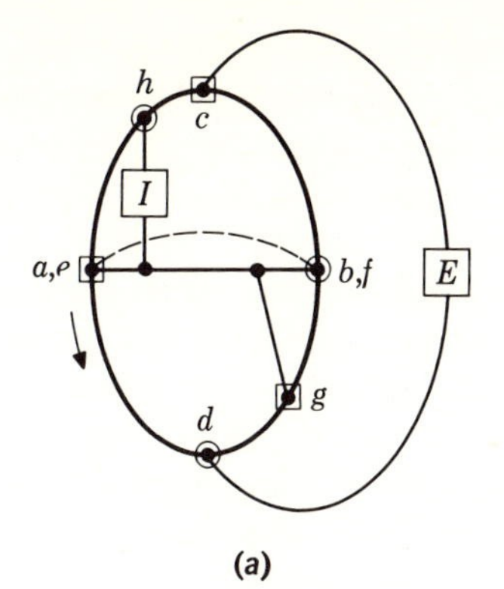

(a)

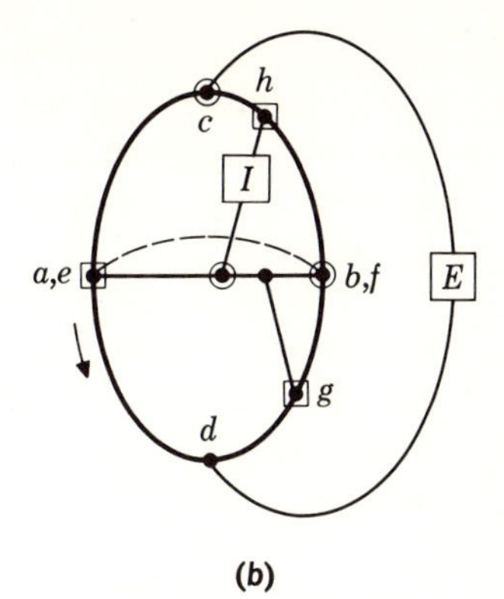

(b)

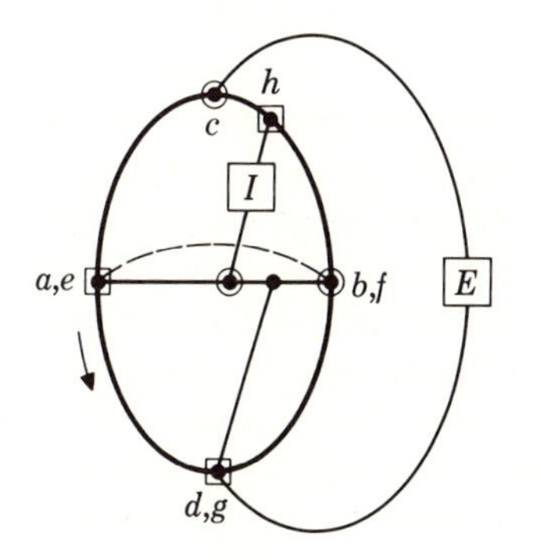

(c)

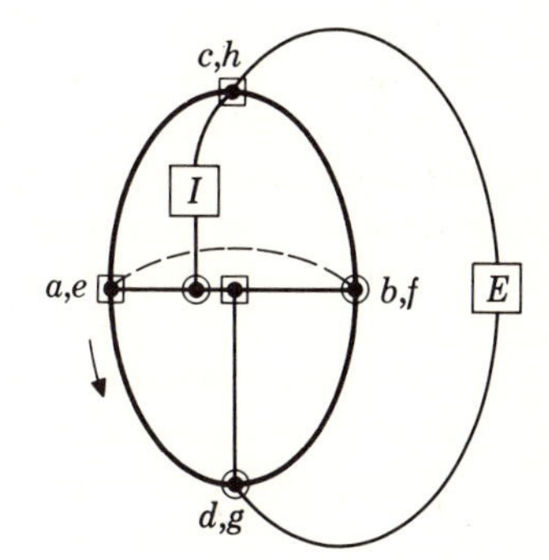

(d)

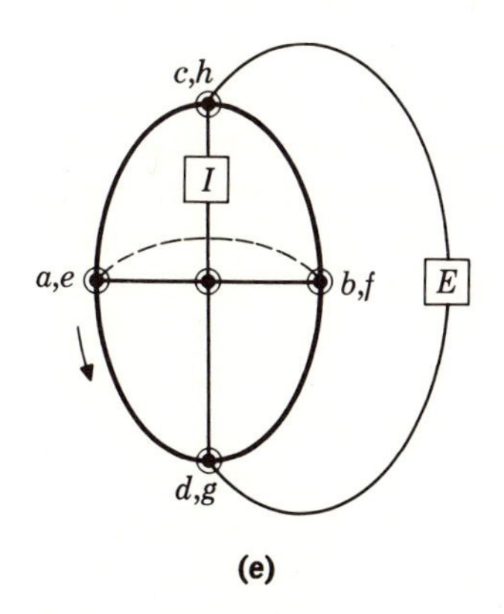

(e)

Figure 8-16

seven or fewer edges because the complete graph with six vertices has a total of 15 edges. Since the star graph has 10 edges and the utility graph has nine edges, one of the two graphs G and $\bar{G}$ must be planar. The same argument can be applied when G has less than six vertices.

Let us examine now the case in which G has seven vertices. Suppose that both G and $\bar{G}$ contain a subgraph that is isomorphic, to within vertices of degree 2, to the star graph. This means that in the complete graph there is a vertex whose degree is equal to or larger than 8 $(= 4 + 4)$, which obviously is impossible. Suppose that one of G and $\bar{G}$ contains a subgraph that is isomorphic, to within vertices of degree 2, to the star graph and the other contains a subgraph that is isomorphic, to within vertices of degree 2, to the utility graph. This means that in the complete graph there is a vertex whose degree is equal to or larger than 7 $(= 4 + 3)$, which is also impossible. Suppose that both G and $\bar{G}$ contain a subgraph that is isomorphic, to within vertices of degree 2, to the utility graph. In the utility graph, two vertices are said to be "on the same side of the street" if both of them are houses or both of them are utilities. We claim that there are two vertices that are on the same side of the street in both the subgraph of G and the subgraph of $\bar{G}$ that are isomorphic, to within vertices of degree 2, to the utility graph. Let a, b, and c denote the houses, and let d, e, and f denote the utilities, in the subgraph of G. If no two of the vertices a, b, and c are on the same side of the street in the subgraph of $\bar{G}$, one is neither a house nor a utility in this subgraph. Thus, two of the vertices d, e, and f must be on the same side of the street in this subgraph. Since two vertices on the same side of the street are not adjacent in the utility graph, there are two vertices in the complete graph whose degrees are equal to or larger than 7 $(= 3 + 3 + 1)$. However, this is impossible. ■

8-4 DUAL GRAPHS

A graph G_2 is said to be a *dual* of a graph G_1 if there is a one-to-one correspondence between the edges in G_1 and the edges in G_2 such that a set of edges in G_1 is a vector in the circuit subspace of G_1 if and only if its corresponding set of edges in G_2 is a vector in the cut-set subspace of G_2. In other words, there is a one-to-one correspondence between the sets of edges in the circuit subspace of G_1 and the sets of edges in the cut-set subspace of G_2. For example, the graph in Fig. 8-17*b* is a dual of the graph in Fig. 8-17*a* where corresponding edges are labeled with the same letter, starred and unstarred. The reader can check that corresponding to the circuit $\{a,d,g,f,b\}$ in the graph in Fig. 8-17a, there is the cut-set $\{a^*,d^*,g^*,f^*,b^*\}$ in the graph in Fig. 8-17b, and so on. Notice that a dual of a linear graph might be a multigraph. To be specific, when a graph contains one or more vertices of degree 2, a dual of the graph will be a multigraph. (This will become evident after we prove Corollary 8-6.1.) As an example, Fig. 8-17d shows a dual of the graph in Fig. 8-17c. The graph in Fig. 8-17c is a linear graph, whereas the graph in Fig. 8-17d is a multigraph.

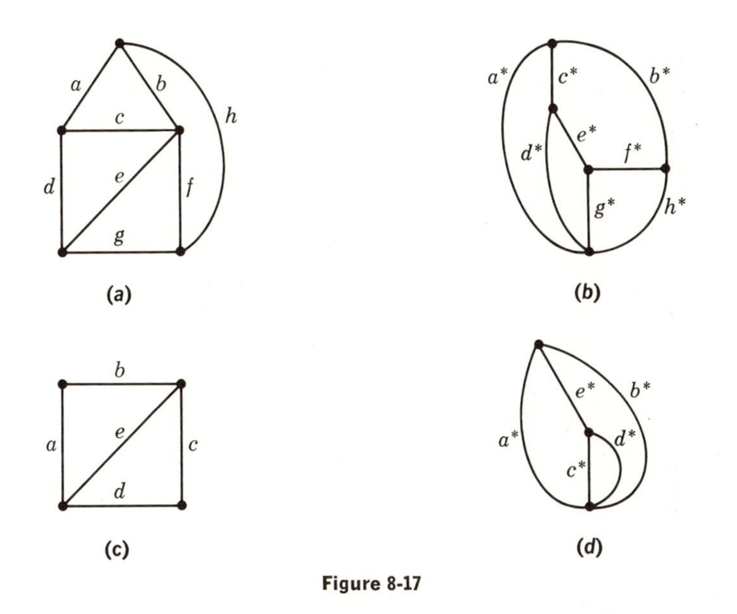

Figure 8-17

Theorems 8-5 and 8-6 and their corollaries show the relationship between the circuit subspace and the cut-set subspace of a graph and that of its dual graph.

Theorem 8-5 *Let G_2 be a dual of G_1. If a set of edges is a circuit in G_1, then its corresponding set of edges is a cut-set in G_2. If a set of edges is a cut-set in G_2, then its corresponding set of edges is a circuit in G_1.*

Proof Let L be a set of edges that is a circuit in G_1. Let L^* denote the corresponding set of edges in G_2. According to the definition of G_2 as a dual graph of G_1, L^* is either a cut-set or an edge-disjoint union of cut-sets in G_2. Suppose that L^* is an edge-disjoint union of the cut-sets L_1^*, L_2^*, ... Let L_1, L_2, . . . be the corresponding sets of edges in G_1. Then each one of L_1, L_2, . . . is a circuit or an edge-disjoint union of circuits in G_1. However, this is impossible because no proper subset of a circuit is a circuit or an edge-disjoint union of circuits. Therefore, L^* must be a cut-set in G_2.

The other half of the theorem can be proved in a similar manner. ■

Theorem 8-6 *Let G_2 be a dual of G_1. A set of edges is a vector in the cut-set subspace of G_1 if and only if its corresponding set of edges is a vector in the circuit subspace of G_2.*

Proof Let C be a vector in the cut-set subspace of G_1. Let C^* be its corresponding set of edges in G_2. Since C has an even number of edges in common with every vector in the circuit subspace of G_1, C^* has an even number of edges in common with every vector in the cut-set subspace of G_2. According to Theorem 7-17, C^* is either a circuit or an edge-disjoint union of circuits in G_2. ■

Corollary 8-6.1 *If a set of edges is a cut-set in G_1, then its corresponding set of edges is a circuit in G_2. If a set of edges is a circuit in G_2, then its corresponding set of edges is a cut-set in G_1.*

Proof Similar to that of Theorem 8-5. ■

Corollary 8-6.2 *If G_2 is a dual of G_1, G_1 is a dual of G_2.*

Therefore, it makes sense to say that two graphs G_1 and G_2 are dual graphs as they are duals of each other.

A somewhat unexpected result is that the concept of duality is closely related to the concept of planarity. As a matter of fact, Theorem 8-7 below can be taken as an alternative definition for the planarity of a graph.

Theorem 8-7 *A graph is planar if and only if it has a dual.*

Proof To prove that every planar graph G_1 has a dual, we show a procedure for constructing a dual G_2 of G_1. For each region in G_1, let there be a corresponding vertex in G_2. For an edge in the boundary of two adjacent regions in G_1, let there be a corresponding edge which joins the two vertices, corresponding to the two adjacent regions, in G_2. We claim that G_2 is a dual of G_1. First of all, the one-to-one correspondence between the edges in G_1 and G_2 is clear. Secondly, the contour of a region in G_1 corresponds to a cut-set which separates the corresponding vertex from the other vertices in G_2. According to Theorem 8-3, the set of contours of the finite regions is a basis of the circuit subspace of G_1. According to Theorem 7-20, the corresponding set of cut-sets is a basis of the cut-set subspace in G_2. Because there is a one-to-one correspondence between the vectors in a basis of the circuit subspace of G_1 and the vectors in a basis of the cut-set subspace of G_2, there is a one-to-one correspondence between the vectors in the circuit subspace of G_1 and the vectors in the cut-set subspace of G_2.

The proof that every graph having a dual is planar is too lengthy to be included here.† We shall just sketch the proof. Let M_L denote a circuit matrix of the graph G_1. The matrix M_L must then be a cut-set matrix of G_2, which is a dual of G_1. The existence of G_2, therefore, amounts to the existence of a graph having M_L as a cut-set matrix. It was proved by Tutte that a graph can be constructed with M_L as a cut-set matrix if and only if M_L is a circuit matrix of a graph that does not contain a subgraph that is isomorphic, to within vertices of degree 2, to either one of the Kuratowski graphs. (Note that the "if" part has already been proved in the preceding paragraph. We need the "only if" part to complete the proof of the theorem.) Therefore, according to Theorem 8-4, we conclude that a graph has a dual if and only if it is planar. ■

The construction procedure of a dual of a given graph in the proof of Theorem 8-7 can be carried out systematically as illustrated in Fig. 8-18. In Fig. 8-18, the edges of a given planar graph G_1 are in solid lines.

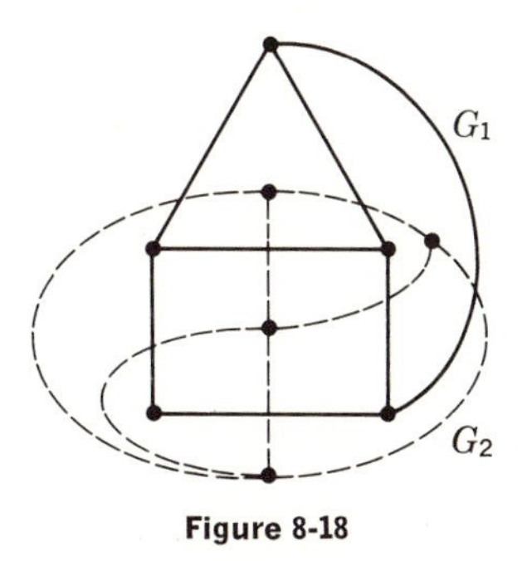

Figure 8-18

A vertex of G_2, a dual of G_1, is placed inside each of the regions of G_1. (Note that there is also a vertex in the infinite region of G_1.) The edges of the dual graph G_2, shown in dashed lines, join vertices that are in adjacent regions of G_1.

The careful reader must have detected by now that we have been using the term "a dual of a graph" instead of "the dual of a graph." Indeed, a given graph might have two dual graphs that are not isomorphic

† See Sec. 8-5.

graphs. For example, both the graphs in Fig. 8-19*b* and Fig. 8-19*c* are dual graphs of the graph in Fig. 8-19*a*. Clearly, they are not isomorphic graphs.

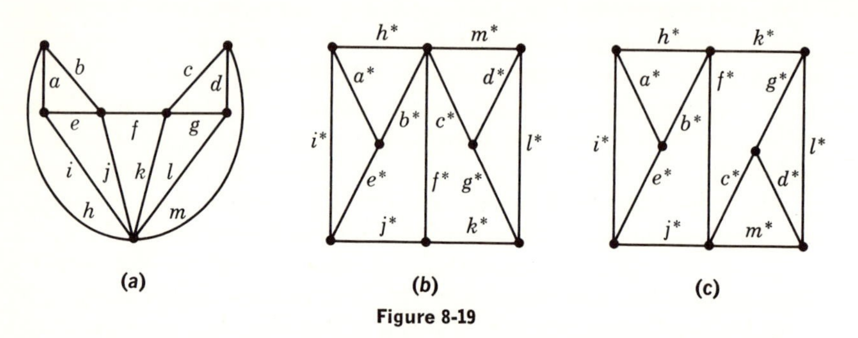

Figure 8-19

To investigate the nonuniqueness of a dual of a graph, we introduce the definition of two graphs being 2-isomorphic. Two graphs G and G' are *2-isomorphic* if they become isomorphic under repeated application of either or both of the following operations:

1. Separation of a graph into components by splitting an articulation point. Figure 8-20*a* illustrates such a separation.

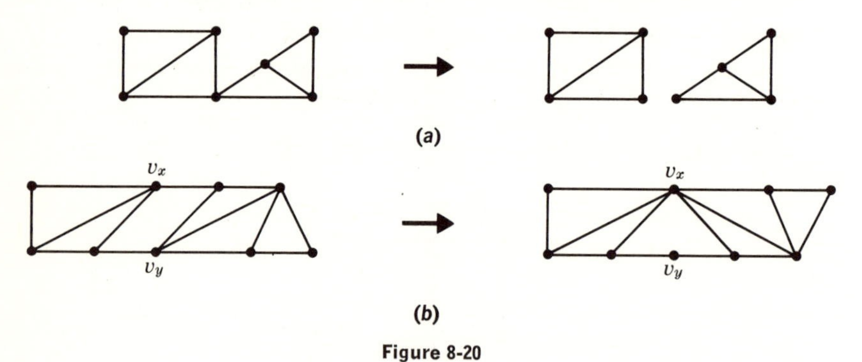

Figure 8-20

2. If a graph can be divided into two disjoint subgraphs that have two vertices in common, the interchange of these two vertices in one of the subgraphs. (Such an operation can also be visualized as splitting the two common vertices, turning one of the graphs around at the two vertices, and then rejoining the two vertices.) Figure 8-20*b* illustrates such an interchange of vertices, where the common vertices are labeled v_x and v_y.

As an example, the reader can check that the two graphs in Fig. 8-21*a* and Fig. 8-21*b* are 2-isomorphic.

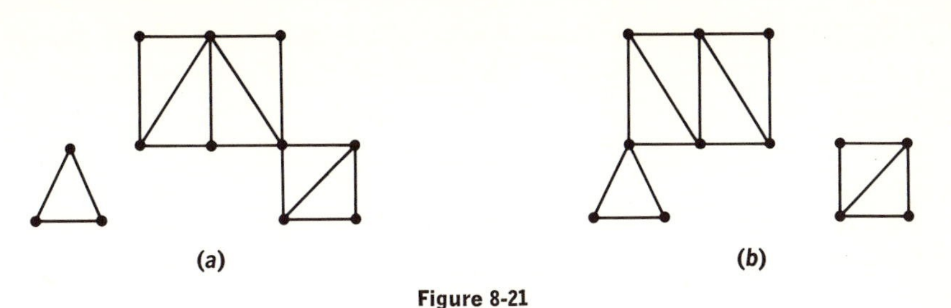

Figure 8-21

As it turns out, the dual graphs of a graph are 2-isomorphic graphs. The result is stated as two theorems.

Theorem 8-8 *If G_2 is a dual of G_1 and G_2' is 2-isomorphic to G_2, then G_2' is also a dual of G_1.*

Proof According to the definition of 2-isomorphism, there is a one-to-one correspondence between the edges in G_2 and the edges in G_2', since the operations of splitting an articulation point and interchanging the two common vertices in two subgraphs do not change the number of edges in the graphs. Moreover, a circuit in G_2 is also a circuit in G_2', and vice versa, because the two operations do not add or remove any circuits to or from the graphs.

It follows that there is a one-to-one correspondence between the edges in G_1 and the edges in G_2'. Moreover, a set of edges in G_2' is a circuit if and only if its corresponding set of edges in G_1 is a cutset because a set of edges in G_2' is a circuit if and only if its corresponding set of edges in G_2 is a circuit. ■

We state the following theorem without proof as the proof is quite lengthy.†

Theorem 8-9 *If two graphs G_2 and G_2' are duals of a graph G_1, then G_2 and G_2' are 2-isomorphic graphs.*

The reader can check that the two graphs in Fig. 8-19*b* and Fig. 8-19*c* are 2-isomorphic, although they are not isomorphic.

Example 8-6 Figure 8-22 shows an electrical network in which there are 10 switches controlling the excitation of the light. Our problem is to design another network, called a complementary network, such that the light it controls will be turned on when the light in the network in Fig. 8-22 is off, and the light it controls will be turned off when the light in the network in Fig. 8-22 is on.

† For a proof, see Whitney [10].

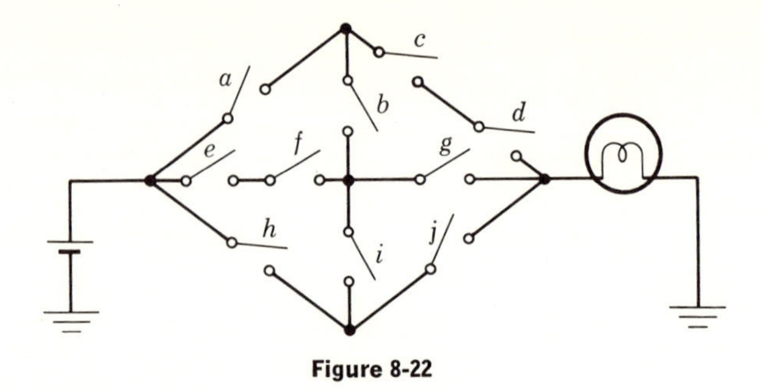

Figure 8-22

An electromagnetic relay consists of an electromagnet and a collection of contacts which can be used as switches in electrical circuits. There are two types of contacts: the normally open contacts and the normally closed contacts. When the electromagnet is deenergized, a normally open contact remains open, and a normally closed contact remains closed. When the electromagnet is energized, a normally open contact will be closed, and a normally closed contact will be opened. We use the symbol in Fig. 8-23*a*

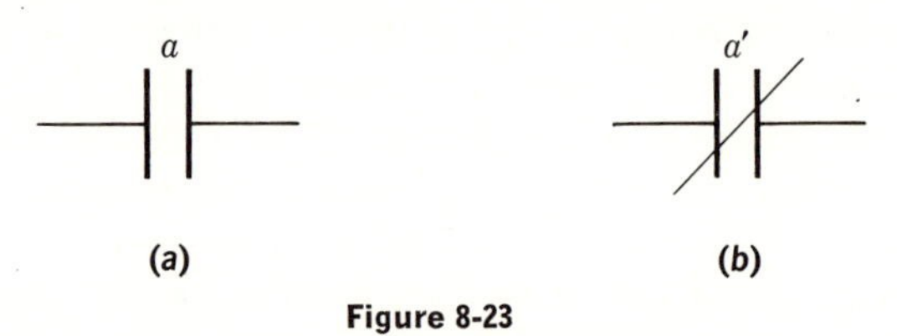

Figure 8-23

to denote a normally open contact and the symbol in Fig. 8-23*b* to denote a normally closed contact. Moreover, when a capital letter is used to denote the electromagnet of a relay, the corresponding lowercase unprimed letter will be used to denote a normally open contact, and the corresponding lowercase primed letter will be used to denote a normally closed contact controlled by the electromagnet. The circuit in Fig. 8-22 can be rebuilt as that in Fig. 8-24*a* where the switches control the electromagnets, which in turn control the contacts. It follows that the normally closed contacts controlled by these electromagnets can be used to build the complementary network.

We define a *one-terminal-pair graph* as a graph with two vertices specially designated as the *terminals* of the graph. A *planar one-terminal-pair graph* is a one-terminal-pair graph that is planar and remains planar when an edge joining the two terminals is added to it. A dual of a planar one-terminal-pair graph is also a planar one-terminal-pair graph which is constructed as follows: An edge joining

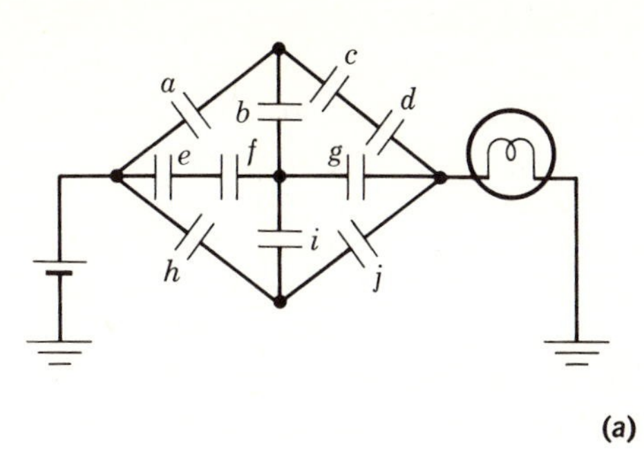

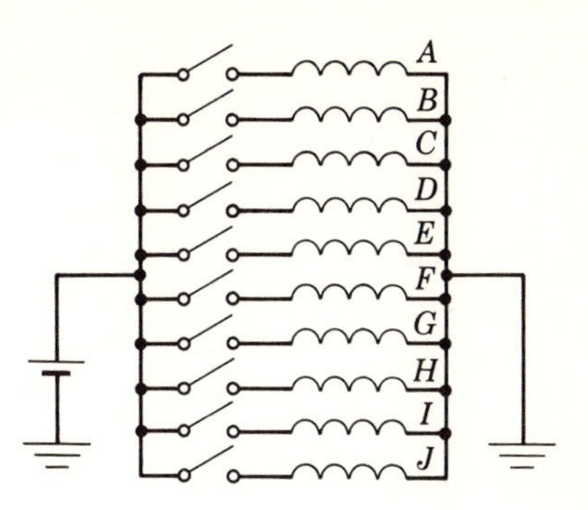

(a)

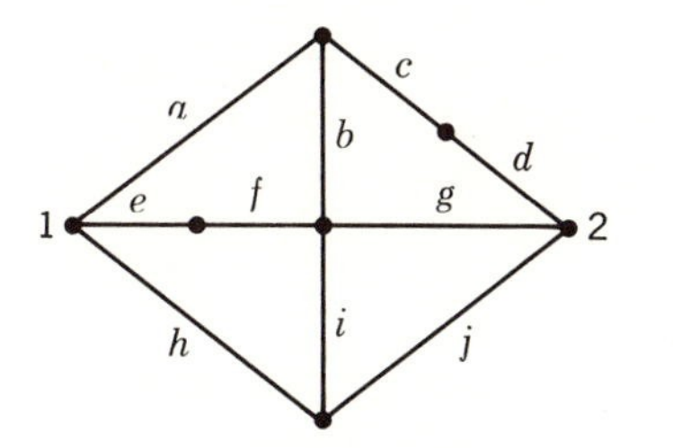

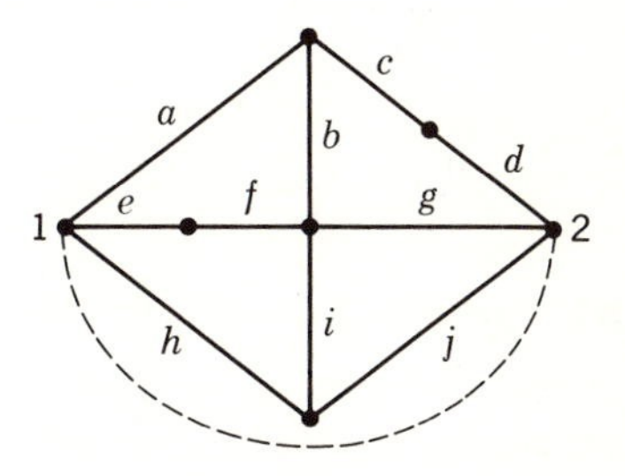

(b)

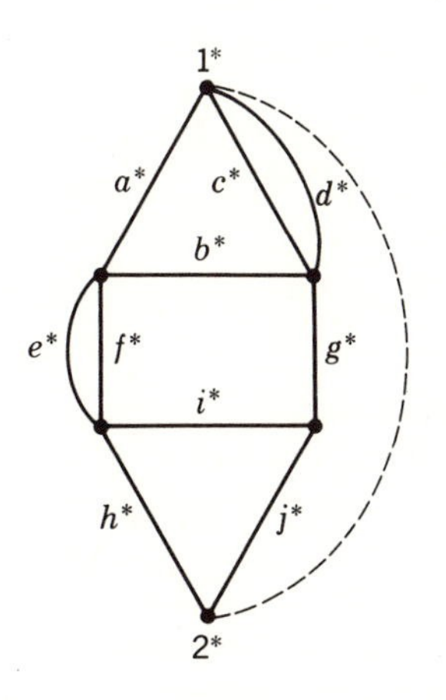

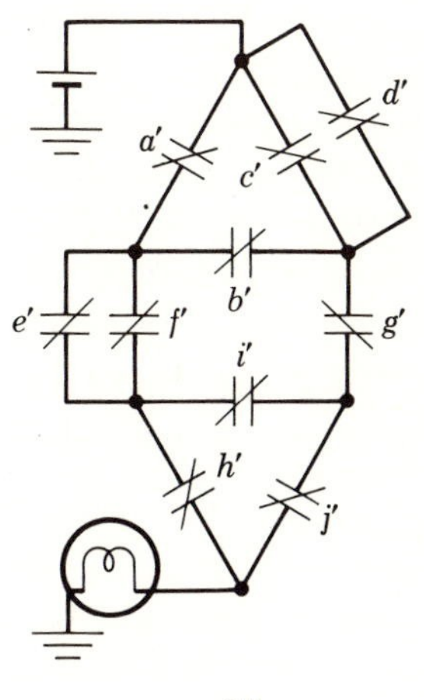

(c)

(d)

Figure 8-24

the two terminals of the planar one-terminal-pair graph is added to the graph. A dual of the resultant graph is constructed. In the dual graph, the edge corresponding to the one joining the two terminals in the original planar one-terminal-pair graph is deleted, and the two vertices joined by the deleted edge are labeled as the terminals of the dual graph.

We note that a path joining the two terminals in a planar one-terminal-pair graph corresponds to a cut-set separating the two terminals in its dual. Therefore, our design problem becomes a very simple one. The network in Fig. 8-24*a* is represented as a one-terminal-pair graph in Fig. 8-24*b* where the two terminals are marked as 1 and 2. A dual of the one-terminal-pair graph is shown in Fig. 8-24*c* where the two terminals are marked as 1* and 2*. The complementary network is shown in Fig. 8-24*d*. Notice that a *closed* path between the battery and the light in the network in Fig. 8-24*a* corresponds to an *opened* cut-set separating the battery and the light in the network in Fig. 8-24*d*. ■

8-5 SUMMARY AND REFERENCES

In this chapter, we studied two important and closely related concepts in the theory of graphs: planarity and duality. Historically, there are three pieces of work of great significance:

1. Kuratowski's theorem (Theorem 8-4). This theorem characterizes a planar graph as one that does not contain a subgraph being isomorphic, to within vertices of degree 2, to either one of the Kuratowski graphs.
2. Whitney's work on dual graphs. Whitney's original definition of duality is that a graph G_2 is a dual of the graph G_1 if there is a one-to-one correspondence between the edges of these two graphs such that if H_1 is any subgraph of G_1 and $\bar{H}_2$ is the complement of the corresponding subgraph of G_2, then

$$\begin{pmatrix}\text{The dimension of the}\\ \text{cut-set subspace of } \bar{H}_2\end{pmatrix} = \begin{pmatrix}\text{the dimension of the}\\ \text{cut-set subspace of } G_2\end{pmatrix} - \begin{pmatrix}\text{the dimension of the}\\ \text{circuit subspace of } H_1\end{pmatrix}$$

   Such a definition is equivalent to ours as will be seen in Prob. 8-11. Proceeding from his original definition, Whitney proved that a graph is planar if and only if it has a dual.
3. Tutte's work on the theory of matroids. Tutte obtained a necessary and sufficient condition for a matrix of 0's and 1's being a cut-set matrix (a circuit matrix) of a graph. According to his result, a

necessary and sufficient condition for the existence of a graph having a cut-set matrix (a circuit matrix) equal to a circuit matrix (a cut-set matrix) of a given graph is that the given graph does not contain a subgraph that is isomorphic, to within vertices of degree 2, to a Kuratowski graph.

Our definition of duality is a natural consequence of our study of the notion of the circuit subspace and the cut-set subspace of a graph. In this way, as was pointed out in the proof of Theorem 8-7, we see an explicit tie between Whitney's result and Tutte's result. As a matter of fact, a proof of Tutte's result will lead directly to a proof of Whitney's result after the equivalence between Whitney's original definition of duality and our definition of duality is shown.

The notion of planarity is a very interesting one. Our definition of planarity, mappable on the plane with no crossing edges, is the most intuitive one. However, either Kuratowski's characterization of a planar graph or Whitney's characterization of a planar graph can also be adopted as the *definition* of a planar graph. (In that case, "mappable on the plane with no crossing edges" becomes a *property* of planar graphs and can be proved from the definition of planarity.) As a matter of fact, MacLane has studied another characterization of planar graphs as follows: A graph is planar if and only if it has a basis of the circuit subspace such that together with an additional circuit, the set of circuits contains each of the edges exactly twice.

The concept of duality is useful in many areas of applications of the theory of graphs, e.g., electrical network analysis and switching-circuit design. Its extension and generalization have been investigated in Whitney's study of the theory of matroids. Also, we shall see in Chap. 12 a very similar concept of duality in linear programming problems.

See Chap. 21 of Berge [1] for a discussion of planar graphs. Our proof of Kuratowski's theorem follows that of Berge. Whitney's work can be found in his classical papers [9, 10, 11]. Chapter 3 of Seshu and Reed [6] covers the topic of dual graphs following Whitney's original development. For the study of the theory of matroids, see Whitney [12], Tutte [7, 8], and Minty [5]. The proof of Fary's theorem appeared in [3] and can also be found in Busacker and Saaty [2]. MacLane's characterization of planar graphs can be found in [4].

1. Berge, C.: "The Theory of Graphs and Its Applications," John Wiley & Sons, Inc., New York, 1962.
2. Busacker, R. G., and T. L. Saaty: "Finite Graphs and Networks: An Introduction with Applications," McGraw-Hill Book Company, New York, 1965.
3. Fary, I.: On Straight Line Representation of Planar Graphs, *Acta Sci. Math.*, **11**(4):229–233 (1948).
4. MacLane, S.: A Combinatorial Condition for Planar Graphs, *Fundamental Math.*, **28**:22–32 (1937).

5. Minty, G. J.: On the Axiomatic Foundations of the Theories of Directed Linear Graphs, Electrical Networks and Network-Programming, *J. Math. and Mech.*, **15**:485–520 (1966).
6. Seshu, S., and M. B. Reed: "Linear Graphs and Electrical Networks," Addison-Wesley Publishing Company, Inc., Reading, Mass., 1961.
7. Tutte, W. T.: A Homotopy Theorem for Matroids, I, II, *Trans. Am. Math. Soc.*, **88**:144–174 (1958).
8. Tutte, W. T.: Matroids and Graphs, *Trans. Am. Math. Soc.*, **90**:527–552 (1959).
9. Whitney, H.: Non-separable and Planar Graphs, *Trans. Am. Math. Soc.*, **34**:339–362 (1932).
10. Whitney, H.: 2-Isomorphic Graphs, *Am. J. Math.*, **55**:245–254 (1933).
11. Whitney, H.: Planar Graphs, *Fundamental Math.*, **21**:73–84 (1933).
12. Whitney, H.: On the Abstract Properties of Linear Dependence, *Am. J. Math.*, **57**:509–533 (1935).

PROBLEMS

8-1. Prove that in a connected planar graph in which every vertex is of at least degree 3, there exists a region with fewer than six edges in the boundary.

8-2. Show that in a planar graph with six vertices and 12 edges, each of the regions is bounded by three edges.

8-3. In this problem, Euler's formula is derived in an alternative way.

(*a*) According to Theorem 8-2, a planar graph can be mapped on the plane in such a way that the edges are all straight-line segments. It follows that the regions are all polygons. Let $n_1, n_2, \ldots, n_{r-1}$ denote the numbers of edges in the boundaries of the $\mathbf{r} - 1$ finite regions. Show that the sum of the interior angles of these $\mathbf{r} - 1$ polygons, T, is equal to

$$[n_1 + n_2 + \cdots + n_{r-1} - 2(\mathbf{r} - 1)] \times 180°$$

(*b*) Let $\mathbf{e}$ denote the number of edges in the graph, and let $\mathbf{e}_x$ denote the number of edges in the boundary of the infinite region. Show that

$$T = [2\mathbf{e} - \mathbf{e}_x - 2(\mathbf{r} - 1)] \times 180°$$

(*c*) The quantity T can be computed in another way. Let $\mathbf{v}$ denote the number of vertices in the graph, and let $\mathbf{v}_x$ denote the number of vertices in the boundary of the infinite region. Show that

$$T = (\mathbf{v} - \mathbf{v}_x) \times 360° + (\mathbf{e}_x - 2) \times 180°$$

(*d*) Using $\mathbf{e}_x = \mathbf{v}_x$, obtain Euler's formula by equating the results in parts (*b*) and (*c*).

8-4. Prove that in a connected planar graph the ring sum of the contours of the finite regions is the contour of the infinite region.

8-5. Prove that if G and $\bar{G}$ are two complementary graphs with 11 or more vertices, then either G or $\bar{G}$ is nonplanar. (As a matter of fact, the statement is true for G and $\bar{G}$ being complementary graphs with nine or more vertices. However, the proof is much harder.)

8-6. Let G be a planar graph the degree of each vertex of which is 3. Prove that a dual of G will have an odd number of finite regions.

8-7. Find two nonisomorphic dual graphs of the graph in Fig. 8P-1.

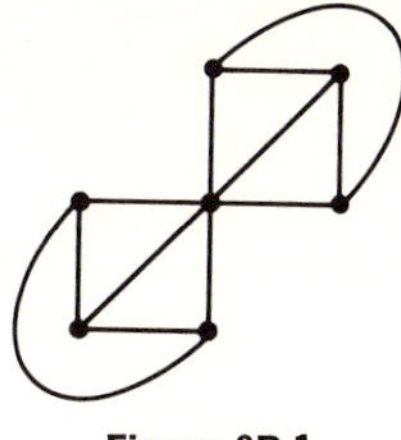

Figure 8P-1

8-8. Find two nonisomorphic dual graphs of the graph in Fig. 8P-2.

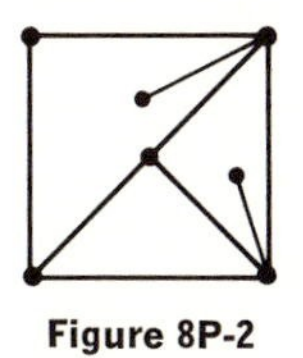

Figure 8P-2

8-9. Show that a dual of a planar biconnected graph containing no loops is also biconnected.

8-10. A series-parallel graph is a one-terminal-pair graph defined recursively as follows:

1. A single edge is a series-parallel graph.
   If G' and G'' are series-parallel then:
2. The series combination of G' and G'' is a series-parallel graph. By the series combination of G' and G'', we mean the joining of one of the terminals of G' with one of the terminals of G'' as illustrated in Fig. 8P-3*a*.
3. The parallel combination of G' and G'' is a series-parallel graph. By the parallel combination of G' and G'', we mean the joining of the two terminals of G' with the two terminals of G'' as illustrated in Fig. 8P-3*b*.

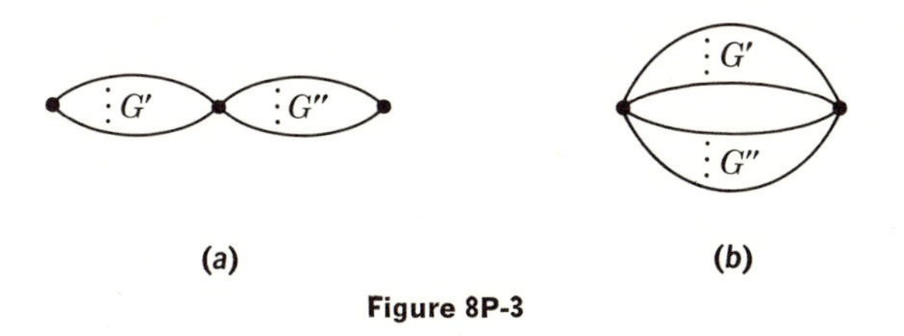

Figure 8P-3

Show that a dual of a graph G is a series-parallel graph if and only if G is series-parallel.

8-11 Let G_1 and G_2 be two graphs with a one-to-one correspondence between their edges. Let H_1 be any subgraph of G_1. Let n_1 denote the dimension of the circuit subspace of H_1. Let $\bar{H}_2$ be the complement of the corresponding subgraph of H_1 in G_2. Prove that

$$r_2 = R_2 - n_1$$

if and only if G_1 and G_2 are dual graphs, where r_2 and R_2 are the dimensions of the cutset subspaces of $\bar{H}_2$ and G_2, respectively.

Chapter 9
Domination, Independence, and Chromatic Numbers

9-1 DOMINATING SETS

A set of vertices in a graph† is said to be a *dominating set* if every vertex not in the set is adjacent to one or more vertices in the set. A *minimal dominating set* is a dominating set such that no proper subset of it is also a dominating set. For the graph in Fig. 9-1, $\{a,c,e,g\}$ is a dominating set,

† Unless otherwise specified, our discussion in this chapter is limited to undirected graphs.

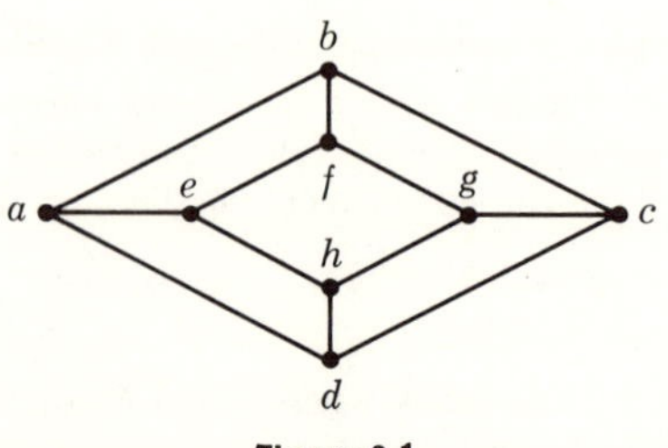

Figure 9-1

and both $\{a,c,f,h\}$ and $\{a,g\}$ are minimal dominating sets. The *domination number* $\alpha(G)$ of a graph G is the size of the smallest minimal dominating set. It can be seen that the domination number of the graph in Fig. 9-1 is 2.

An interesting problem based on the concept of dominating sets is to try to place five queens on an 8×8 chessboard such that they will dominate each of the 64 cells. Figure 9-2 shows one of the solutions to

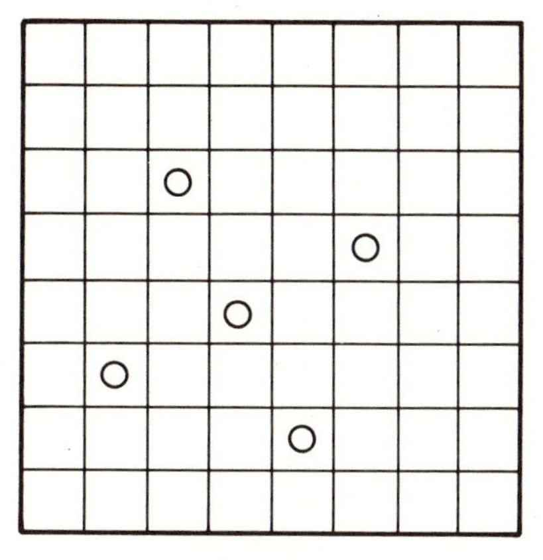

Figure 9-2

the problem. Since no fewer queens can dominate all the cells on the board, the domination number of the graph which has 64 vertices (corresponding to the 64 cells) and represents the dominance relation of the queen piece is equal to 5.

Example 9-1 Communication links are set up between cities, and transmitting stations are to be built in some of them so that every city can receive messages from at least one of the transmitting stations through the links. The problem of selecting the sites for the transmitting stations is exactly that of finding a minimal dominating set of the graph having the cities as vertices and the communication links as edges.

We want to have two groups of transmitting stations so that when one group of stations breaks down, the other group can take over and transmit. The question is as follows: Is it possible to arrange the transmitting stations in such a way that there will not be two stations in the same city? In other words, can there be two disjoint minimal dominating sets in a given graph? The answer is affirmative for any graph having no isolated vertices, as we shall show presently. As an example, for the graph in Fig. 9-1, $\{a,g\}$ and

$\{c,e\}$ are two disjoint minimal dominating sets. We now show that this is true in general.

We want to show first that if D is a minimal dominating set in a graph that has no isolated vertices, then $V - D$, the set of vertices not in D, is a dominating set. Suppose that there is a vertex d in D that is not adjacent to any one of the vertices in $V - D$. Since d is not an isolated vertex, d must be adjacent to at least one of the vertices in D. It follows that $D - \{d\}$ is also a dominating set. However, this is impossible because D is a minimal dominating set. Since every vertex in D is adjacent to one or more of the vertices in $V - D$, $V - D$ is a dominating set.

It is quite clear that any dominating set contains a minimal dominating set, because we can always remove from the dominating set some of the vertices until it becomes a minimal dominating set. We have thus proved that in a graph that contains no isolated vertices, for any given minimal dominating set, there is a disjoint minimal dominating set. Therefore, it is possible to set up two groups of transmitting stations such that no city will have two transmitting stations.

A direct consequence of our result is that for any graph having no isolated vertices, the domination number is less than or equal to half of the number of vertices in the graph, that is, $\alpha(G) \leq \mathbf{v}/2$, where $\mathbf{v}$ is the number of vertices in G. ∎

The minimal dominating sets of a graph can be found by an algebraic method which is based on the notion of generating functions discussed in Chap. 2. We shall illustrate the method by an example. For the graph in Fig. 9-1, in order that the vertex a will be "dominated," either a or b or d or e must be included in a dominating set. This can be represented symbolically as $a + b + d + e$. Similarly, so that the vertex b will be dominated, either b or a or c or f must be included in a dominating set. Again, this can be represented symbolically as $b + a + c + f$. It follows that for the domination of all the vertices in the graph we have the expression

$$\begin{aligned}(a + b + d + e)&(b + a + c + f)(c + b + d + g)(d + a + c + h)\\ \times\ (e + a + f + h)&(f + b + e + g)(g + c + f + h)(h + d + e + g)\\ &= a^4g^4 + b^4h^4 + a^4c^2fh + b^3fh^4 + \cdots\end{aligned}$$

The term a^4g^4 corresponds to the minimal dominating set $\{a,g\}$, the term b^4h^4 corresponds to the minimal dominating set $\{b,h\}$, the term a^4c^2fh corresponds to the minimal dominating set $\{a,c,f,h\}$, and so on. It should be noted that every term in the expression corresponds to a dominating set, but not necessarily to a minimal dominating set. For example, the term b^3fh^4 corresponds to the dominating set $\{b,f,h\}$, which is not a minimal set. Clearly, a set corresponding to a term in the expression is a minimal dominating set if and only if no subset of it has a corresponding term in the expression. (For those readers who are familiar with the subject of Boolean algebras, it can be observed that if multiplication and addition are carried out as "logical and" and "logical or," respectively, then a minimal sum of products form of the expression will give only the minimal dominating sets.)

Example 9-2 The notion of dominating sets in a graph can be extended in two ways. First, we may want to find a set of vertices that dominate only a subset of the vertices in the graph. Second, we also want to define the notion of dominating sets in directed graphs. In a directed graph, a dominating set is a set of vertices such that every vertex not in the set is adjacent from one or more of the vertices in the set. Let us consider the problem of stationing police patrol cars in some areas of the city so that they can be dispatched to other areas in the case of emergency. The street maps of the patrolled areas are represented by the directed graph in Fig. 9-3, since the streets are all one-way streets. (Notice

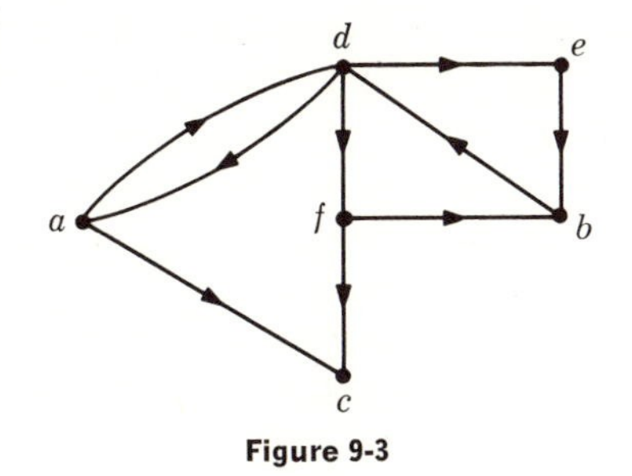

Figure 9-3

that a two-way street can be represented by two edges oriented in both directions.) Suppose that the areas represented by vertices a, b, c, and d have higher crime rates, and we want to have a police car stationed either in such an area or in an adjacent area. The problem of choosing strategic points for stationing the minimal number of police cars is that of finding a minimal dominating set which dominates the vertices a, b, c, and d. Thus, proceeding as before, from Fig. 9-3 we obtain the expression

$$(a + d)(b + e + f)(c + a + f)(d + a + b)$$
$$= a^3b + a^3e + a^3f + b^2cd + bcd^2 + \cdots$$

where the term $a + d$ means that a patrol car should be stationed in either area a or area d to protect area a, the term $b + e + f$ means that a patrol car should be stationed in either area b or area e or area f to protect area b, and so on. Therefore, the term a^3e in the expansion means that a police car in area a and a police car in area e will be sufficient. Moreover, the car in area a will be responsible to calls from three of the four areas, whereas the car in area e will be responsible to calls from only one area. Similarly, although the two terms b^2cd and bcd^2 mean the same way of stationing the cars, there is a difference in the assignment of responsibilities. Corresponding to the term b^2cd, the car in area b is responsible to calls from both areas b and d; corresponding to the term bcd^2, the car in area d is responsible to calls from both areas a and d. ■

9-2 INDEPENDENT SETS

A set of vertices in a graph is said to be an *independent set* if no two vertices in it are adjacent. A set is *dependent* if at least two of the vertices in it are adjacent. A *maximal independent set* is an independent set which becomes dependent when any vertex is added to the set. For the graph in Fig. 9-1, $\{b,e,g\}$ is an independent set, whereas $\{b,h\}$ and $\{b,d,e,g\}$ are maximal independent sets. The *independence number* $\beta(G)$ of a graph G is the size of the largest maximal independent set. For instance, the independence number of the graph in Fig. 9-1 is 4. As an example, we see that to invite a group of guests to dinner with no two of them being friends is to find an independent set of the graph that represents the acquaintance relation. Other examples are to place nontaking rooks or nontaking queens on a chessboard. As a matter of fact, up to eight nontaking queens can be placed on an 8×8 chessboard. Figure 9-4 shows one way of doing so. It follows that for the graph G

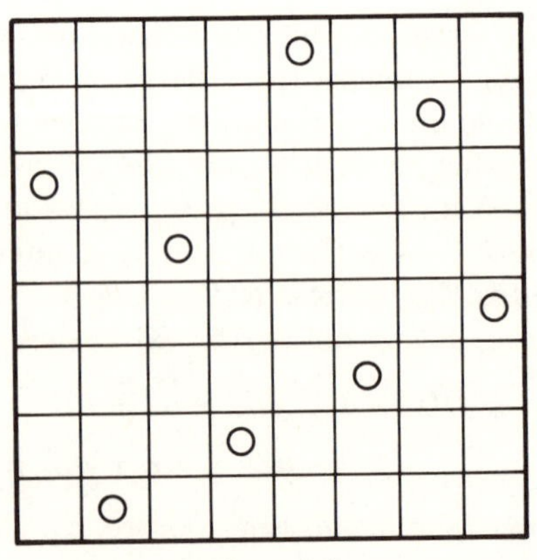

Figure 9-4

which represents the dominance relation of the queen piece, $\beta(G)$ is equal to 8.

As the reader may suspect, there is a close relationship between the dominating sets and the independent sets of a graph, as shall be seen in the following example.

Example 9-3 One major feature of a computerized automatic library system is the cross-reference facility which can supply a user of the library system with a list of books or articles that are related to a particular book in which the user is interested. In the system, such cross-reference information can be represented by an undirected graph. The vertices of the graph correspond to the books in the library, and the edges indicate the cross-reference relationship between the books; i.e., two vertices are joined by an edge if the corresponding books are related. Suppose that we want to select a subset of the books such that the selected books do not cross-refer one another, whereas an unselected book is related to at least one that is selected. In a sense, such a selection covers all the categories of books in the library (a book is either selected or related to a selected one) without duplication in the selection (no two books in the selection are related).

The problem of making such a selection is that of finding an independent set in a graph which is also a dominating set. We shall show that in any arbitrary graph *an independent set is also a dominating set if and only if it is maximal.* In showing this, we not only prove that such a selection is always possible but also reduce the selection problem to that of finding a maximal independent set. Since any vertex not in a maximal independent set is adjacent to one or more vertices in the set, a maximal independent set is also a dominating set. On the other hand, an independent set that is also a dominating set must be maximal since any vertex not in a dominating set is adjacent to one or more vertices in the set.

Furthermore, we conclude that for any given graph, the independence number is always larger than or equal to the domination number. An example is to place queens on an 8×8 chessboard. As we have seen, eight nontaking queens can be placed on a chessboard [$\beta(G) = 8$], whereas only five queens will be enough to dominate all the cells [$\alpha(G) = 5$]. ■

Example 9-4 Let us consider a problem in information theory. The product of two graphs G and H, $G \times H$, is a graph whose set of vertices is the cartesian product of the sets of vertices of G and H. Let $\{a_1, a_2, \ldots\}$ be the set of vertices of G, and let $\{b_1, b_2, \ldots\}$ be the set of vertices of H. In the graph $G \times H$:

1. There is an edge joining the vertices (a_1,b_1) and (a_1,b_2) if there is an edge joining the vertices b_1 and b_2 in H.
2. There is an edge joining the vertices (a_1,b_1) and (a_2,b_1) if there is an edge joining the vertices a_1 and a_2 in G.
3. There is an edge joining the vertices (a_1,b_1) and (a_2,b_2) if there is an edge joining the vertices a_1 and a_2 in G *and* an edge joining the vertices b_1 and b_2 in H.

As an example, Fig. 9-5*c* shows the product of graph G in Fig. 9-5*a* and graph H in Fig. 9-5*b*.

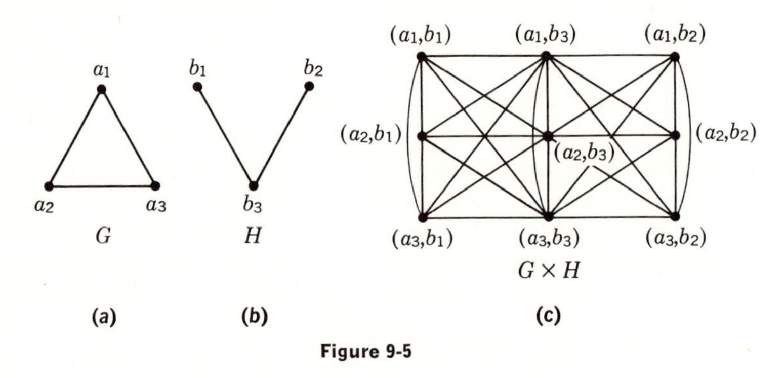

Figure 9-5

Consider a communication channel represented by the graph in Fig. 9-6*a*. At the transmitting end, five letters a, b, c, d, and e can

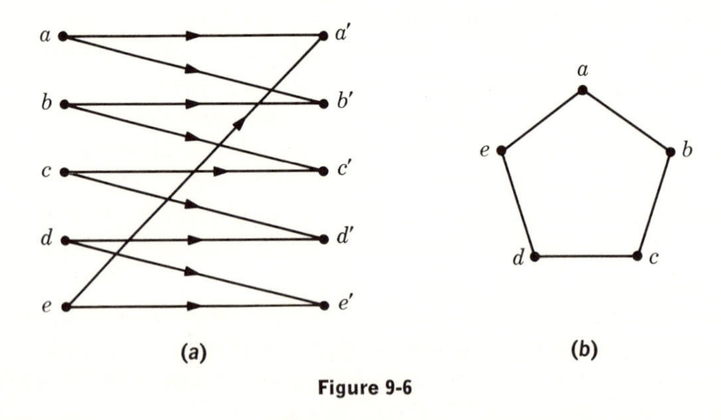

Figure 9-6

be transmitted. Five corresponding letters a', b', c', d', and e' will be received at the receiving end. However, because of disturbances in the channel, the letter a will either be received as a' or be received as b', the letter b will either be received as b' or be received as c', and so on, as shown in Fig. 9-6*a*. In order that distinct messages can be transmitted unambiguously through the channel, we shall use only some of the five letters to represent the messages. For example, if only the letters a and c are to be used, we can recognize that the transmitted letter is a when either a' or b' is received and that the transmitted letter is c when either c' or d' is received. The problem of selecting a subset of the letters for error-free reception is that of finding an independent set of the graph in Fig. 9-6*b* in which the vertices are the letters at the transmitting end and in which two vertices are joined by an edge if they might be confused as the same letter at the receiving end. Since the independence number of the graph is 2, we know that only two of the five letters can be used. For example, we can use the sets $\{a,c\}$ or $\{a,d\}$ or $\{b,e\}$ because they are all maximal independent sets.

When more than two distinct messages are to be transmitted, we must represent the messages by "words" made up of several letters. For example, when four messages are to be transmitted, they can be represented by *aa*, *ac*, *ca*, and *cc*. Because the letters a and c will always be recognized unambiguously at the receiving end, the four messages will also be recognized unambiguously. However, when we are to encode the messages as words of two letters, we can actually accommodate five messages by using the set of words *aa*, *bc*, *ce*, *db*, and *ed*, since, as the following table shows, there will be no confusion at the receiving end.

Word transmitted	*Words received*			
aa	$a'a'$	$a'b'$	$b'a'$	$b'b'$
bc	$b'c'$	$b'd'$	$c'c'$	$c'd'$
ce	$c'e'$	$c'a'$	$d'e'$	$d'a'$
db	$d'b'$	$d'c'$	$e'b'$	$e'c'$
ed	$e'd'$	$e'e'$	$a'd'$	$a'e'$

The problem of finding a largest possible set of two-letter words for error-free reception is that of finding a maximal independent set of the graph $G \times G$, where G is the graph of Fig. 9-6*b*. The vertices in $G \times G$ are words made up of two letters, and two vertices are joined by an edge if the two words might be confused when they are transmitted. As a matter of fact, it can be shown that $\beta(G \times G)$ is equal to 5. This brings up a very interesting point. The average number of messages per letter transmitted is greater when two-letter words are used instead of single letters.

Would one obtain a still greater average if words of three letters, and in general words of n letters, were used? To measure the richness of words that can be used, we define the capacity of a graph G, $\theta(G)$, as

$$\theta(G) = \sup_n \sqrt[n]{\beta(G^n)}$$

where G^n denotes the product of n G's. (The notation $\sup_n \sqrt[n]{\beta(G^n)}$ means the maximal value of $\sqrt[n]{\beta(G^n)}$ for all possible values of n.) This definition is quite natural. Since $\beta(G^n)$ is the number of n-letter words that can be used for error-free communication, $\sqrt[n]{\beta(G^n)}$ is the "number" of single letters that can be received unambiguously at the receiving end. (Recall the simple fact that from k letters, k^n n-letter words can be composed.) However, for a given graph G, there is no general way of determining its capacity. As a matter of fact, the capacity of the graph in Fig. 9-6*b* has not yet been determined. ■

9-3 CHROMATIC NUMBERS

By *coloring* a graph we mean to paint the vertices of the graph with one or more distinct colors. By *properly coloring* a graph we mean to paint the vertices of the graph in such a way that no two adjacent vertices are painted with the same color. The *chromatic number* $\gamma(G)$ of a graph G is the least number of distinct colors that can be used to color the graph properly. For example, the chromatic number of the graph in Fig. 9-7*a* is 3. One way of properly coloring the graph is shown in

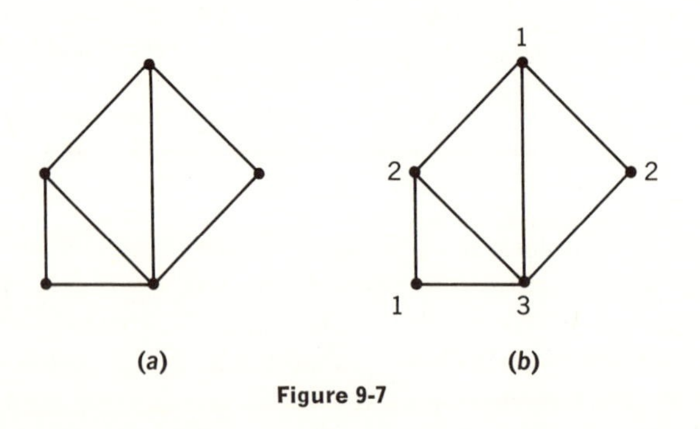

Figure 9-7

Fig. 9-7*b* where the vertices are labeled with the three colors 1, 2, and 3. Moreover, it is quite obvious that the graph cannot be colored properly with two colors.

In map making, one wants to color the regions of a map in such a way that no two adjacent regions are of the same color. In this connection we shall also talk about properly coloring all the regions, including the infinite region of a planar graph. (As was pointed out in Chap. 8, it is not meaningful to talk about the regions of a nonplanar graph.) However, the problem of coloring the regions of a planar graph is the same as that of coloring the vertices of a dual of the graph. For example, to color the regions of the map in Fig. 9-8*a* is the same as to color the vertices of the graph in Fig. 9-8*b*.

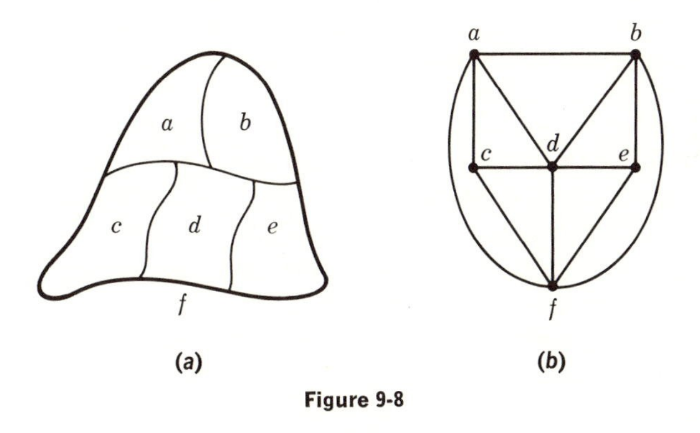

Figure 9-8

A variation of the coloring problem is to color the edges of a graph such that all the edges incident with one vertex are colored distinctly. Again, this problem is equivalent to that of coloring the vertices of another graph which has a vertex corresponding to every edge in the original graph and has an edge between two vertices when their corresponding edges in the original graph are incident with the same vertex. For example, the problem of coloring the edges of the graph in Fig. 9-9*a*

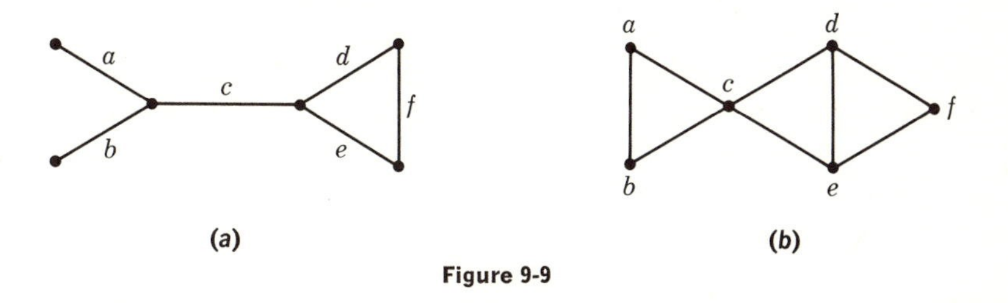

Figure 9-9

is equivalent to that of coloring the vertices of the graph in Fig. 9-9*b*. The edges of the graph in Fig. 9-9*a* and their corresponding vertices in Fig. 9-9*b* are labeled with the same letters to exhibit the correspondence.

The following theorem gives us an upper bound on the chromatic number of a graph.

Theorem 9-1 *If the maximal degree of the vertices in a graph is k, the chromatic number of the graph is equal to or less than $k + 1$.*

Proof We shall show that such a graph can always be properly colored with $k + 1$ colors. We start by picking an arbitrary vertex and painting it with one of the $k + 1$ colors. We then pick any unpainted vertex and paint it with a color that has not been used to paint those vertices that are adjacent to it. This is always possible because at most k colors were used to paint the adjacent vertices. Such a procedure can be repeated until all the vertices in the graph are painted. ■

It is clear that for the complete graph with $k + 1$ vertices (that is, the degrees of the vertices are all equal to k), $k + 1$ colors are needed to properly color the graph. A more interesting result,† which we shall state without proof, is the following: If the maximal degree of the vertices in a graph is k and if no connected component of the graph is a complete graph with $k + 1$ vertices, then the chromatic number of the graph is equal to or less than k.

Theorem 9-2 *A graph can be properly colored with two colors if and only if it contains no circuits of odd length.*

Proof For a graph that contains no circuits of odd length, we pick an arbitrary vertex and paint it with one of the two colors, say red. We then paint the vertices that are adjacent to a red vertex with the other color, say blue, and paint the vertices that are adjacent to a blue vertex with the color red. This procedure is repeated until all the vertices are painted. In this way, no vertex is ever painted with both colors, red and blue, since this will happen only when there are two elementary paths between two vertices, one of the paths having odd length, the other having even length. However, this implies the existence of a circuit of odd length.

On the other hand, if a graph can be properly colored with two colors, we shall find the colors of the vertices alternate when a circuit is traversed. Hence, the length of the circuit must be even. ■

Corollary 9-2.1 *The regions of a planar graph can be colored properly with two colors if and only if the degrees of the vertices are all even numbers.*

† This result is due to Brooks [4].

Proof If the degrees of the vertices in a planar graph are all even numbers, the contours of all the regions of a dual of the graph will contain even numbers of edges. According to Theorem 8-3, any circuit can be expressed as a linear combination of the contours of the finite regions. Since the ring sum of two sets that have even numbers of elements also has an even number of elements, we conclude that all the circuits in the dual graph are of even length. ■

Example 9-5 We want to decompose the graph in Fig. 9-10*a* (in the way it is drawn) into several planar graphs. Problems like this arise in

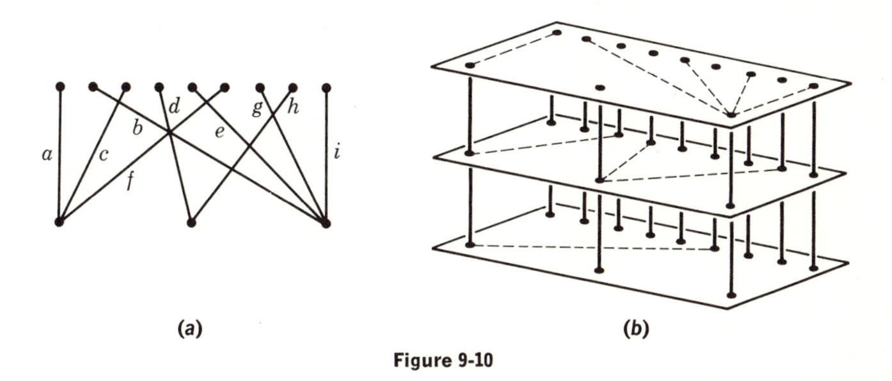

(*a*) (*b*)

Figure 9-10

the design of printed circuits that are made by printing connecting wires on the surface of nonconducting material. When the junctions (the vertices of the graph) are restricted to fixed positions and are to be joined by straight connecting wires (the edges of the graph), some of the wires might unavoidably cross one another. Since printed wires are not insulated, no two of them should meet one another except at the junctions. If the wires are printed on several planes such that the connecting wires on every plane only meet at the junctions, as illustrated in Fig. 9-10*b*, no crossing wires will result. The corresponding junctions in the planes are then tied together by conductors. (In Fig. 9-10*b*, the connecting wires are in dashed lines, and the conductors joining the junctions are in heavy solid lines.)

To decompose the graph in Fig. 9-10*a*, we construct the graph in Fig. 9-11 where there is a vertex corresponding to each of the edges of the graph in Fig. 9-10*a*. Moreover, if two edges cross each other in the graph in Fig. 9-10*a*, there will be an edge joining their corresponding vertices in the graph in Fig. 9-11. Since the chromatic number of the graph in Fig. 9-11 is 3, the connecting wires must be separated into three planes. Moreover, according to the

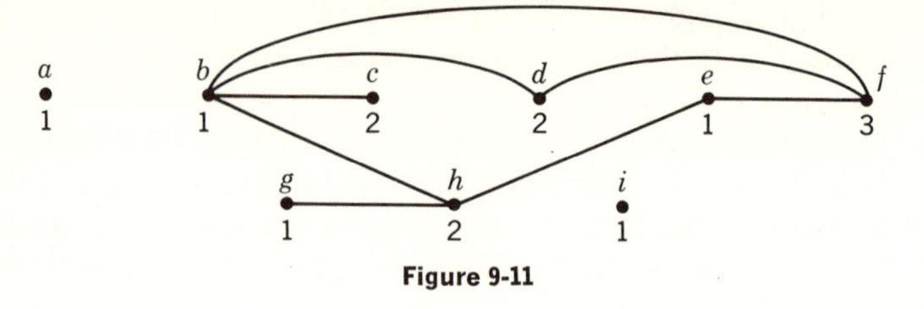

Figure 9-11

coloring scheme shown in Fig. 9-11 (the vertices are labeled with the three colors 1, 2, and 3), we can print the edges that are painted with the same color on one plane as illustrated in Fig. 9-12. ■

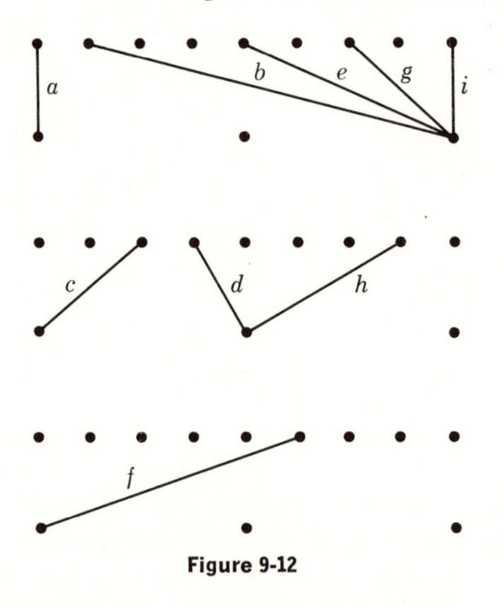

Figure 9-12

Example 9-6 Show that the regions into which a plane is divided by circles drawn on the plane, as illustrated in Fig. 9-13, can be properly colored with two colors. We first ignore those circles that do not intersect the other circles, like circles A and B in Fig. 9-13. Clearly, every vertex of the graph formed by the arcs of the circles is of even degree. According to Corollary 9-2.1, the regions can be

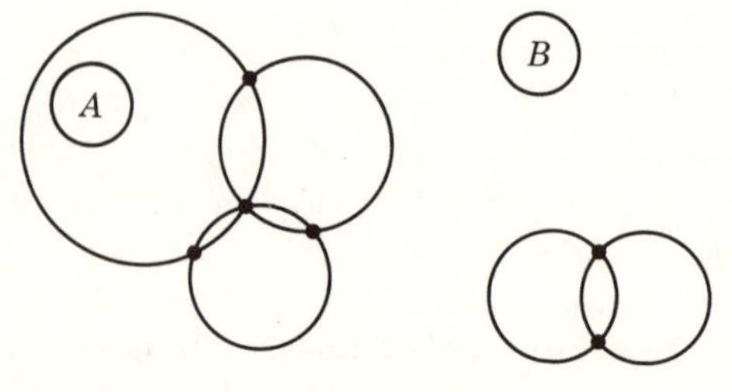

Figure 9-13

properly colored with two colors. For circles like A and B which are completely embedded in a region, we can simply color the interior of such a circle with the color that was not used to color the exterior region. ■

Example 9-7 Show that a linear graph, with no loops, is a nonplanar graph if it has seven vertices and the degrees of the vertices are all equal to 4.

We prove first a useful general result: The regions of a planar graph cannot be properly colored with two colors if the lengths of all but one of the contours are divisible by an integer d ($d > 1$). Suppose that the regions can be properly colored with two colors. Each of the edges in the graph must be on the contours of two distinctly colored regions. (We can disregard those edges that are connected to a vertex of degree 1 as they do not affect our argument.) Let p be the number of regions that are painted with the first color, and let $n_1', n_2', \ldots, n_p'$ be the numbers of edges in the contours of these regions. Also, let q be the number of regions that are painted with the second color, and let $n_1'', n_2'', \ldots, n_q''$ be the numbers of edges in the contours of these regions. We have

$$n_1' + n_2' + \cdots + n_p' = \text{total number of edges in the graph}$$
$$n_1'' + n_2'' + \cdots + n_q'' = \text{total number of edges in the graph}$$

and thus

$$n_1' + n_2' + \cdots + n_p' = n_1'' + n_2'' + \cdots + n_q''$$

However, this equation cannot possibly hold if all but one of $n_1', n_2', \ldots, n_p', n_1'', n_2'', \ldots, n_q''$ are divisible by d.

Now, assume that a loop-free linear graph having seven vertices, the degrees of all of which are equal to 4, is a planar graph. The number of edges in the graph, $\mathbf{e}$, must equal $(7 \times 4)/2 = 14$. According to Euler's formula the number of regions is

$$\mathbf{r} = \mathbf{e} - \mathbf{v} + 2 = 14 - 7 + 2 = 9$$

Let $n_1, n_2, n_3, \ldots, n_9$ denote the number of edges in the contours of the regions. Since the graph is a linear graph, the n's are all equal to or larger than 3. Thus

$$n_1 + n_2 + n_3 + \cdots + n_9 \geq 27$$

Therefore, we have

$$n_1 + n_2 + n_3 + \cdots + n_9 = 2\mathbf{e} = 28$$

It follows that eight of the nine n's are equal to 3, and the remaining n is equal to 4. According to the result we have just proved, since all but one of the nine n's are divisible by 3, the regions of the graph cannot be properly colored with two colors.

However, since the degrees of the vertices of the graph are all equal to 4, according to Corollary 9-2.1, the region of the graph can be colored properly with two colors. We have, therefore, a contradiction, and can conclude that the graph is nonplanar. ■

★9-4 THE CHROMATIC POLYNOMIALS

For a given graph with n vertices, let $P(\lambda)$ denote the number of ways of properly coloring the graph with λ or fewer distinct colors. First of all, note that $P(\lambda)$ is a polynomial of degree n. Let m_i denote the number of ways of properly coloring the graph with exactly i distinct colors. (For example, m_1 equals 0 unless the graph contains only isolated vertices, and m_n equals $n!$.) It follows that there are $m_i \binom{\lambda}{i}$ ways of properly coloring the graph with i of the λ distinct colors. Therefore we have

$$P(\lambda) = \frac{m_1}{1!}\lambda + \frac{m_2}{2!}\lambda(\lambda - 1) + \frac{m_3}{3!}\lambda(\lambda - 1)(\lambda - 2) + \cdots + \frac{m_i}{i!}\lambda(\lambda - 1) \cdots (\lambda - i + 1) + \cdots + \frac{m_n}{n!}\lambda(\lambda - 1) \cdots (\lambda - n + 1)$$

which is a polynomial of degree n. The polynomial $P(\lambda)$ is called the *chromatic polynomial* of the graph. As an example, for the graph in Fig. 9-14a, since $m_1 = 0$, $m_2 = 0$, and $m_3 = 3!$, the chromatic polynomial is

$$P(\lambda) = \frac{3!}{3!}\lambda(\lambda - 1)(\lambda - 2) = \lambda(\lambda - 1)(\lambda - 2)$$

In general, the chromatic polynomial of a graph can be found more expeditiously by the application of the principle of inclusion and exclusion.

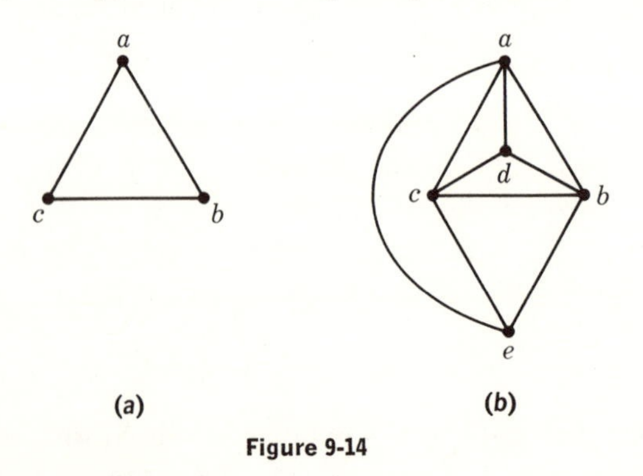

Figure 9-14

As an example, for the graph in Fig. 9-14*a* we consider all the possible ways, proper and improper, of coloring the vertices. Let a_1 be the property that vertices a and b are painted with the same color. Similarly, let a_2 and a_3 be the properties that vertices a and c and vertices b and c are painted with the same color, respectively. Therefore,

$$N = \lambda^3$$
$$N(a_1) = N(a_2) = N(a_3) = \lambda^2$$
$$N(a_1a_2) = N(a_1a_3) = N(a_2a_3) = \lambda$$
$$N(a_1a_2a_3) = \lambda$$

and

$$P(\lambda) = N(a_1'a_2'a_3') = \lambda^3 - 3\lambda^2 + 3\lambda - \lambda = \lambda(\lambda - 1)(\lambda - 2)$$

As another example, for the graph in Fig. 9-14*b* let the properties be defined as in the following table:

Property	*Vertices that are painted the same color*
a_1	a, b
a_2	a, c
a_3	a, d
a_4	a, e
a_5	b, c
a_6	b, d
a_7	b, e
a_8	c, d
a_9	c, e

We then have

$$s_1 = 9\lambda^4 \qquad s_2 = 36\lambda^3 \qquad s_3 = 7\lambda^3 + 77\lambda^2$$
$$s_4 = 51\lambda^2 + 75\lambda \qquad s_5 = 15\lambda^2 + 111\lambda \qquad s_6 = 2\lambda^2 + 82\lambda$$
$$s_7 = 36\lambda \qquad s_8 = 9\lambda \qquad s_9 = \lambda$$

That $s_1 = 9\lambda^4$ is quite obvious, because

$$N(a_1) = N(a_2) = N(a_3) = \cdots = N(a_9) = \lambda^4$$

Similarly, we have $s_2 = 36\lambda^3$, because

$$N(a_1a_2) = N(a_1a_3) = \cdots = N(a_8a_9) = \lambda^3$$

To compute s_3, we compute the values of the $\binom{9}{3} = 84$ terms which are of the form $N(a_ia_ja_k)$. Note that seven of these 84 combinations of three properties will have the vertices divided into three color-groups. For example, the properties a_1, a_2, and a_5 mean that the vertices a, b, and c are painted with one color, the vertex d is painted with one color, and the vertex e is painted with one color. The remaining 77 of these 84 combinations will have the vertices divided into two color-groups. For example, the properties a_1, a_3, and a_9 mean that the vertices a, b, and d are painted with one color and the vertices c and e are painted with one color. The other s's are computed similarly. Thus,

$$\begin{aligned} P(\lambda) &= N(a_1'a_2'a_3' \cdots a_9') \\ &= \lambda^5 - 9\lambda^4 + 36\lambda^3 - (7\lambda^3 + 77\lambda^2) \\ &\qquad + (51\lambda^2 + 75\lambda) - (15\lambda^2 + 111\lambda) \\ &\qquad\qquad + (2\lambda^2 + 82\lambda) - 36\lambda + 9\lambda - \lambda \\ &= \lambda^5 - 9\lambda^4 + 29\lambda^3 - 39\lambda^2 + 18\lambda \\ &= \lambda(\lambda - 1)(\lambda - 2)(\lambda - 3)^2 \end{aligned}$$

Incidentally, because of the factors $(\lambda - 1)$, $(\lambda - 2)$, and $(\lambda - 3)$ in $P(\lambda)$, we can conclude that at least four distinct colors are needed to properly color the graph.

9-5 THE FOUR-COLOR PROBLEM

A famous and as yet unsolved problem in the theory of graphs is to determine the smallest number of distinct colors that are sufficient to properly color any planar graph. The graph in Fig. 9-15 shows that

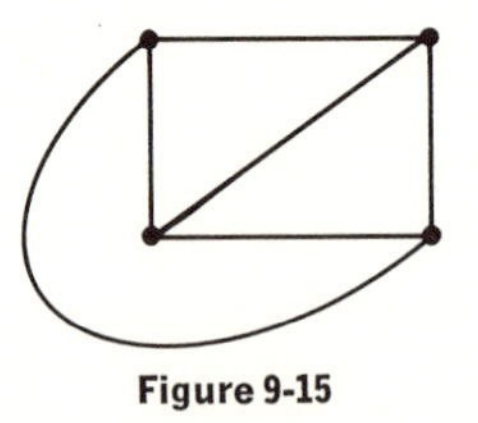

Figure 9-15

three colors are not enough, and Theorem 9-3 below shows that five colors are always sufficient. The question is as follows: Are four colors always sufficient? The most widely accepted conjecture is that four colors are always sufficient. However, for over 100 years, the conjecture has not been proved or disproved. This problem is known as the "four-color problem."

Theorem 9-3 *Five colors are sufficient to color any planar graph properly.*

Proof We shall prove the theorem by induction on the number of vertices in a graph. As the basis of induction, it is obvious that any graph with five or fewer vertices can be properly colored with five colors. As the induction hypothesis, suppose that any planar graph with $n - 1$ vertices can be properly colored with five colors. Let there be a planar graph with n vertices. According to the result in Example 8-4, there is, in the graph, a vertex the degree of which is 5 or less. We shall denote this vertex by x. If we delete from the graph the vertex x and the edges that are incident with it, we have a subgraph with $n - 1$ vertices. According to the induction hypothesis, the subgraph can be properly colored with five colors. If x is adjacent to four or fewer vertices or if x is adjacent to five vertices but two or more of them were colored with the same color, we can pick a remaining color for x. The only case that needs a closer look is that shown in Fig. 9-16a where x is adjacent to five

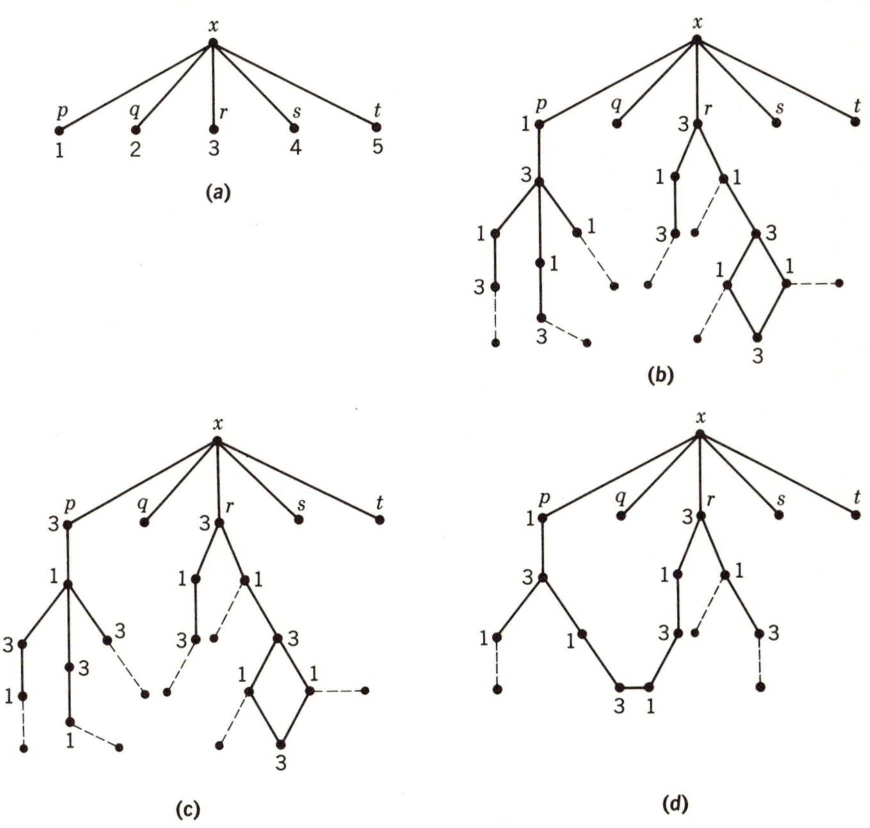

Figure 9-16

vertices p, q, r, s, and t which are colored with the five colors 1, 2, 3, 4, and 5. There are two possible cases:

1. There is no path connecting the vertices p and r such that all the vertices on the path are painted with the two colors 1 and 3. This case is illustrated in Fig. 9-16*b*, where the dashed lines are edges that are incident with vertices that are colored with colors other than 1 and 3. Starting from the vertex p, we interchange the colors 1 and 3 of the vertices that are connected to p by a sequence of edges that are incident with vertices painted with the colors 1 and 3 as shown in Fig. 9-16*c*. Since p is now painted with the color 3, whereas r is still painted with the color 3, x can be painted with the color 1.
2. There is a path connecting the vertices p and r such that all the vertices on the path are painted with the two colors 1 and 3 as illustrated in Fig. 9-16*d*. Since the graph is planar, there is no path connecting the vertices q and s such that all the vertices on the path are painted with the two colors 2 and 4. Using the same argument as in (1), we interchange the colors of the vertices that are painted with the colors 2 and 4 and are connected to q. The vertex x can then be painted with the color 2. ∎

Since we cannot prove that four colors are enough for the proper coloring of any planar graph, it will be interesting to find out some sufficient conditions under which a planar graph can be properly colored with four colors. We define a *triangular graph* to be a planar graph in which each region (including the infinite region) is bounded by exactly three edges. We shall call a region bounded by three edges a *triangular region*. A region that is not bounded by exactly three edges in a graph can always be divided into triangular regions by adding a vertex inside of the region and joining this vertex with all the vertices on the boundary by edges. The *triangular transformation* of a graph is the graph obtained by dividing all the nontriangular regions of a graph into triangular

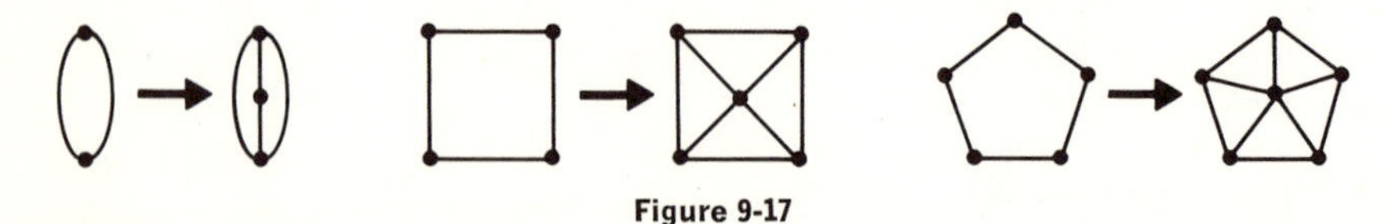

Figure 9-17

regions. Typical transformations are illustrated in Fig. 9-17. Clearly, if the triangular transformation of a graph can be properly colored with four colors, then the graph can also be properly colored with four colors, because we can always delete those vertices added to triangularize the graph after the triangular transformation of the graph is properly colored. We shall therefore limit our discussion to the coloring of triangular graphs.

Theorem 9-4 *A triangular graph can be properly colored with four colors if and only if the edges of the graph can be colored with three colors such that the edges in the contour of every region are colored distinctly.*

Proof Suppose the edges of a triangular graph are colored with three colors α, β, and γ such that the edges in the contour of every region are colored distinctly. In the subgraph which contains those edges that are painted with the colors α and β, all the circuits must be of even length. This is so because the contour of each of the regions of the subgraph must contain exactly four edges, as illustrated in Fig. 9-18. Therefore, the vertices of the subgraph

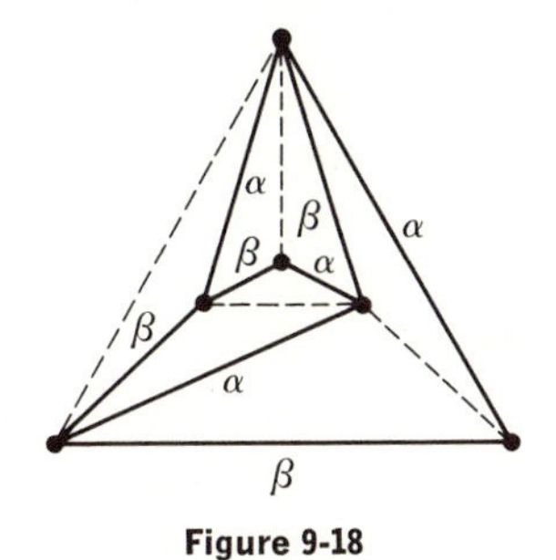

Figure 9-18

can be properly colored with two colors, say A and B. Notice that every vertex in the given graph, except the isolated vertices and those vertices of degree 1 that are incident with an edge painted with color γ, will have been painted with one of the two colors A and B. Similarly, the vertices in the subgraph containing those edges painted with the colors α and γ can be painted with two colors, say u and v. If we superimpose these two subgraphs, we find that each vertex is painted in one of the four ways Au, Bu, Av, and Bv. Moreover, any two adjacent vertices are painted distinctly because they are distinguished either by the colors A and B or by the colors u and v. If we associate a distinct color with each of these four combinations, we have properly colored the vertices of the graph. The isolated vertices and vertices of degree 1 can then be colored after the other vertices are properly colored.

On the other hand, suppose that the vertices of the graph can be properly colored with four colors 1, 2, 3, and 4. We shall paint those edges that join a 1 vertex and a 2 vertex, or a 3 vertex and a 4 vertex, with the color α; we shall paint those edges that join a 1 vertex and a 3 vertex, or a 2 vertex and a 4 vertex, with color β; and we shall paint those edges that join a 1 vertex and a 4 vertex, or a 2 vertex and a 3 vertex, with color γ. We see that the edges on the contour of any triangular region are painted distinctly. Since the three vertices of the region are painted with three of the four distinct colors 1, 2, 3, and 4, every two edges in the contour (which are incident with one common vertex) must be painted with two distinct colors. ■

The result in Theorem 9-4 gives us a *sufficient* condition on a graph being properly colorable with four colors; namely, the edges of the triangular transformation of the graph can be colored with three colors such that the edges in the contour of every region are colored distinctly. The following theorem, in turn, gives us a condition on the possibility of so coloring the edges of the graph.

Theorem 9-5 *The edges of a triangular graph can be colored with three colors such that the edges in the contour of every region are colored distinctly if and only if a coefficient equal to* 1 *or* 2 *can be assigned to each region such that the sum of the coefficients of the regions that have a common vertex is equal to a multiple of* 3.

Proof Assume that such an assignment of coefficients has been made. We start by picking an arbitrary edge and painting it with color α. If the coefficient of a finite region is 1 or if the coefficient of the infinite region is 2, then the edges in the contour of the region will be colored such that in a clockwise direction around the contour the colors of the edges read α, β, and γ. If the coefficient of a finite region is 2 or if the coefficient of the infinite region is 1, the edges in the contour of the region will be colored such that in a clockwise direction around the contour the colors of the edges read α, γ, and β. Such colorings are illustrated in Fig. 9-19. Going around a vertex

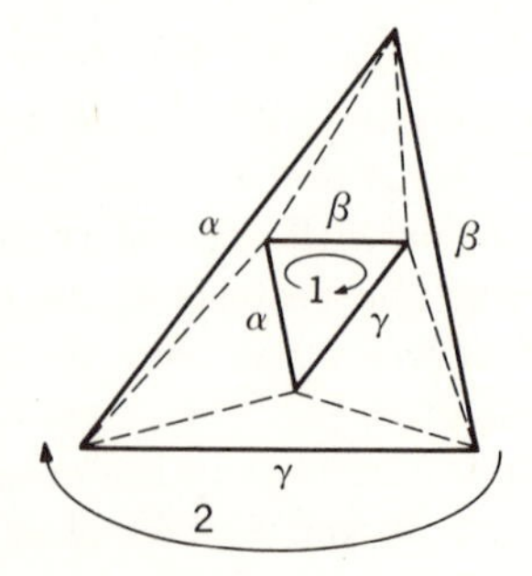

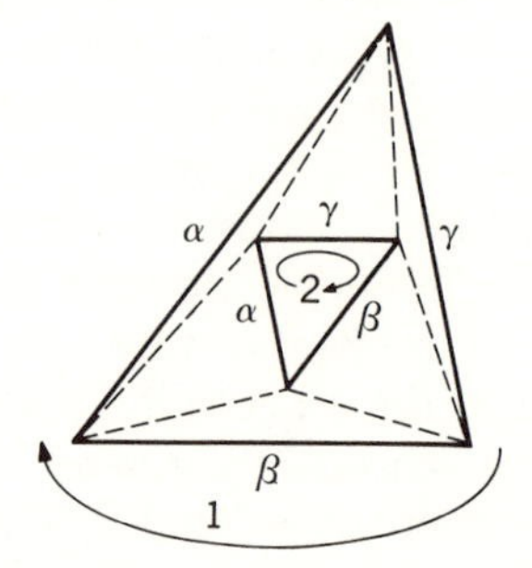

Figure 9-19

in a clockwise direction, we see that such a coloring scheme always colors two edges with the same color if they are separated by regions (finite as well as infinite) the sum of whose coefficients is a multiple of 3. Various cases are illustrated in Fig. 9-20.

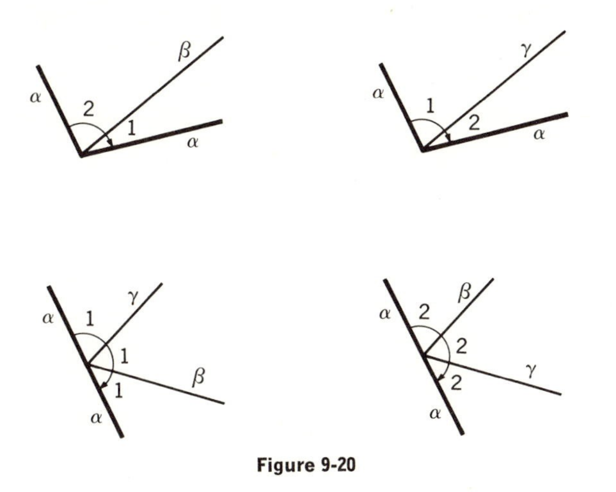

Figure 9-20

Therefore, the condition that the sum of the coefficients of the regions that have a common vertex is equal to a multiple of 3 assures that no conflicts will arise in our coloring scheme.

On the other hand, suppose that the edges in the contour of every region are colored distinctly with three colors α, β, and γ. We shall follow the edges around the contour of each of the regions in a clockwise direction. For each finite region, if the order of the colors reads α, β, and γ, we let the coefficient of the region be 1; if the order of the colors reads α, γ, and β, we let the coefficient of the region be 2. For the infinite region, if the order of the colors reads α, β, and γ, we let the coefficient of the region be 2; if the order of the colors reads α, γ, and β, we let the coefficient of the region be 1. For such an assignment of coefficients, we observe that going around a vertex in a clockwise direction, the sum of coefficients of those regions separating two edges of the same color is always a multiple of 3, using the same argument that was presented above. ■

Corollary 9-5.1 *A triangular graph can be properly colored with four colors if the degree of each vertex is a multiple of* 3.

Proof Assign the coefficient 1 to every region. ■

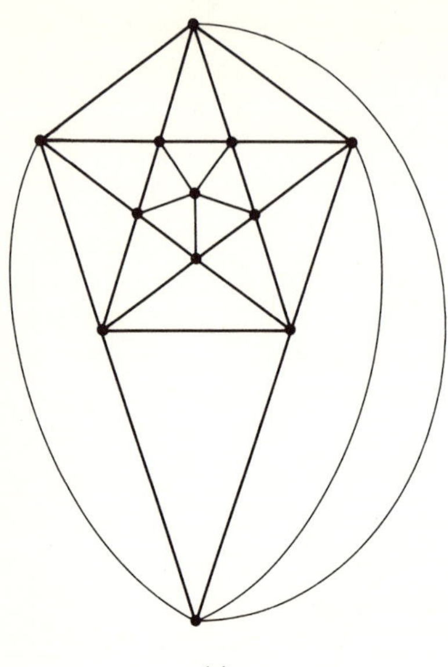

(a)

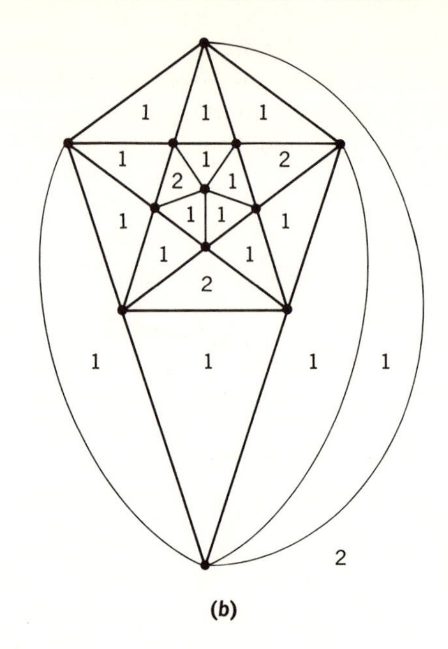

(b)

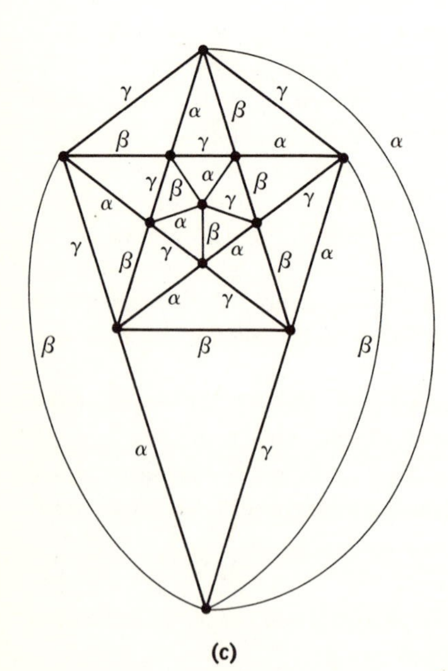

(c)

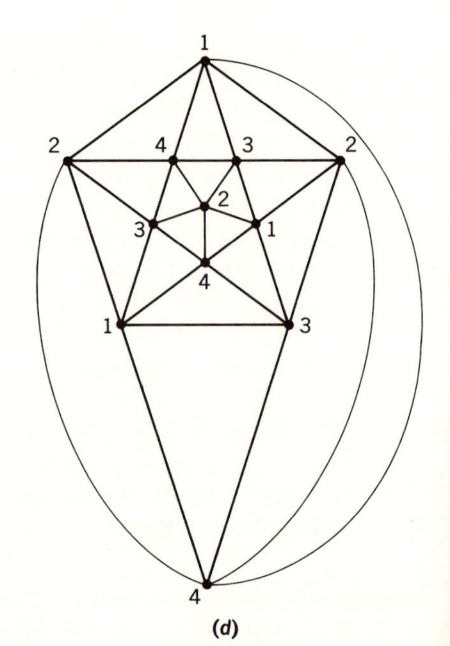

(d)

Figure 9-21

Corollary 9-5.2 *A triangular graph can be properly colored with four colors if the degree of each vertex is a multiple of* 2.

Proof Assign the coefficients 1 and 2 alternately to the regions around a common vertex in a clockwise direction. ∎

Example 9-8 Properly color the graph in Fig. 9-21*a*. To illustrate the application of the results in Theorems 9-4 and 9-5, we assign to each region a coefficient equal to 1 or 2 as shown in Fig. 9-21*b*. We then color the edges with three colors α, β, and γ as shown in Fig. 9-21*c*. Finally, the vertices of the graph are colored with the colors 1, 2, 3, and 4 as shown in Fig. 9-21*d*. ∎

9-6 SUMMARY AND REFERENCES

For the subject matter in this chapter, see Chaps. 13 and 14 of Ore [7] and Chap. 4 of Berge [2]. In Berge [2], the term *externally stable set* is used for dominating set and the term *internally stable set* is used for independent set. Further discussion on the capacity of a graph can be found in Shannon [8] and Chap. 4 of Berge [2]. Ball [1] has an excellent account of the historical notes of the four-color problem. The little book by Dynkin and Uspenski [6] is a good elementary introduction to the subject of map coloring. For coloring planar graphs and chromatic polynomials, see Chap. 21 of Berge [2] and Chap. 4 of Busacker and Saaty [5]. The original paper on chromatic polynomials is by Whitney [9]. An alternative way of finding the chromatic polynomial was presented in a paper by Birkhoff [3].

1. Ball, W. W. R.: "Mathematical Recreation and Essays," The Macmillan Company, New York, 1960.
2. Berge, C.: "The Theory of Graphs and Its Applications," John Wiley & Sons, Inc., New York, 1962.
3. Birkhoff, G. D.: A Determinant Formula for the Number of Ways of Coloring a Map, *Ann. Math.*, **2**:42–46 (1912).
4. Brooks, R. L.: On Coloring the Nodes of a Network, *Proc. Cambridge Phil. Soc.*, **37**:194–197 (1941).
5. Busacker, R. G., and T. L. Saaty: "Finite Graphs and Networks: An Introduction with Applications," McGraw-Hill Book Company, New York, 1965.
6. Dynkin, E. B., and W. A. Uspenski: "Multicolor Problems," D. C. Heath and Company, Boston, 1952.
7. Ore, O.: "Theory of Graphs," American Mathematical Society, Providence, R.I., 1962.
8. Shannon, C. E.: The Zero Error Capacity of a Noisy Channel, *IRE Trans. Inform. Theory*, **IT-2**:8–19 (September, 1956).
9. Whitney, H.: The Coloring of Graphs, *Ann. Math.*, **2**:688–718 (1932).

PROBLEMS

9-1. Let G be a graph with n vertices. Show that $\beta(G)\gamma(G) \geq n$.

9-2. For any two graphs G and H, show that $\gamma(G \times H) \leq \gamma(G)\gamma(H)$.

9-3. For any two graphs G and H, show that $\beta(G \times H) \geq \beta(G)\beta(H)$.

9-4. Show that the regions of a planar graph cannot be properly colored with two colors if the graph has nine vertices and 17 edges and the degree of each vertex is at least 3.

9-5. Show that a linear planar graph with 17 edges and 10 vertices cannot be properly colored with two colors.

9-6. Show that two colors are sufficient to color the regions into which the plane is divided by infinite straight lines drawn in an arbitrary manner.

9-7. A knight has made n moves on an 8×8 chessboard and has returned to the square from which it started. Prove that n must be an even number.

9-8. n circles are drawn on the plane. In each circle a chord is drawn so that chords in two different circles have at most one point in common. Show that the regions into which the plane is divided can be properly colored with three colors.

Hint: Prove this by induction on the number of circles.

9-9. (*a*) Show that a linear planar graph with less than 30 edges has a vertex of degree 4 or less.

(*b*) Show that a linear planar graph with less than 30 edges can be properly colored with four colors.

9-10. (*a*) Show that if a planar graph has less than 12 regions and the degree of each vertex is at least 3, then there is a region bounded by four or fewer edges.

(*b*) Prove that four colors are sufficient to color the regions of such a graph.

9-11. Show that no planar graph with n vertices has a chromatic polynomial that is of the form

$$P(\lambda) = \lambda(\lambda^2 - 6\lambda + 5)Q(\lambda)$$

where $Q(\lambda)$ is a polynomial of degree $n - 3$.

9-12. Find the chromatic polynomial of the graph in Fig. 9P-1.

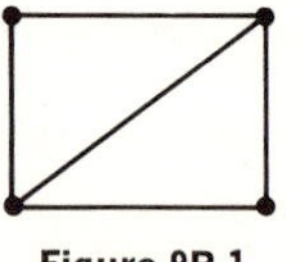

Figure 9P-1

9-13. Show that the chromatic polynomial of a tree of n vertices is $P(\lambda) = \lambda(\lambda - 1)^{n-1}$.

9-14. (*a*) Find the chromatic polynomial of a circuit of n vertices, where n is an even number.

(*b*) Find the chromatic polynomial of a circuit of n vertices, where n is an odd number.

Check your results by noticing that two colors are sufficient to color the vertices in a circuit of even length but are not sufficient to color the vertices in a circuit of odd length.

Chapter 10
Transport Networks

10-1 INTRODUCTION

A directed graph that is connected and contains no loops is said to be a *transport network* if in the graph the following conditions are satisfied:

1. There is one and only one vertex that has no incoming edges; it is called the *source* and is denoted by a.
2. There is one and only one vertex that has no outgoing edges; it is called the *sink* and is denoted by z.
3. There is a nonnegative number associated with each edge; it is called the *capacity* of the edge. The capacity of the edge (i,j) is denoted by $\alpha(i,j)$.

Clearly, a transport network represents a general model for the transportation of material from the origin of supply to the destination through shipping routes, where there are upper limits on the amount of material that can be shipped through the routes. Figure 10-1a shows an example of a transport network.

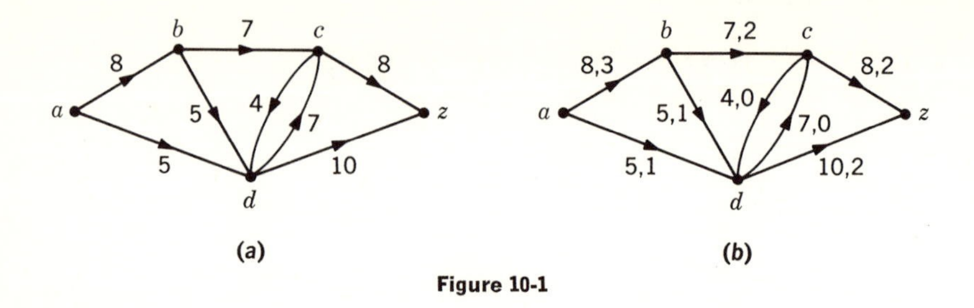

Figure 10-1

A *flow* in a transport network, ϕ, is an assignment of a nonnegative number $\phi(i,j)$ to each edge (i,j) such that the following conditions are satisfied:

1. $\phi(i,j) \leq \alpha(i,j)$ for each edge (i,j).
2. $\sum_{\text{all } i} \phi(i,j) = \sum_{\text{all } k} \phi(j,k)$ for each vertex j except the source a and the sink z.†

In terms of the transportation of material, $\phi(i,j)$ is the amount of material to be shipped through the route (i,j). Condition 1 means that the amount of material to be shipped through a route cannot exceed the capacity of the route. Condition 2 means that, except at the source and at the sink, the amount of material flowing into a vertex must equal the amount of material flowing out of the vertex. For example, Fig. 10-1b shows a flow in the transport network in Fig. 10-1a. The first number associated with an edge is the capacity of the edge, and the second number associated with an edge is the flow in the edge. The quantity $\sum_{\text{all } i} \phi(a,i)$ is said to be the *value of the flow* ϕ and is denoted by ϕ_v. Intuitively, it is clear that

$$\phi_v = \sum_{\text{all } i} \phi(a,i) = \sum_{\text{all } k} \phi(k,z)$$

that is, the total outgoing flow at the source is equal to the total incoming flow at the sink. This result is proved rigorously in the following section (Corollary 10-1.1). For a given flow, an edge (i,j) is said to be saturated if $\phi(i,j) = \alpha(i,j)$ and is said to be unsaturated if $\phi(i,j) < \alpha(i,j)$. A *maximal flow* in a transport network is a flow that achieves the largest possible value. It is conceivable that there might be more than one maximal flow in a transport network. In other words, there might be a number of different flows, all of which attain the largest possible value.

A variation of our model of a transport network is a network having

† We define $\phi(i,j)$ to be zero if there is no edge from i to j.

several sources, each of which can supply a certain amount of material, and several sinks, each of which demands a certain amount of material. Figure 10-2a shows an example, where a_1, a_2, and a_3 are the

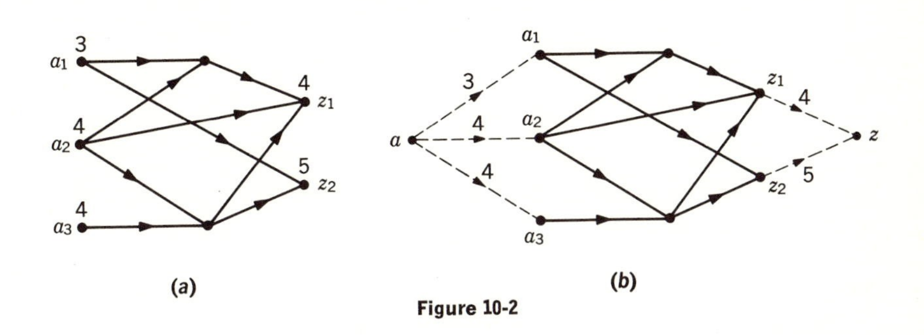

Figure 10-2

sources and z_1 and z_2 are the sinks; the numbers associated with the sources and sinks are the supplies and demands at the sources and the sinks, respectively. Such a network can be augmented to form a network with a single source and a single sink as shown in Fig. 10-2b, namely, a new source a is added together with an edge from a to each of the sources a_1, a_2, and a_3. The capacity of the edge (a,a_i), $i = 1, 2, 3$, is equal to the amount of material that the source a_i can supply. Also, a new sink z is added together with an edge from each of the sinks z_1 and z_2 to z. The capacity of the edge (z_i,z), $i = 1, 2$, is equal to the demand at the sink z_i.

10-2 CUTS

A *cut* in a transport network is a cut-set of the undirected graph, obtained from the transport network by ignoring the direction of the edges, that separates the source from the sink. The notation $(P,\bar{P})$ is used to denote a cut that divides the vertices into two subsets P and $\bar{P}$, where the subset P contains the source and the subset $\bar{P}$ contains the sink. The *capacity of a cut,* denoted by $\alpha(P,\bar{P})$, is defined to be the sum of the capacities of those edges incident from the vertices in P to the vertices in $\bar{P}$; that is,

$$\alpha(P,\bar{P}) = \sum_{i \in P; j \in \bar{P}} \alpha(i,j)$$

For example, the dashed line in Fig. 10-3 shows a cut that separates the subset of vertices $P = \{a,d\}$ from the subset of vertices $\bar{P} = \{b,c,z\}$. The capacity of this cut is equal to $8 + 7 + 10 = 25$.

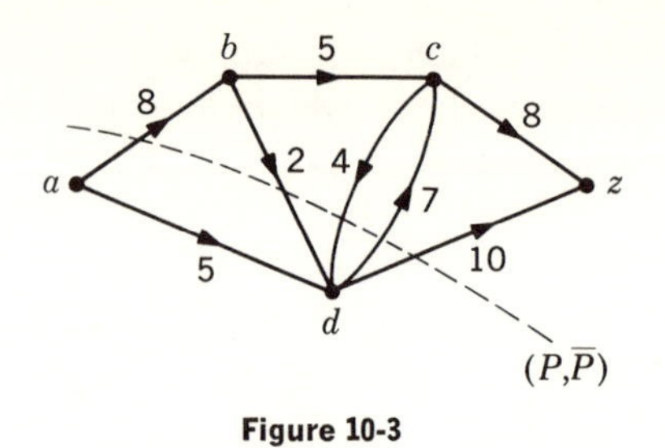

Figure 10-3

The following theorem gives an upper bound on the values of flows in a transport network.

Theorem 10-1 *The value of any flow in a given transport network is less than or equal to the capacity of any cut in the network.*

Proof Let ϕ be a flow and $(P,\bar{P})$ be a cut in a transport network. For the source a,

$$\sum_{\text{all } i} \phi(a,i) - \sum_{\text{all } j} \phi(j,a) = \sum_{\text{all } i} \phi(a,i) = \phi_v \tag{10-1}$$

since $\phi(j,a) = 0$ for any j. For a vertex p other than a, in P,

$$\sum_{\text{all } i} \phi(p,i) - \sum_{\text{all } j} \phi(j,p) = 0 \tag{10-2}$$

Combining Eqs. (10-1) and (10-2), we have

$$\begin{aligned}
\phi_v &= \sum_{p\in P} \Big[\sum_{\text{all } i} \phi(p,i) - \sum_{\text{all } j} \phi(j,p)\Big] \\
&= \sum_{p\in P;\text{all } i} \phi(p,i) - \sum_{p\in P;\text{all } j} \phi(j,p) \\
&= \Big[\sum_{p\in P;i\in P} \phi(p,i) + \sum_{p\in P;i\in \bar{P}} \phi(p,i)\Big] \\
&\qquad - \Big[\sum_{p\in P;j\in P} \phi(j,p) + \sum_{p\in P;j\in \bar{P}} \phi(j,p)\Big]
\end{aligned} \tag{10-3}$$

Note that

$$\sum_{p\in P;i\in P} \phi(p,i) = \sum_{p\in P;j\in P} \phi(j,p)$$

because both sums run through all the vertices in P. Thus, Eq. (10-3) becomes

$$\phi_v = \sum_{p\in P;i\in \bar{P}} \phi(p,i) - \sum_{p\in P;j\in \bar{P}} \phi(j,p) \tag{10-4}$$

But, since $\sum_{p \in P; j \in \bar{P}} \phi(j,p)$ is always a nonnegative quantity, we have

$$\phi_v \leq \sum_{p \in P; i \in \bar{P}} \phi(p,i) \leq \sum_{p \in P; i \in \bar{P}} \alpha(p,i) = \alpha(P,\bar{P}) \quad \blacksquare$$

For example, the value of any flow in the transport network in Fig. 10-3 cannot exceed 12 because the capacity of the cut consisting of the edges (b,c), (b,d), and (a,d) is 12.

Equation (10-4) is a useful result which is stated as the following corollary.

Corollary 10-1.1 *The value of a flow in a transport network is equal to the sum of the flows in the edges from the vertices in P to the vertices in $\bar{P}$ minus the sum of the flows in the edges from the vertices in $\bar{P}$ to the vertices in P for any cut $(P,\bar{P})$.*

10-3 THE MAX-FLOW MIN-CUT THEOREM

One frequently wants to determine the largest amount of material that can be shipped from the source of a given transport network to its sink. Moreover, it is desirable to have an algorithm for the construction of a flow in the network that achieves the largest possible value. To these ends, we prove the following theorem.

Theorem 10-2 *In a transport network, the maximal value that a flow can achieve is equal to the minimal value of the capacities of all the cuts in the network.*

Proof In view of Theorem 10-1, we have only to show that there exists a flow ϕ having a value equal to the capacity of some cut $(P,\bar{P})$. Observe that ϕ must be a maximal flow; if there were a larger flow, then its value would exceed the capacity of the cut $(P,\bar{P})$. Similarly, the capacity of the cut $(P,\bar{P})$ must be a minimal one; if there were a cut with smaller capacity, then ϕ_v would exceed the capacity of that cut.

The existence of such a flow is proved by giving a construction procedure that is known as the *labeling procedure*. To start the procedure, we must construct an initial flow ϕ in the network. However, such construction poses no problem as we can always start, trivially, with zero flow in every edge.

At first, the source a is labeled $(-,\infty)$. (The significance of such a label will become clear later.) Next, all the vertices that are adjacent from a are scanned. A vertex b that is adjacent from a

is labeled $(a^+,\Delta(b))$, where $\Delta(b)$ is equal to $\alpha(a,b) - \phi(a,b)$, if $\alpha(a,b) > \phi(a,b)$; it is not labeled if $\alpha(a,b) = \phi(a,b)$. After all the vertices that are adjacent from the source a are scanned and labeled (if possible), those vertices that are adjacent to or from the labeled vertices are scanned. Let b be a labeled vertex, and let q be a vertex that is adjacent *from* b. The vertex q is labeled $(b^+,\Delta(q))$, where $\Delta(q)$ is equal to the smaller of the two quantities $\Delta(b)$ and $[\alpha(b,q) - \phi(b,q)]$ if $\alpha(b,q) > \phi(b,q)$. The vertex q is not labeled if $\alpha(b,q) = \phi(b,q)$. Let b be a labeled vertex, and let q be a vertex that is adjacent *to* b. The vertex q is labeled $(b^-,\Delta(q))$, where $\Delta(q)$ is equal to the smaller of the two quantities $\Delta(b)$ and $\phi(q,b)$ if $\phi(q,b) > 0$. The vertex q is not labeled if $\phi(q,b) = 0$. Such a labeling procedure is not necessarily unique. The vertex q might be adjacent to or from more than one labeled vertex. Also, there might even be an edge incident from b to q as well as an edge incident from q to b. In any case, when a vertex can be labeled in more than one way, an arbitrary choice of these ways is made.

Let us examine the meanings of these labels before proceeding with the presentation of the remaining steps in the procedure. For a vertex that is adjacent from the source (like the vertex b), the label $(a^+,\Delta(b))$ means that the flow into b can be increased by an amount equal to $\Delta(b)$. Moreover, such an increment can be drawn from the source a. Similarly, for a vertex q that is adjacent from a labeled vertex b, the label $(b^+,\Delta(q))$ means that by drawing the increment from the vertex b, the total incoming flow into q from the labeled vertices can be increased by $\Delta(q)$. For a vertex q that is adjacent to a labeled vertex b, the label $(b^-,\Delta(q))$ means that by decreasing the flow from q to b, the total outgoing flow from q to the labeled vertices can be decreased by $\Delta(q)$. In either of these cases, an increase in the flow equal to $\Delta(q)$ from the vertex q to the unlabeled vertices is assured. The meaning of the label of the source, $(-,\infty)$, should also become clear now. It means that (out from nowhere) the source can supply an infinite amount of material to the other vertices.

If we repeat the procedure of labeling the vertices that are adjacent to or from the labeled vertices, one of the following two cases shall arise:

Case 1 The sink z is labeled, say, with a label $(y^+,\Delta(z))$. [Of course, z will never have a label like $(y^-,\Delta(z))$.] We can increase the flow in the edge (y,z) from $\phi(y,z)$ to $\phi(y,z) + \Delta(z)$, as the increment is guaranteed by the vertex y. If y is labeled $(q^+,\Delta(y))$, we shall in turn draw the increment from the vertex q by increasing the flow in the edge (q,y) from $\phi(q,y)$ to $\phi(q,y) + \Delta(z)$. On the other hand, if y is labeled $(q^-,\Delta(y))$, we shall decrease the flow in the edge (y,q) from $\phi(y,q)$ to $\phi(y,q) - \Delta(z)$ so that the increment $\Delta(z)$ from y to z is compensated. The process is continued back to the source a, and the value of the flow in the transport network will be increased by the amount $\Delta(z)$. The labeling procedure can now be started all over again to further increase the value of the flow in the network.

Case 2 The sink z is not labeled. Let us denote all the labeled vertices by P and all the unlabeled vertices by $\bar{P}$. The fact that the sink z is not labeled means that the flow in each of the edges incident from the vertices in P to the vertices in $\bar{P}$ is equal to the capacity of that edge, and that the flow in each of the edges incident from the vertices in $\bar{P}$ to the vertices in P is equal to zero. According to Corollary 10-1.1, we have thus obtained a flow, the value of which is equal to the capacity of the cut $(P,\bar{P})$. The flow, therefore, is a maximal flow. ∎

Corollary 10-2.1 *If the capacities of the edges in a transport network are all integral values, there is a maximal flow such that the flows in the edges are also integral values.*

Proof If we start with zero flows in the edges and use the labeling procedure to construct a maximal flow, the increments in the flows in the edges are always integral values. ∎

Consider the following illustrative example. For the transport network in Fig. 10-4a, we start with zero flow in every edge. (The first number associated with an edge is its capacity, and the second number is the flow in the edge.) Figure 10-4b shows the first pass of the labeling procedure. Notice that the sink z can be labeled either with $(d^+,3)$ or with $(b^+,2)$. We choose arbitrarily the label $(d^+,3)$. Figure 10-4c shows the second pass, and Fig. 10-4d shows the third pass. Notice that in Fig. 10-4d the vertex b is labeled $(c^+,6)$ and the vertex d is labeled $(c^+,4)$; that is, the vertex c has guaranteed a total flow of 10 to the vertices b and d although $\Delta(c)$ is only equal to 9. However, since the vertex c would have to supply either the increment of flow at b *or* the increment of flow at d, but not at both, in the augmentation step, no difficulty will arise. Figure 10-4e shows the last pass of the labeling procedure, which yields a maximal flow of 13.

Example 10-1 In the graph shown in Fig. 10-5a, the numbers associated with the edges are the distances between vertices. Find a shortest path between the vertices a and z. (It is possible that there are

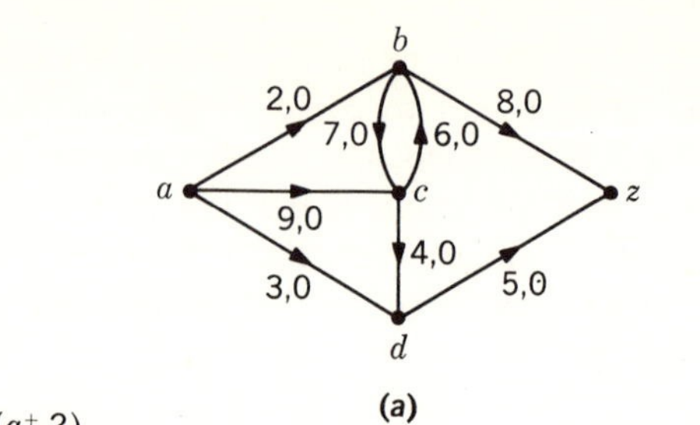

(a)

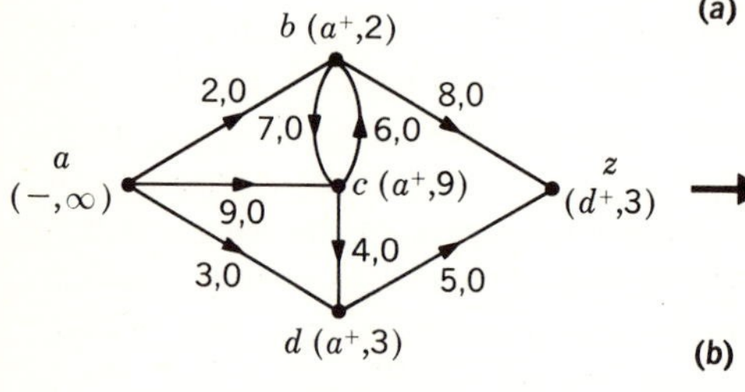

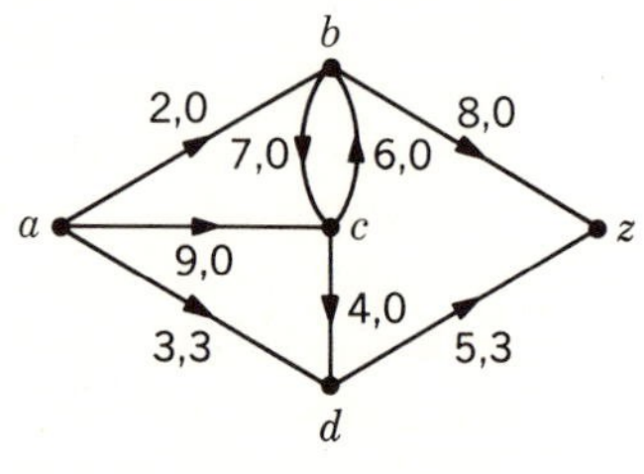

(b)

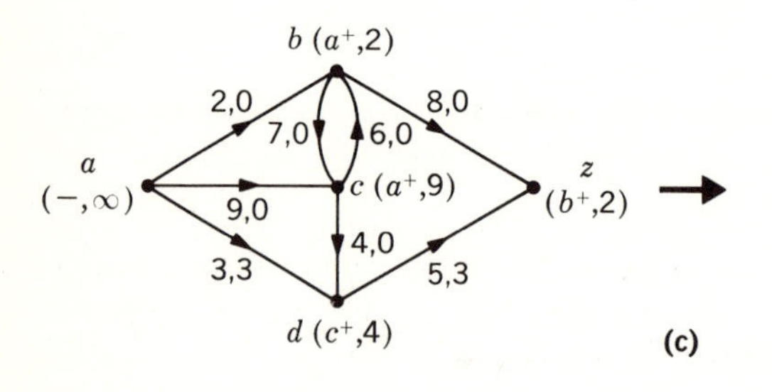

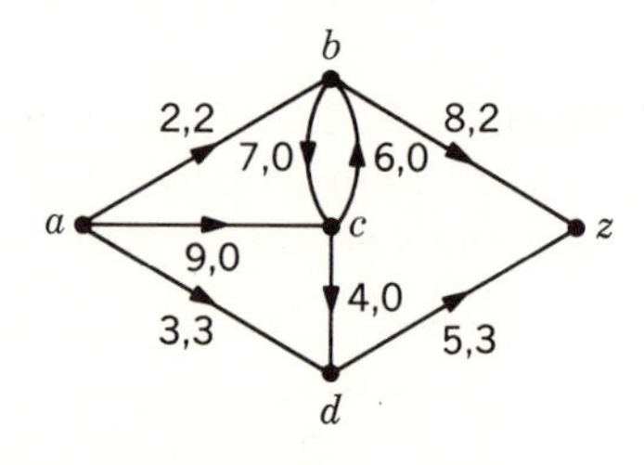

(c)

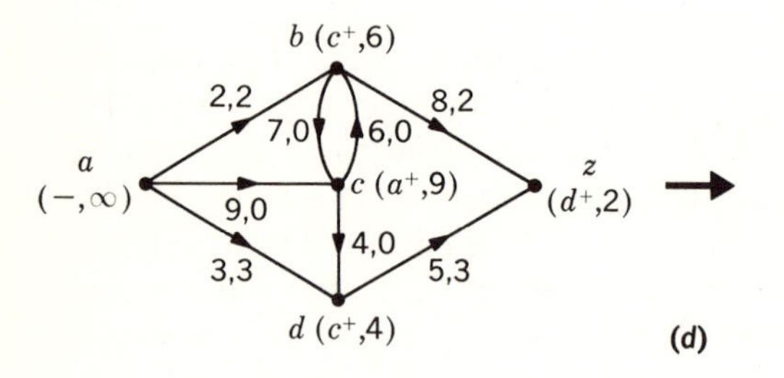

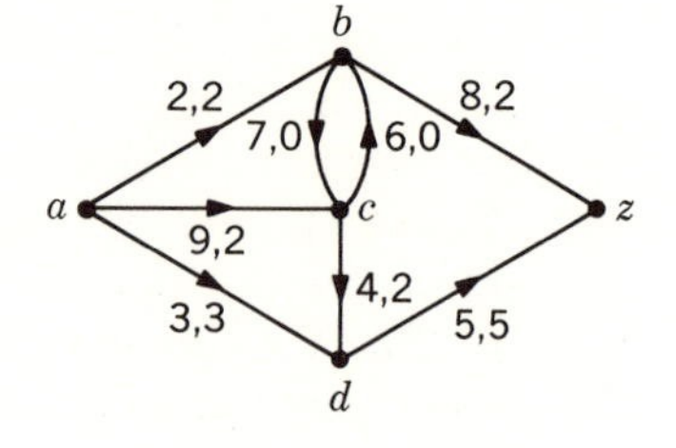

(d)

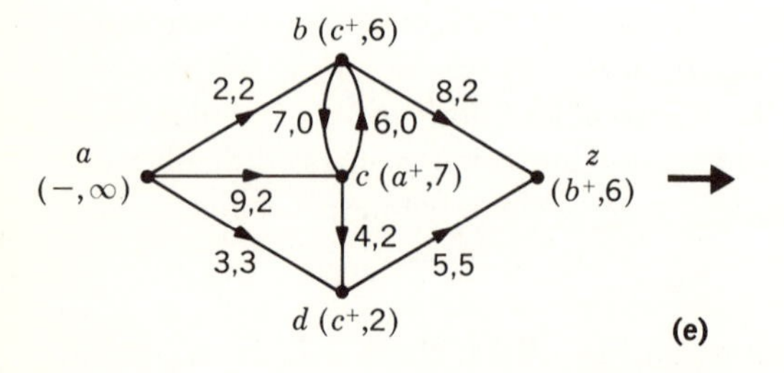

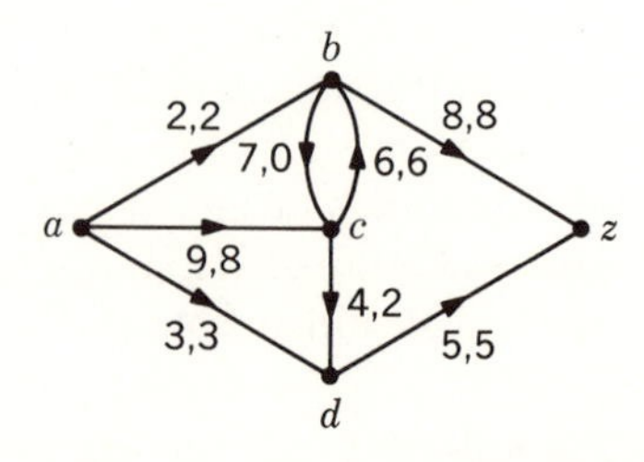

(e)

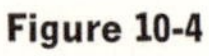

Figure 10-4

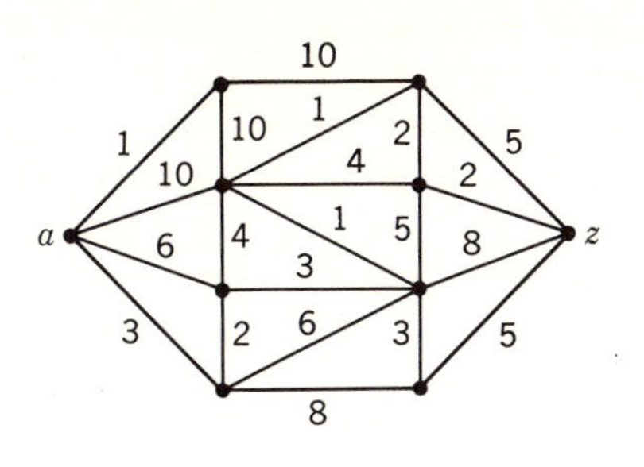

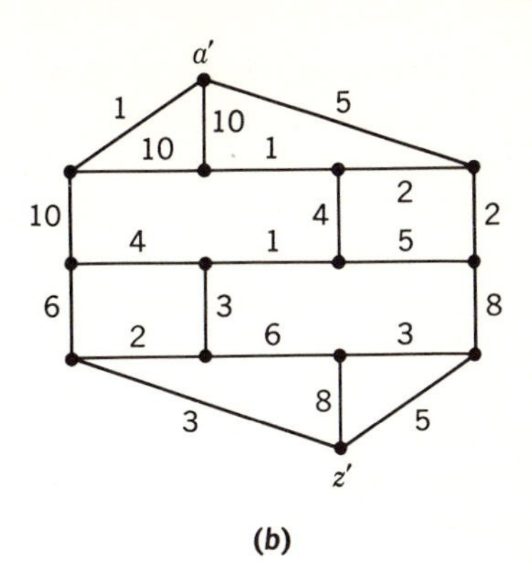

Figure 10-5

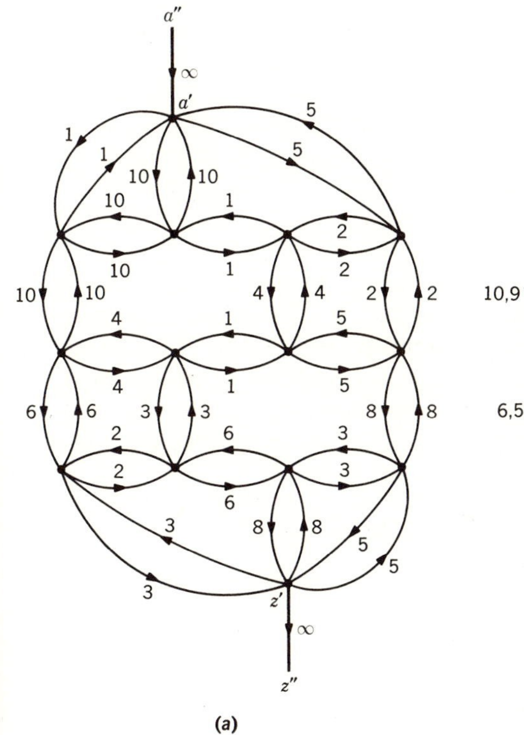

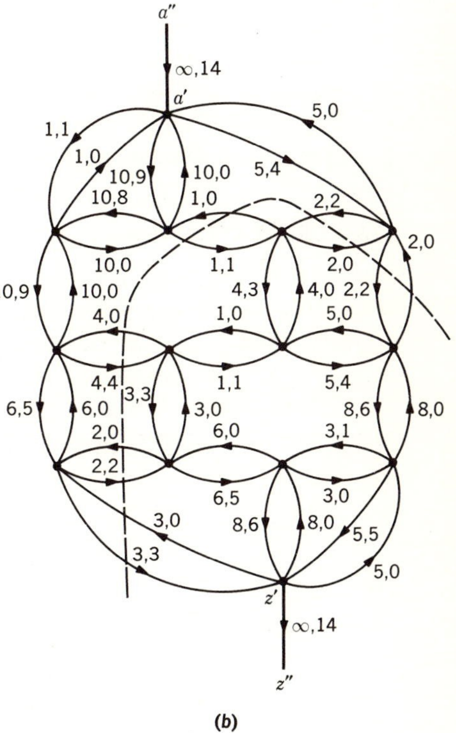

Figure 10-6

several paths of minimal distance.) Consider the graph in Fig. 10-5a as a one-terminal-pair planar graph with a and z as its terminals. We construct a dual of the graph, which is shown in Fig. 10-5b. We also assign to the edges of the graph in Fig. 10-5b the same numbers that were assigned to their corresponding edges in Fig. 10-5a and interpret these numbers as the capacities of the edges. As shown in Example 8-6, a path between the vertices a and z in the graph in Fig. 10-5a corresponds to a cut-set separating the vertices a' and z' in the graph in Fig. 10-5b. Therefore, the problem of finding a shortest path between a and z in Fig. 10-5a is equivalent to that of finding a cut-set separating a' and z' in Fig. 10-5b such that the sum of the capacities of the edges in the cut-set is minimal. To find such a cut-set, the transport network in Fig. 10-6a is constructed by replacing each edge in the graph in Fig. 10-5b with two edges that are of the same capacity and are directed in both directions. Also, edges of infinite capacity are added between a'' and a' and between z' and z'' so that a'' and z'' will be the source and the sink of the transport network, respectively. Applying the labeling procedure to find a maximal flow in the transport network, we find a cut of minimal capacity as shown in Fig. 10-6b when the procedure terminates. This minimal cut corresponds to a shortest path of length 14 in the graph in Fig. 10-5a. ■

★10-4 AN EXTENSION

Suppose for every edge in the transport network there is an upper bound on the flow in the edge (the capacity) as well as a lower bound; that is, to every edge (i,j) two nonnegative numbers $\alpha(i,j)$ and $\beta(i,j)$, with $\alpha(i,j) \geq \beta(i,j)$, are assigned. A flow ϕ in the network is said to be a *feasible flow* if for each edge (i,j), $\phi(i,j)$ is less than or equal to $\alpha(i,j)$ and at the same time larger than or equal to $\beta(i,j)$. In some transport net-

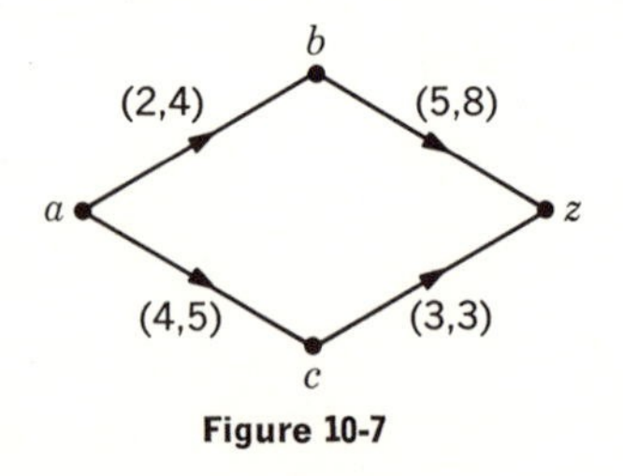

Figure 10-7

work having such limitations, no feasible flow may exist. Figure 10-7 shows one such network. We have, therefore, for a given transport network, two questions: Can there be feasible flows in the network?

How does one find a maximal feasible flow in the network? Theorems 10-3 and 10-4, which follow, provide answers to these questions.

Theorem 10-3 *For any feasible flow* ϕ *and for any cut* $(P,\bar{P})$ *in a given transport network,*

$$\beta(P,\bar{P}) - \alpha(\bar{P},P) \leq \phi_v \leq \alpha(P,\bar{P}) - \beta(\bar{P},P)$$

where

$$\alpha(P,\bar{P}) = \sum_{i\in P;j\in\bar{P}} \alpha(i,j) \qquad \text{and} \qquad \alpha(\bar{P},P) = \sum_{i\in P;j\in\bar{P}} \alpha(j,i)$$

$$\beta(P,\bar{P}) = \sum_{i\in P;j\in\bar{P}} \beta(i,j) \qquad \text{and} \qquad \beta(\bar{P},P) = \sum_{i\in P;j\in\bar{P}} \beta(j,i)$$

Proof This theorem is a direct extension of Theorem 10-1. According to Eq. (10-4) in the proof of Theorem 10-1, we have

$$\phi_v = \sum_{p\in P;i\in\bar{P}} \phi(p,i) - \sum_{p\in P;j\in\bar{P}} \phi(j,p)$$

Thus,

$$\phi_v \leq \sum_{p\in P;i\in\bar{P}} \alpha(p,i) - \sum_{p\in P;j\in\bar{P}} \beta(j,p) = \alpha(P,\bar{P}) - \beta(\bar{P},P)$$

and

$$\phi_v \geq \sum_{p\in P;i\in\bar{P}} \beta(p,i) - \sum_{p\in P;j\in\bar{P}} \alpha(j,p) = \beta(P,\bar{P}) - \alpha(\bar{P},P) \quad \blacksquare$$

Although the determination of the conditions for the existence of feasible flows in the transport network is not the main purpose of proving this theorem, the inequalities we have obtained do impose a necessary condition on the existence of feasible flows in a transport network. Since for any two cuts $(P,\bar{P})$ and $(Q,\bar{Q})$, we must have

$$\beta(P,\bar{P}) - \alpha(\bar{P},P) \leq \phi_v \leq \alpha(P,\bar{P}) - \beta(\bar{P},P)$$

and

$$\beta(Q,\bar{Q}) - \alpha(\bar{Q},Q) \leq \phi_v \leq \alpha(Q,\bar{Q}) - \beta(\bar{Q},Q)$$

that either

$$\beta(P,\bar{P}) - \alpha(\bar{P},P) > \alpha(Q,\bar{Q}) - \beta(\bar{Q},Q)$$

or

$$\beta(Q,\bar{Q}) - \alpha(\bar{Q},Q) > \alpha(P,\bar{P}) - \beta(\bar{P},P)$$

will mean the nonexistence of feasible flows in the network. As an example, for the transport network and the two cuts $(P,\bar{P})$ and $(Q,\bar{Q})$

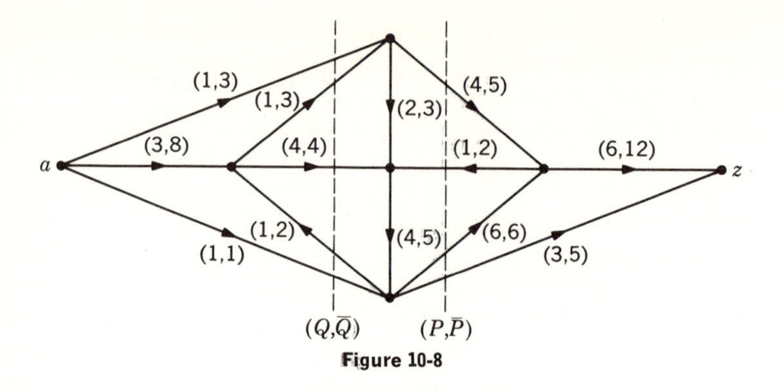

Figure 10-8

shown in Fig. 10-8, we have

$$\beta(P,\bar{P}) - \alpha(\bar{P},P) = 4 + 6 + 3 - 2 = 11$$

and

$$\alpha(Q,\bar{Q}) - \beta(\bar{Q},Q) = 3 + 3 + 4 + 1 - 1 = 10$$

Therefore, we conclude that no feasible flow can exist in the network.

Theorem 10-4 *In a transport network where there are both upper bounds and lower bounds on the flows in the edges, the value of the maximal feasible flow is equal to the minimal value of $\alpha(P,\bar{P}) - \beta(\bar{P},P)$ for all the cuts $(P,\bar{P})$ in the network.*

Proof The proof of this theorem is also similar to that of Theorem 10-2. In view of Theorem 10-3, we need only show the existence of a feasible flow, the value of which is equal to $\alpha(P,\bar{P}) - \beta(\bar{P},P)$ for a cut $(P,\bar{P})$. The labeling procedure presented in the proof of Theorem 10-2 is modified as follows:

1. A vertex b that is adjacent to the source a is labeled $(a^+,\Delta(b))$, with $\Delta(b) = \alpha(a,b) - \phi(a,b)$, if $\alpha(a,b) > \phi(a,b)$; it is not labeled if $\alpha(a,b) = \phi(a,b)$.
2. If an unlabeled vertex q is adjacent from a labeled vertex b, then q is labeled $(b^+,\Delta(q))$, with $\Delta(q)$ equal to the smaller one of the two quantities $\Delta(b)$ and $\alpha(b,q) - \phi(b,q)$, if $\alpha(b,q) > \phi(b,q)$; however, q is not labeled if $\alpha(b,q) = \phi(b,q)$.
3. If an unlabeled vertex q is adjacent to a labeled vertex b, then q is labeled $(b^-,\Delta(q))$, with $\Delta(q)$ equal to the smaller one of the two quantities $\Delta(b)$ and $\phi(q,b) - \beta(q,b)$, if $\phi(q,b) > \beta(q,b)$; however, q is not labeled if $\phi(q,b) = \beta(q,b)$.

It is quite clear that at the termination of such a labeling procedure the value of the flow in the transport network is equal to $\alpha(P,\bar{P}) - \beta(\bar{P},P)$, where P is the set of labeled vertices and $\bar{P}$ is the set of unlabeled vertices. ∎

The labeling procedure presented here can be used to construct a maximal feasible flow in a transport network, provided that we start the procedure with an *initial* feasible flow in the network. Unlike the case where there are only upper bounds on the flows in the edges, the existence of an initial feasible flow is not at all obvious, because zero flows in all the edges might not constitute a feasible flow in the network. Theorem 10-5, which follows shortly, will settle the issues of (1) the existence of a feasible flow in a transport network and (2) the construction of an initial feasible flow.

Given a transport network G with upper and lower bounds on the flows in the edges, we construct a corresponding network $\hat{G}$ with only upper bounds on the flows in the edges. The capacity of an edge (i,j) in $\hat{G}$ is denoted by $\hat{\alpha}(i,j)$.

1. Network $\hat{G}$ contains all the vertices in G with two additional vertices $\hat{a}$ and $\hat{z}$, which are the source and the sink of $\hat{G}$.
2. Network $\hat{G}$ contains all the edges in G. The capacity $\hat{\alpha}(i,j)$ of an edge (i,j) is equal to $\alpha(i,j) - \beta(i,j)$.
3. Corresponding to each edge (i,j) in G, the source $\hat{a}$ is joined to the vertex j by an edge $(\hat{a},j)$ of capacity $\hat{\alpha}(\hat{a},j) = \beta(i,j)$, and the vertex i is joined to the sink $\hat{z}$ by an edge $(i,\hat{z})$ of capacity $\hat{\alpha}(i,\hat{z}) = \beta(i,j)$.
4. The vertex a and the vertex z (the source and the sink of G) are joined by an edge of capacity $\hat{\alpha}(z,a) = \infty$.

As an example, for the transport network G in Fig. 10-9*a* the corresponding network $\hat{G}$ is shown in Fig. 10-9*b*, where heavy lines are used to indicate those edges that were edges in G.

Theorem 10-5 *A feasible flow exists in G if and only if there is a (maximal) flow in $\hat{G}$ such that all the edges incident into the sink $\hat{z}$ are saturated.*

Proof Suppose that there is a feasible flow ϕ in G. Let a flow $\hat{\phi}$ in $\hat{G}$ be constructed as follows:

$$\begin{aligned}\hat{\phi}(i,j) &= \phi(i,j) - \beta(i,j)\\ \hat{\phi}(i,\hat{z}) &= \hat{\alpha}(i,\hat{z})\\ \hat{\phi}(\hat{a},j) &= \hat{\alpha}(\hat{a},j)\\ \hat{\phi}(z,a) &= \phi_v\end{aligned} \tag{10-5}$$

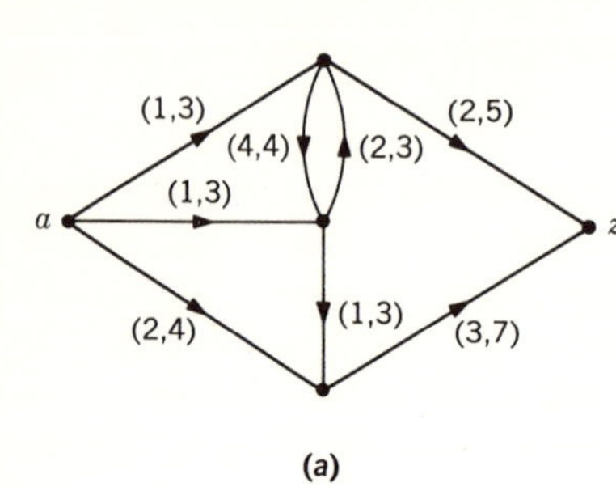

(a)

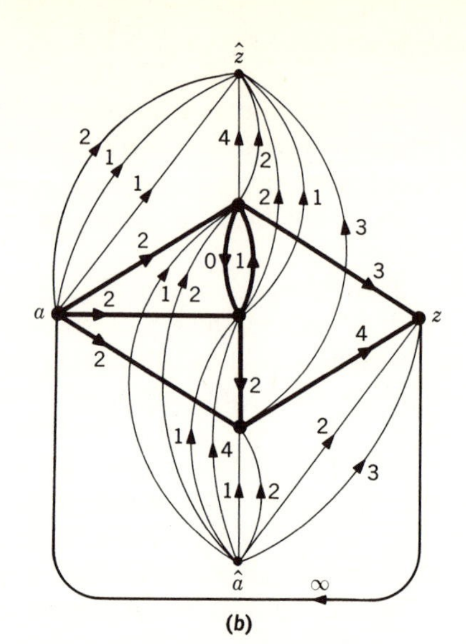

(b)

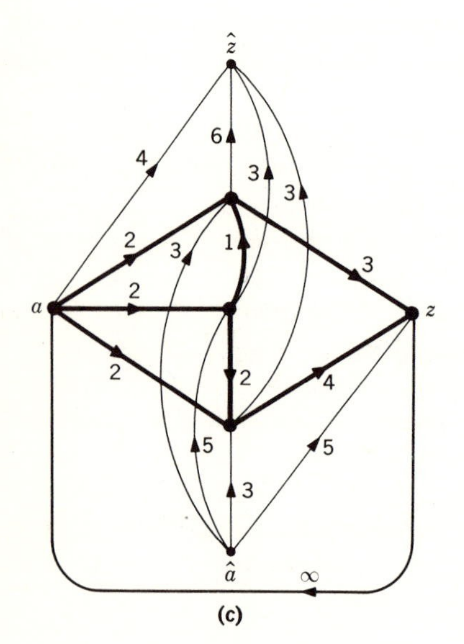

(c)

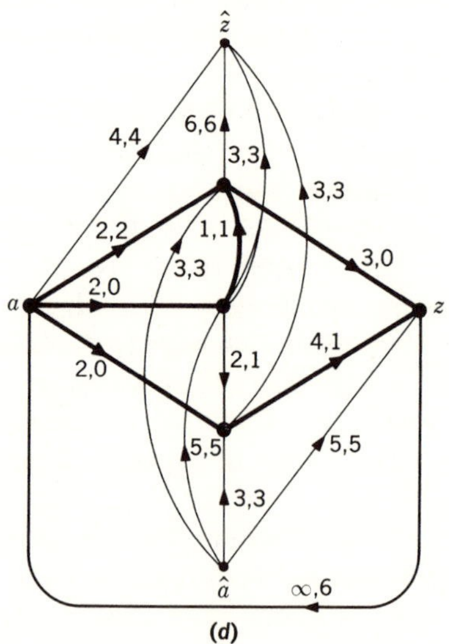

(d)

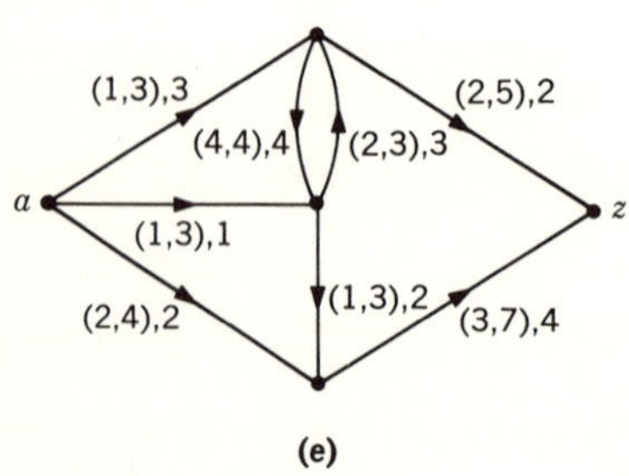

(e)

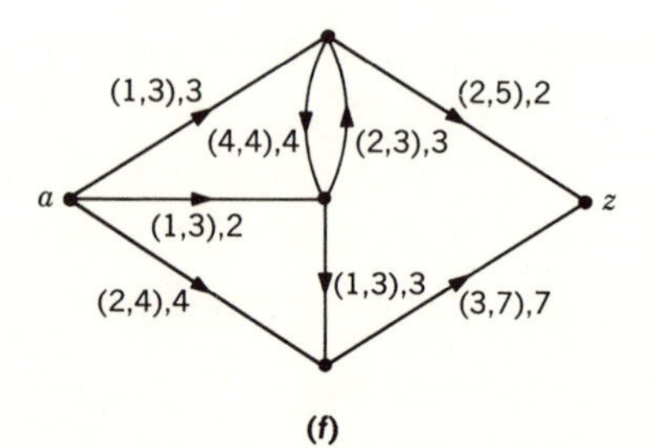

(f)

Figure 10-9

We show that $\hat{\phi}$ is indeed a flow in the network $\hat{G}$. First, it is clear that

$$\hat{\phi}(i,j) = \phi(i,j) - \beta(i,j) \geq \beta(i,j) - \beta(i,j) = 0$$

and

$$\hat{\phi}(i,j) = \phi(i,j) - \beta(i,j) \leq \alpha(i,j) - \beta(i,j) = \hat{\alpha}(i,j)$$

Next, it can be shown that for every j other than $\hat{a}$ and $\hat{z}$,

$$\sum_{\substack{\text{all } i \\ \text{in } \hat{G}}} \hat{\phi}(i,j) - \sum_{\substack{\text{all } k \\ \text{in } \hat{G}}} \hat{\phi}(j,k) = 0$$

Notice that

$$\sum_{\substack{\text{all } i \\ \text{in } \hat{G}}} \hat{\phi}(i,j) = \sum_{\substack{\text{all } i \\ \text{in } G}} \hat{\phi}(i,j) + \sum_{\substack{\text{all} \\ \text{edges from} \\ \hat{a} \text{ to } j}} \hat{\phi}(\hat{a},j)\dagger$$

Since

$$\sum_{\substack{\text{all} \\ \text{edges from} \\ \hat{a} \text{ to } j}} \hat{\phi}(\hat{a},j) = \sum_{\substack{\text{all } i \\ \text{in } G}} \beta(i,j)$$

we have

$$\begin{aligned}\sum_{\substack{\text{all } i \\ \text{in } \hat{G}}} \hat{\phi}(i,j) &= \sum_{\substack{\text{all } i \\ \text{in } G}} \hat{\phi}(i,j) + \sum_{\substack{\text{all } i \\ \text{in } G}} \beta(i,j) \\ &= \sum_{\substack{\text{all } i \\ \text{in } G}} \phi(i,j) - \sum_{\substack{\text{all } i \\ \text{in } G}} \beta(i,j) + \sum_{\substack{\text{all } i \\ \text{in } G}} \beta(i,j) \\ &= \sum_{\substack{\text{all } i \\ \text{in } G}} \phi(i,j)\end{aligned}$$

Similarly, we have

$$\sum_{\substack{\text{all } k \\ \text{in } \hat{G}}} \hat{\phi}(j,k) = \sum_{\substack{\text{all } k \\ \text{in } G}} \phi(j,k) - \sum_{\substack{\text{all } k \\ \text{in } G}} \beta(j,k) + \sum_{\substack{\text{all } k \\ \text{in } G}} \beta(j,k) = \sum_{\substack{\text{all } k \\ \text{in } G}} \phi(j,k)$$

It follows that

$$\sum_{\substack{\text{all } i \\ \text{in } \hat{G}}} \hat{\phi}(i,j) - \sum_{\substack{\text{all } k \\ \text{in } \hat{G}}} \hat{\phi}(j,k) = \sum_{\substack{\text{all } i \\ \text{in } G}} \phi(i,j) - \sum_{\substack{\text{all } k \\ \text{in } G}} \phi(j,k) = 0$$

Therefore, we conclude that $\hat{\phi}$ is a flow in the network $\hat{G}$. Moreover, according to Eq. (10-5), $\hat{\phi}$ is a flow such that all the edges incident into $\hat{z}$ are saturated.

† There might be more than one edge from $\hat{a}$ to j.

On the other hand, suppose that there is a flow $\hat{\phi}$ in $\hat{G}$ such that all the edges incident into $\hat{z}$ are saturated. Let a flow ϕ in G be constructed as follows:

$$\phi(i,j) = \hat{\phi}(i,j) + \beta(i,j)$$

We show that ϕ is indeed a feasible flow in the network G. We notice that

$$\phi(i,j) = \hat{\phi}(i,j) + \beta(i,j) \geq \beta(i,j)$$

and

$$\phi(i,j) = \hat{\phi}(i,j) + \beta(i,j) \leq \hat{\alpha}(i,j) + \beta(i,j) = \alpha(i,j) - \beta(i,j) + \beta(i,j) = \alpha(i,j)$$

Next, it can be shown that for every j other than a and z,

$$\sum_{\substack{\text{all } i\\ \text{in } G}} \phi(i,j) - \sum_{\substack{\text{all } k\\ \text{in } G}} \phi(j,k) = 0$$

Since the saturation of all the edges incident into $\hat{z}$ means the saturation of all the edges incident from $\hat{a}$, we have

$$\sum_{\substack{\text{all } i\\ \text{in } G}} \beta(i,j) = \sum_{\substack{\text{all edges}\\ \text{from } \hat{a} \text{ to } j}} \hat{\phi}(\hat{a},j)$$

and

$$\sum_{\substack{\text{all } k\\ \text{in } G}} \beta(j,k) = \sum_{\substack{\text{all edges}\\ \text{from } j \text{ to } \hat{z}}} \hat{\phi}(j,\hat{z})$$

It follows that

$$\begin{aligned}
\sum_{\substack{\text{all } i\\ \text{in } G}} \phi(i,j) - \sum_{\substack{\text{all } k\\ \text{in } G}} \phi(j,k) &= \Big[\sum_{\substack{\text{all } i\\ \text{in } G}} \hat{\phi}(i,j) + \sum_{\substack{\text{all } i\\ \text{in } G}} \beta(i,j)\Big] - \Big[\sum_{\substack{\text{all } k\\ \text{in } G}} \hat{\phi}(j,k) + \sum_{\substack{\text{all } k\\ \text{in } G}} \beta(j,k)\Big] \\
&= \Big[\sum_{\substack{\text{all } i\\ \text{in } G}} \hat{\phi}(i,j) + \sum_{\substack{\text{all edges}\\ \text{from } \hat{a} \text{ to } j}} \hat{\phi}(\hat{a},j)\Big] - \Big[\sum_{\substack{\text{all } k\\ \text{in } G}} \hat{\phi}(j,k) + \sum_{\substack{\text{all edges}\\ \text{from } j \text{ to } \hat{z}}} \hat{\phi}(j,\hat{z})\Big] \\
&= \sum_{\substack{\text{all } i\\ \text{in } \hat{G}}} \hat{\phi}(i,j) - \sum_{\substack{\text{all } k\\ \text{in } \hat{G}}} \hat{\phi}(j,k) = 0
\end{aligned}$$

Thus, we conclude that ϕ is a feasible flow in the network G. ■

For a given transport network G in which both upper and lower bounds on the flow in each edge are specified, we can find an initial feasible flow in G by constructing the corresponding transport network $\hat{G}$ and finding a maximal flow $\hat{\phi}$ in $\hat{G}$. If the maximal flow in $\hat{G}$ does not saturate all the edges incident into $\hat{z}$, we conclude that no feasible flow can exist in the network G. If the maximal flow in $\hat{G}$ does saturate all the edges leading to $\hat{z}$, we can construct an initial feasible flow in G as shown in the proof of Theorem 10-5. Once an initial feasible flow has been found in G, we can find a maximal feasible flow in G by carrying out the labeling procedure presented in the proof of Theorem 10-4.

As an example, to find a maximal feasible flow in the network G in Fig. 10-9*a*, we construct the corresponding network $\hat{G}$ in Fig. 10-9*b*, which is, in turn, simplified by combining edges between vertices as in Fig. 10-9*c*. A maximal flow in $\hat{G}$ is constructed as shown in Fig. 10-9*d*. An initial feasible flow in G can then be constructed as shown in Fig. 10-9*e*. Finally, a maximal feasible flow in G is constructed as shown in Fig. 10-9*f*.

10-5 SUMMARY AND REFERENCES

As was pointed out earlier, problems concerning the shipment of merchandise and the allocation of resources can be formulated as network flow problems. Moreover, there are other applications of the result as was seen in Example 10-1 and as will also be seen in Chap. 11 when we study the theory of matching.

The model of a transport network can be modified in several ways as discussed in Probs. 10-4, 10-7, 10-9, and 10-15. A further extension is to have a set of positive numbers associated with each of the edges in a transport network and to require that the flow in an edge be equal to one of the numbers associated with that edge. The problem is then to construct a feasible flow as well as to construct a maximal feasible flow in the network. Another extension is the problem of multiple-commodity flow. Suppose that in a transport network there are k sources, each supplying one of k commodities, and there are k sinks, each demanding one of the k commodities. The multiple-commodity flow problem is to construct k simultaneous flows for the k commodities such that the demands at the k sinks are satisfied and the sum of the values of the k flows in each edge does not exceed the capacity of that edge.

The book by Ford and Fulkerson [3] has an excellent coverage of the subject material. Also see Chap. 8 of Berge [1] and Chap. 7 of Busacker and Saaty [2].

1. Berge, C.: "The Theory of Graphs and Its Applications," John Wiley & Sons, Inc., New York, 1962.

2. Busacker, R. G., and T. L. Saaty: "Finite Graphs and Networks: An Introduction with Applications," McGraw-Hill Book Company, New York, 1965.
3. Ford, L. R., Jr., and D. R. Fulkerson: "Flows in Networks," Princeton University Press, Princeton, N.J., 1962.

PROBLEMS

10-1. Use the labeling procedure to find a maximal flow in the transport network in Fig. 10P-1.

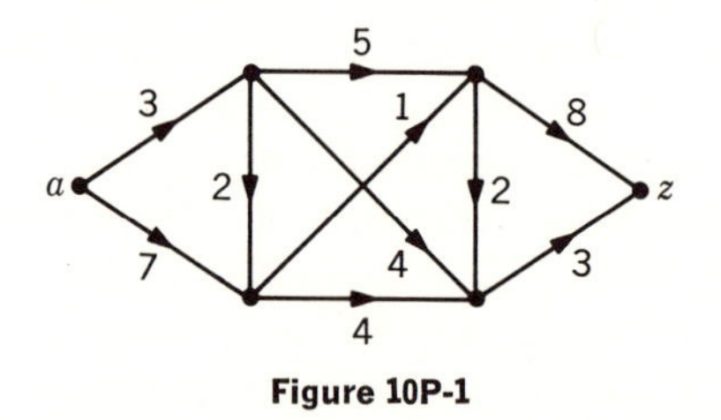

Figure 10P-1

10-2. Find a maximal flow in the transport network in Fig. 10P-2.

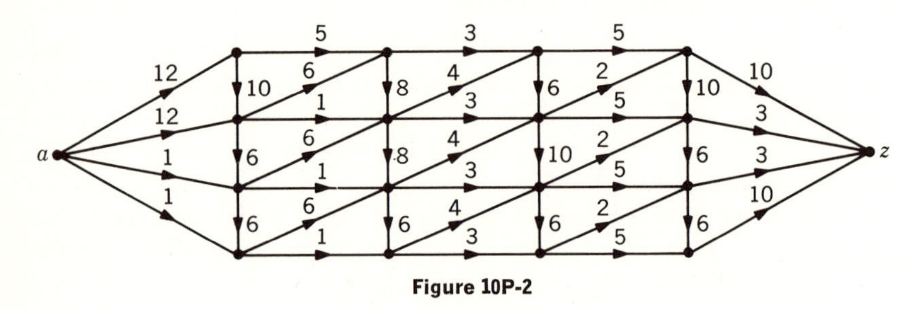

Figure 10P-2

10-3. Equipment is manufactured at three factories x_1, x_2, and x_3 and is to be shipped to three depots y_1, y_2, and y_3 through the transport network shown in Fig. 10P-3.

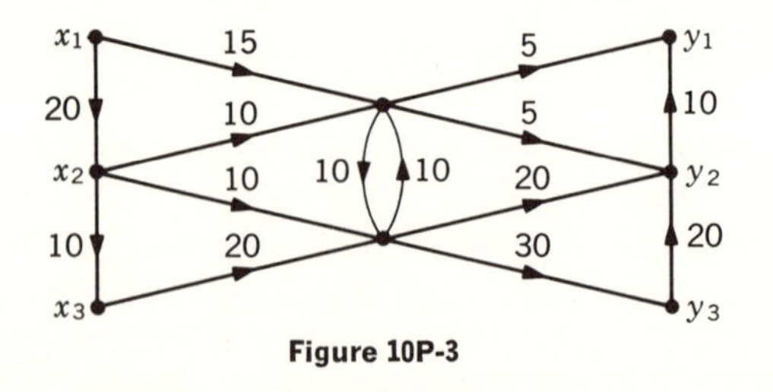

Figure 10P-3

Factory x_1 can make 40 units, factory x_2 can make 20 units, and factory x_3 can make 10 units. Depot y_1 needs 15 units, depot y_2 needs 25 units, and depot y_3 needs 10 units. How many units should each factory make so that they can be transported to the depots?

10-4. Find a maximal flow in the transport network in Fig. 10P-4 in which flows in the unoriented edges can be in either direction.

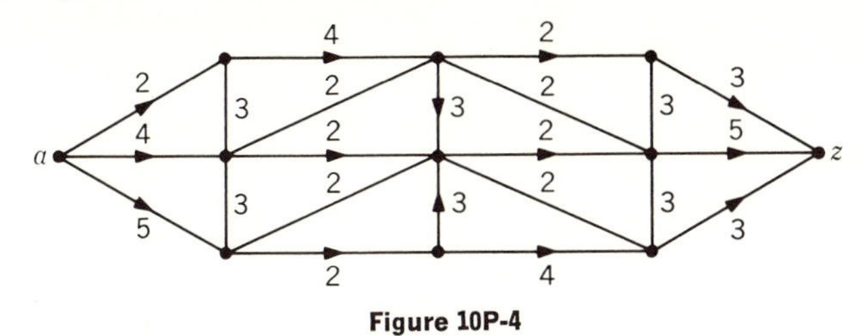

Figure 10P-4

10-5. (*a*) Seven kinds of military equipment are to be flown to a destination by five cargo planes. There are four units of each kind, and the five planes can carry eight, eight, five, four, and four units, respectively. Can the equipment be loaded in such a way that no two units of the same kind are on one plane?

(*b*) Give a solution to part (*a*) when the capacities of the planes are seven, seven, six, four, and four units, respectively.

10-6. Construct a directed 2-graph with four vertices such that the outgoing and incoming degrees of the vertices are (5,4), (3,3), (1,2), and (2,2), respectively, by solving the corresponding network flow problem.

10-7. In the transport network in Fig. 10P-5, not only are there upper bounds on the flows in the edges, but there are also upper bounds on the total flows into the intermediate vertices. Find a maximal flow in the network.

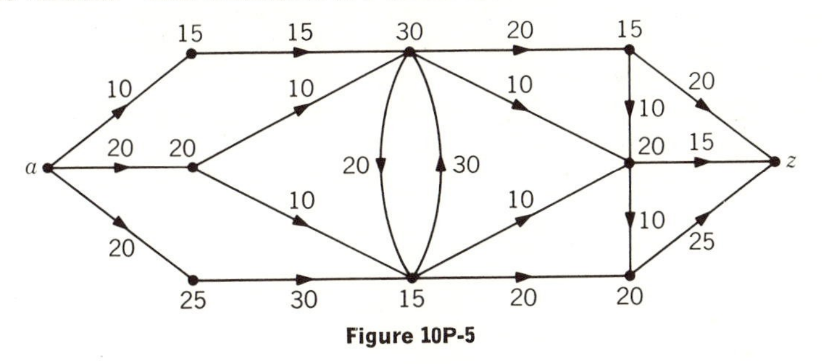

Figure 10P-5

10-8. In the network in Fig. 10P-6 the numbers associated with the edges are the distances between vertices. Find a shortest path between *a* and *z*.

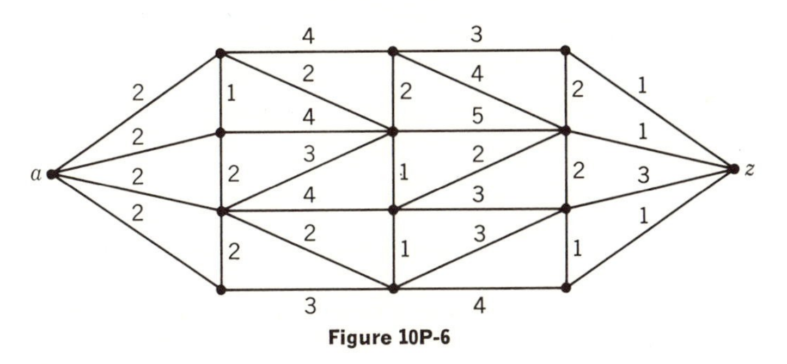

Figure 10P-6

10-9. Given a transport network, we wish to find a flow such that the flow in each edge is larger than or equal to the capacity of the edge and the value of the flow is minimal.

(*a*) Define the "capacity of a cut," and state the minimal-flow maximal-cut theorem which is analogous to the maximal-flow minimal-cut theorem proved in Sec. 10-3.

(*b*) Prove the minimal-flow maximal-cut theorem by designing an algorithm to find a minimal flow.

10-10. Engineers and technicians are to be hired by a company to participate in three projects. The personnel requirements of these three projects are listed in the following table:

	Minimal number of people needed in each project	*Minimal number in each category*			
		Mechanical engineers	*Mechanical technicians*	*Electrical engineers*	*Electrical technicians*
Project I	40	5	10	10	5
Project II	40	10	5	15	5
Project III	20	5	0	10	5

Moreover, to prepare for later expansion, the company wants to hire at least 30 mechanical engineers, 20 mechanical technicians, 20 electrical engineers, and 20 electrical technicians. What is a minimal number of persons in each category that the company should hire, and how should they be allocated to the three projects?

10-11. In the graph in Fig. 10P-7 a minimal set of edges is to be selected such that every vertex is incident with at least one of the edges in the set. Solve this problem as a minimal flow problem associated with a transport network.

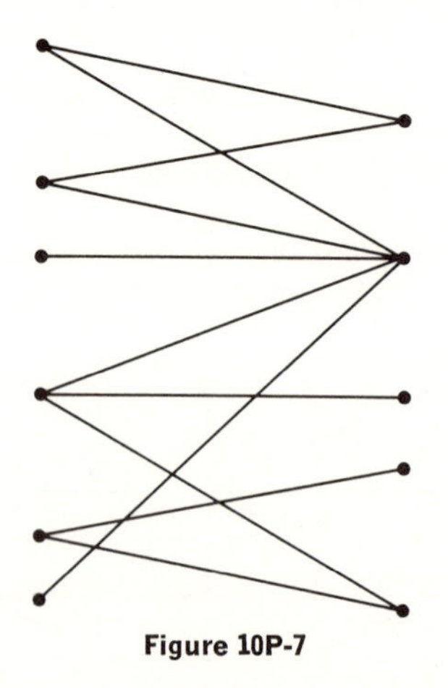

Figure 10P-7

10-12. Three candidates x_1, x_2, and x_3 have been promised minimal amounts of campaign money of \$40,000, \$23,000, and \$50,000, respectively. Each candidate in turn has promised three campaign areas at least the amounts of money shown in the table below. In addition, each candidate will need at least \$5,000 for his own expenses.

	Campaign area		
Candidate	C_1	C_2	C_3
x_1	\$20,000	\$10,000	\$10,000
x_2	10,000	5,000	2,000
x_3	5,000	10,000	20,000

If the three campaign areas C_1, C_2, and C_3 require a minimum of \$30,000, \$25,000, and \$50,000, respectively, to conduct a thorough campaign, what is a minimal amount of campaign money each candidate must obtain and how should it be distributed?

10-13. Show that no feasible flow exists in the transport network G in Fig. 10P-8 by

(*a*) using an exhaustive argument

(*b*) finding two cuts $(P,\bar{P})$ and $(Q,\bar{Q})$ such that

$$\beta(Q,\bar{Q}) - \alpha(\bar{Q},Q) > \alpha(P,\bar{P}) - \beta(\bar{P},P)$$

(*c*) constructing $\hat{G}$ and finding a maximal flow in $\hat{G}$

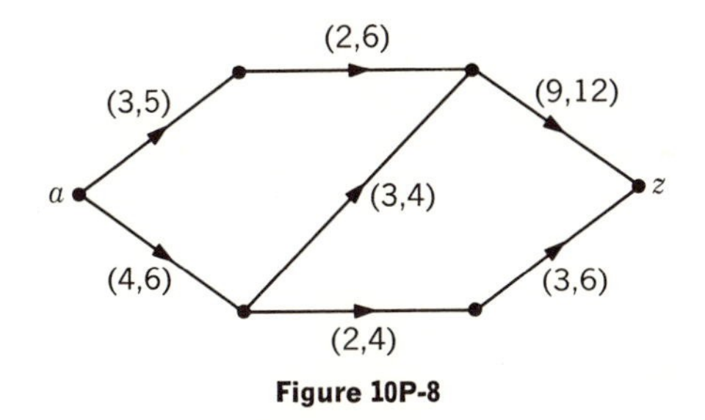

Figure 10P-8

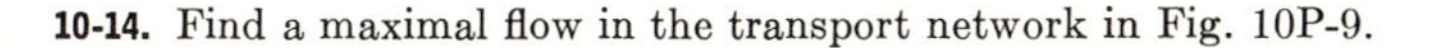

10-14. Find a maximal flow in the transport network in Fig. 10P-9.

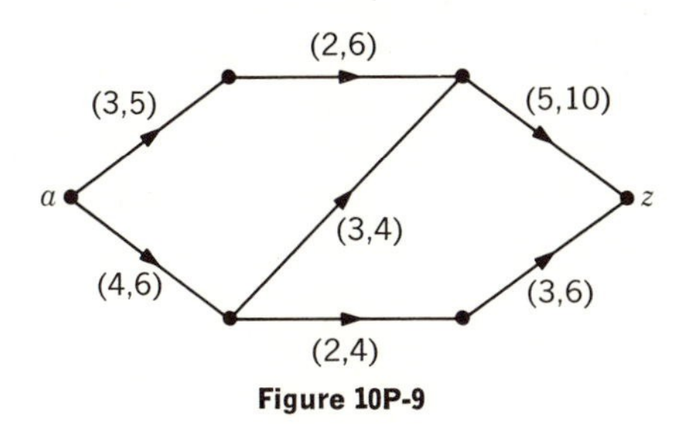

Figure 10P-9

10-15. (*a*) State a procedure for finding a minimal flow in a transport network where there is a lower bound as well as an upper bound on the flow in every edge.

(*b*) Find a minimal flow in the transport network in Fig. 10P-9.

Chapter 11
Matching Theory

11-1 INTRODUCTION

Four workmen x_1, x_2, x_3, and x_4 are available to fill five jobs y_1, y_2, y_3, y_4, and y_5. x_1 is qualified for the jobs y_1 and y_2; x_2 is qualified for the jobs y_1 and y_3; x_3 is qualified for the job y_4; and x_4 is qualified for the jobs y_2, y_3, and y_5. The assignment problem is concerned with the following questions: Can each workman be assigned to a job for which he is qualified? If so, how should the assignment be made? If not, at most, how many of them can be assigned? It can be observed that this problem is one of permutations with forbidden positions, a topic discussed in Chap. 4. In the present context, there are five possible positions for the workmen with the forbidden positions for each workman being the jobs for which he is not qualified. Figure 11-1*a* shows the forbidden positions in tabular form. Applying the principle of inclusion and exclusion, we can find the hit polynomial and thus determine the number of ways of assigning the four workmen to jobs for which they are qualified. (In the case that not all of the workmen can be assigned to jobs for

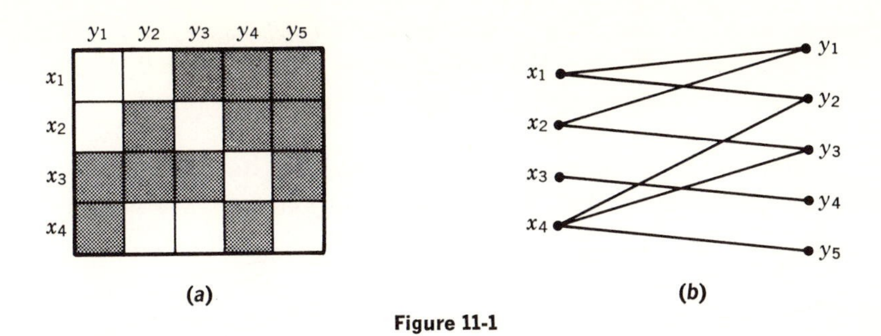

Figure 11-1

which they are qualified, the hit polynomial also gives the largest number of workmen that can be assigned as well as the number of ways of making the assignment.) However, such a computation gives no information about how the assignments should be made.

In this chapter, the assignment problem is solved in a different way. First of all, it is observed that the qualifications of these men can be represented by a graph. The vertices in the graph are x_1, x_2, x_3, x_4, y_1, y_2, y_3, y_4, and y_5 corresponding to the workmen and the jobs, respectively. There is an edge joining x_i and y_j if workman x_i is qualified for the job y_j. Such a representation is shown in Fig. 11-1*b*. The problem of assigning the men to jobs for which they are qualified is then equivalent to the problem of selecting a subset of the edges such that each x will be connected to exactly one y by one of these edges.

A *bipartite* graph is defined as a graph, the vertices of which can be divided into two disjoint subsets such that no vertex in a subset is adjacent to vertices in the same subset. The graph in Fig. 11-1*b* is a bipartite graph. In a bipartite graph G, let X and Y denote the two disjoint subsets of vertices. A *matching* in a bipartite graph is a selection of edges such that no two edges in the selection are incident with the same vertex (in X or in Y); that is, a matching defines a one-to-one correspondence between the vertices in a subset of X and the vertices in a subset of Y. *A complete matching of X into Y* in a bipartite graph is a matching such that there is an edge incident with every vertex in X. In other words, a one-to-one correspondence is defined between all the vertices in X and the vertices in a subset of Y.

11-2 COMPLETE MATCHING

For a given bipartite graph, we want to know whether there is a complete matching of the set of vertices X into the set of vertices Y. Theorem 11-1 gives a necessary and sufficient condition for the existence of a complete matching in a bipartite graph.

Theorem 11-1 *In a bipartite graph a complete matching of X into Y exists if and only if $|A| \leq |R(A)|$ for every subset A of X, where $R(A)$ denotes the set of vertices in Y that are adjacent to the vertices in A.*

Before proving the theorem, let us illustrate its application. For the bipartite graph in Fig. 11-2*a*, there is a complete matching of X

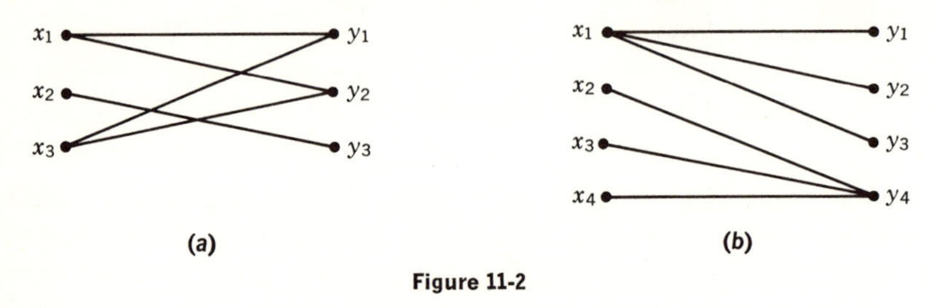

(*a*) (*b*)

Figure 11-2

into Y because, as shown in the Table 11-1, the condition $|A| \leq |R(A)|$ is satisfied for every subset A of X.

Table 11-1

A	$\|A\|$	$R(A)$	$\|R(A)\|$
ϕ	0	ϕ	0
$\{x_1\}$	1	$\{y_1,y_2\}$	2
$\{x_2\}$	1	$\{y_3\}$	1
$\{x_3\}$	1	$\{y_1,y_2\}$	2
$\{x_1,x_2\}$	2	$\{y_1,y_2,y_3\}$	3
$\{x_1,x_3\}$	2	$\{y_1,y_2\}$	2
$\{x_2,x_3\}$	2	$\{y_1,y_2,y_3\}$	3
$\{x_1,x_2,x_3\}$	3	$\{y_1,y_2,y_3\}$	3

For the bipartite graph in Fig. 11-2*b*, there is no complete matching of X into Y, since $|A| > |R(A)|$ for the set $A = \{x_2,x_3\}$.

Proof of Theorem 11-1 That $|A| \leq |R(A)|$ for every subset A of X is a necessary condition for the existence of a complete matching of X into Y is quite obvious. Because, if there is a subset A_0 of X such that $|A_0| > |R(A_0)|$, then it is impossible to define a one-to-one correspondence between the vertices in A_0 and the vertices in a subset of Y.

That $|A| \leq |R(A)|$ for every subset A of X is a sufficient condition for the existence of a complete matching of X into Y will be proved by induction on the number of vertices in X. As the basis of induction, the condition is clearly a sufficient one for any bipartite graph having only one vertex in X, because this vertex must be adjacent to at least one vertex in Y. As the induction hypothesis, assume that the condition is a sufficient one for any bipartite graph having $m - 1$ or less vertices in X. Now, suppose that there is a bipartite graph G having m vertices in X such that $|A| \leq |R(A)|$ for every subset A of X. We shall show that there is a complete matching of X into Y by examining the following cases:

Case 1 For every nonempty proper subset A of X, $|A| < |R(A)|$. We pick an arbitrary vertex x_0 in X and match it with a vertex y_0 in Y. This is possible because x_0 must be adjacent to more than one vertex in Y. Let G' denote the graph obtained from G by removing the vertices x_0 and y_0 and all the edges that are incident with them. In the bipartite graph G', because the vertices in every subset A of $X - \{x_0\}$ are adjacent to $|A|$ or more vertices in $Y - \{y_0\}$, according to the induction hypothesis there is a complete matching of $X - \{x_0\}$ into $Y - \{y_0\}$. Therefore, there is a complete matching of X into Y in the graph G.

Case 2 There is a nonempty proper subset A_0 of X such that $|A_0| = |R(A_0)|$. Let G' denote the subgraph of G containing the set of vertices A_0, the set of vertices $R(A_0)$ that are adjacent to the vertices in A_0, and all the edges joining the vertices in these two sets of vertices. Because the vertices in every subset A of A_0 are adjacent to $|A|$ or more vertices in $R(A_0)$, according to the induction hypothesis there is a complete matching of A_0 into $R(A_0)$ in the graph G'. Let G'' denote the subgraph of G containing the sets of vertices $X - A_0$ and $Y - R(A_0)$ and the edges joining the vertices in these two sets. We claim that there is a complete matching of $X - A_0$ into $Y - R(A_0)$ in the graph G''. Suppose that this is not the case. According to the induction hypothesis, there is a subset of vertices A_1 in $X - A_0$ that are adjacent to less than $|A_1|$ vertices in $Y - R(A_0)$ in the graph G''. However, this means that in the graph G the vertices in the set $A_0 \cup A_1$ are then adjacent to less than $|A_0 \cup A_1|$ vertices, which is a contradiction to the assumption that $|A| \leq |R(A)|$ for every subset A of X. Therefore, there is a complete matching of $X - A_0$ into $Y - R(A_0)$ in the graph G''. It follows that there is a complete matching of X into Y in the graph G. ■

Corollary 11-1.1 below gives a sufficient condition on the existence of a complete matching in a special case.

Corollary 11-1.1 *In a bipartite graph, there exists a complete matching of X into Y if every vertex in X is adjacent to k or more vertices in Y and if every vertex in Y is adjacent to k or less vertices in X.*

Proof For a subset A of X, there are $k|A|$ or more edges incident with the vertices in A. Since these edges must be incident with $|A|$ or more vertices in Y, we have $|A| \leq |R(A)|$. ■

Example 11-1 A set of words, $\{ace,bc,dab,df,fe\}$, is to be transmitted as messages. We want to investigate the possibility of representing each word by one of the letters in the word such that the words will be represented uniquely. If such a representation is possible, we can transmit a single letter instead of a complete word for a message we want to send. We define a bipartite graph G in which the set X is the set of the five words and the set Y is the set of the six letters, $\{a,b,c,d,e,f\}$. There is an edge joining a word in X and a letter in Y if and only if the letter appears in the word. The graph G is shown in Fig. 11-3*a*. The problem is to determine the existence

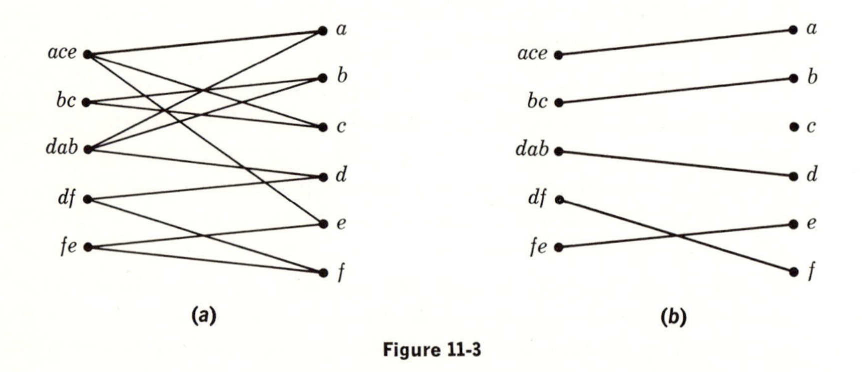

Figure 11-3

of a complete matching of X into Y. As illustrated in Fig. 11-3*b*, a complete matching, and therefore a unique representation of the words by single letters, does exist. (In Sec. 11-4, it will be shown that the method of constructing a maximal flow in a transport network presented in Sec. 10-2 is directly applicable to finding a complete matching in a bipartite graph.) ■

Example 11-2 A *doubly stochastic matrix* is a square matrix in which the following conditions are satisfied:

1. The entries are all nonnegative.
2. The sum of the entries in each row is equal to 1.
3. The sum of the entries in each column is equal to 1.

A *permutation matrix* is a square matrix in which the following conditions are satisfied:

1. There is exactly one entry that is equal to 1 in each row.
2. There is exactly one entry that is equal to 1 in each column.
3. All the other entries are 0's.

Given an $n \times n$ doubly stochastic matrix D, we want to show that D can be expressed as

$$D = c_1P_1 + c_2P_2 + \cdots + c_kP_k$$

where $P_1, P_2, \ldots, P_k$ are $n \times n$ permutation matrices, $c_1, c_2, \ldots, c_k$ are positive numbers, and $c_1 + c_2 + \cdots + c_k = 1$. For instance, the doubly stochastic matrix

$$\begin{bmatrix} 0 & 0.5 & 0.5 \\ 0.8 & 0.2 & 0 \\ 0.2 & 0.3 & 0.5 \end{bmatrix}$$

can be decomposed as

$$0.5\begin{bmatrix} 0 & 1 & 0 \\ 1 & 0 & 0 \\ 0 & 0 & 1 \end{bmatrix} + 0.2\begin{bmatrix} 0 & 0 & 1 \\ 0 & 1 & 0 \\ 1 & 0 & 0 \end{bmatrix} + 0.3\begin{bmatrix} 0 & 0 & 1 \\ 1 & 0 & 0 \\ 0 & 1 & 0 \end{bmatrix}$$

Let us define a bipartite graph that has the vertices $x_1, x_2, \ldots, x_n$ in X, corresponding to the rows of D, and the vertices $y_1, y_2, \ldots, y_n$ in Y, corresponding to the columns of D. In the bipartite graph, there is an edge joining x_i and y_j if and only if the entry in the ith row and the jth column of D is nonzero. We claim that there exists a complete matching of X into Y. Suppose this is not the case. According to Theorem 11-1, if there is no complete matching of X into Y, then there is a set of i rows which have nonzero entries only in j columns, while $j < i$. The sum of the entries in these rows is equal to i when we sum them row by row. The sum of entries in these rows is equal to or less than j when we sum them column by column. This is clearly impossible. The situation is shown schematically in Fig. 11-4 where the dots denote nonzero entries and the circles denote zero entries.

The existence of a complete matching of X into Y means that one can select n nonzero entries in D such that there is exactly one of them in each row and exactly one of them in each column. Let c_1 be the smallest of these n nonzero entries. Clearly, we can write

$$D = c_1P_1 + R$$

where P_1 is an $n \times n$ permutation matrix that has 1's in the positions corresponding to the nonzero entries we have selected.

Since the sum of the entries in each row and the sum of the entries in each column are both equal to $1 - c_1$ in R, the argument

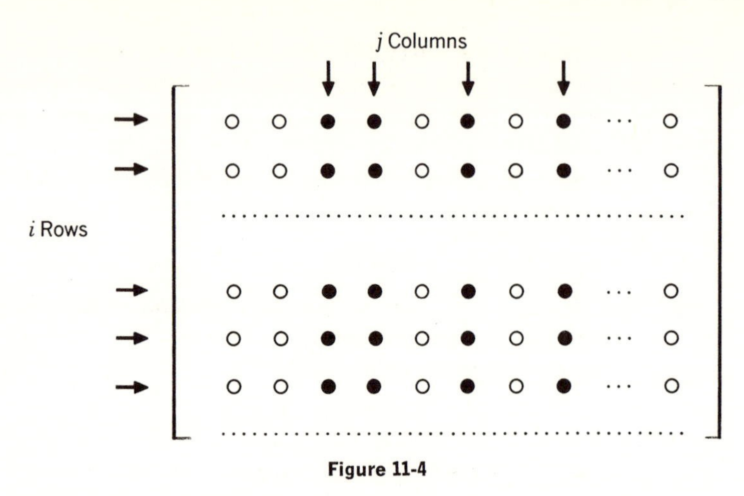

Figure 11-4

above can be repeated to decompose R as a sum of permutation matrices. The repetition terminates when all the entries in R become zeros.

That $c_1 + c_2 + \cdots + c_k = 1$ is a direct consequence of the fact that the sum of the entries in every row of D is equal to 1. ■

Example 11-3 If every girl in the school has k boyfriends and every boy in the school has k girlfriends, is it possible for each girl to go to the dance with one of her boyfriends and for each boy to go to the dance with one of his girlfriends? According to Corollary 11-1.1, the answer to this question is always affirmative regardless of the number of girls and boys in the school. ■

Example 11-4 An $r \times n$ *Latin rectangle* $(r < n)$ is an $r \times n$ matrix that has the numbers $1, 2, 3, \ldots, n$ as entries such that no number appears more than once in the same row or the same column. We shall show that it is always possible to append $n - r$ rows to an $r \times n$ Latin rectangle to form an $n \times n$ Latin square. We define a bipartite graph with $x_1, x_2, \ldots, x_n$ in X corresponding to the n columns in the $r \times n$ Latin rectangle and with $Y = \{1,2, \ldots ,n\}$. In the bipartite graph, there is an edge joining x_i and the number k if and only if k *does not* appear in the ith column of the rectangle. There is a complete matching of X into Y because each x in X is adjacent to exactly $n - r$ numbers in Y (r of the n numbers have appeared in a column), and each number in Y is adjacent to exactly $n - r$ x's in X (each number has not appeared in exactly $n - r$ columns). The existence of a complete matching means that we can assign the n numbers to the n columns such that each column will be assigned a number that has not appeared in the column before. Therefore, the $r \times n$ Latin rectangle can be augmented into an $(r + 1) \times n$ Latin rectangle. This argument can be repeated until an $n \times n$ Latin square is formed. ∎

11-3 MAXIMAL MATCHING

If there is no complete matching of X into Y in a bipartite graph G, then we may wish to match as many of the vertices in X with the vertices in Y as possible. We define a *maximal matching* of X into Y in a bipartite graph to be one that matches the largest number of vertices in X with the vertices in Y.

Let A be a subset of X. The *deficiency* of A, denoted by $\delta(A)$, is defined as $|A| - |R(A)|$. Notice that the deficiency of a set can be a positive number as well as a negative number. We define the *deficiency of a bipartite graph* G, denoted by $\delta(G)$, as the maximal value of the deficiencies of the subsets of X. Since the deficiency of the empty set is always zero, the deficiency of a graph is always larger than or equal to zero. We shall prove two lemmas and a theorem relating the deficiency of a bipartite graph to the maximal number of vertices in X that can be matched into Y.

Lemma 11-1 *Let A_1 and A_2 be two subsets of X.* Then

$$\delta(A_1 \cup A_2) + \delta(A_1 \cap A_2) \geq \delta(A_1) + \delta(A_2)$$

Proof As demonstrated in Fig. 11-5,

$$|A_1 \cup A_2| + |A_1 \cap A_2| = |A_1| + |A_2| \tag{11-1}$$

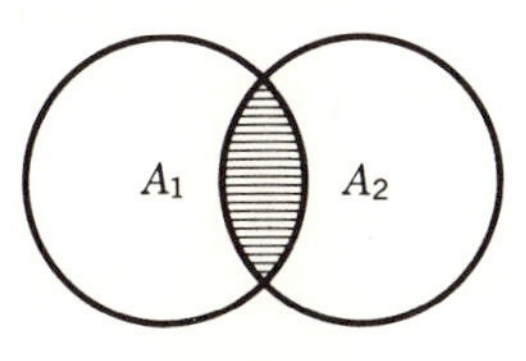

Figure 11-5

Since

$$|R(A_1 \cup A_2)| = |R(A_1) \cup R(A_2)|$$

and

$$|R(A_1 \cap A_2)| \leq |R(A_1) \cap R(A_2)|$$

we have

$$|R(A_1 \cup A_2)| + |R(A_1 \cap A_2)| \leq |R(A_1) \cup R(A_2)| + |R(A_1) \cap R(A_2)| \quad (11\text{-}2)$$

Similar to Eq. (11-1),

$$|R(A_1) \cup R(A_2)| + |R(A_1) \cap R(A_2)| = |R(A_1)| + |R(A_2)|$$

Equation (11-2) can then be rewritten as

$$|R(A_1 \cup A_2)| + |R(A_1 \cap A_2)| \leq |R(A_1)| + |R(A_2)| \quad (11\text{-}3)$$

Subtracting (11-3) from (11-1), we obtain

$$|A_1 \cup A_2| - |R(A_1 \cup A_2)| + |A_1 \cap A_2| - |R(A_1 \cap A_2)| \geq |A_1| - |R(A_1)| + |A_2| - |R(A_2)|$$

That is,

$$\delta(A_1 \cup A_2) + \delta(A_1 \cap A_2) \geq \delta(A_1) + \delta(A_2) \quad \blacksquare$$

Lemma 11-2 *Let A_1 and A_2 be two subsets of X which are such that $\delta(A_1) = \delta(A_2) = \delta(G)$. Then $\delta(A_1 \cap A_2)$ is also equal to $\delta(G)$.*

Proof According to Lemma 11-1,

$$\delta(A_1 \cup A_2) + \delta(A_1 \cap A_2) \geq \delta(A_1) + \delta(A_2) = 2\delta(G)$$

It follows that

$$\delta(A_1 \cup A_2) = \delta(A_1 \cap A_2) = \delta(G)$$

since neither $\delta(A_1 \cup A_2)$ nor $\delta(A_1 \cap A_2)$ can be larger than $\delta(G)$. $\blacksquare$

Theorem 11-2 *In a bipartite graph G the maximal number of vertices in X that can be matched into Y is equal to $|X| - \delta(G)$.*

Proof For the case $\delta(G) = 0$, the theorem is reduced to Theorem 11-1.

For the case $\delta(G) > 0$, let $A_1, A_2, \ldots, A_k$ be all the subsets of X, the deficiencies of which are equal to $\delta(G)$. According to Lemma 11-2, $\delta(A_1 \cap A_2 \cap \cdots \cap A_k) = \delta(G)$, which means that the set $A_1 \cap A_2 \cap \cdots \cap A_k$ is nonempty. Let A_0 denote the set $A_1 \cap A_2 \cap \cdots \cap A_k$. Let x_0 be a vertex in the set A_0, and let G' be the remaining subgraph after x_0 and all the edges incident with it are deleted from G. Since $X - \{x_0\}$ contains no subset with a deficiency equal to $\delta(G)$, we conclude that $\delta(G') < \delta(G)$.

Let A_0' denote the set $A_0 - \{x_0\}$. We have then

$$\delta(A_0') = |A_0'| - |R(A_0')| = |A_0| - 1 - |R(A_0')|$$

Since $\delta(A_0')$ is less than $\delta(G)$, which is equal to $|A_0| - |R(A_0)|$,

$$|A_0| - 1 - |R(A_0')| < |A_0| - |R(A_0)|$$

that is,

$$|R(A_0')| + 1 > |R(A_0)|$$

or

$$|R(A_0')| \geq |R(A_0)| \tag{11-4}$$

On the other hand, since A_0' is a subset of A_0,

$$|R(A_0)| \geq |R(A_0')| \tag{11-5}$$

Combining (11-4) and (11-5), we obtain

$$|R(A_0)| = |R(A_0')|$$

It follows that

$$\delta(A_0') = |A_0| - 1 - |R(A_0)| = \delta(G) - 1$$

Since $\delta(G') < \delta(G)$, we conclude that the deficiency of the graph G' is equal to $\delta(G) - 1$.

Repeating this argument, we see that after the deletion of $\delta(G)$ vertices from the set X, together with all edges incident with these vertices, the deficiency of the remaining subgraph becomes zero and there will be a complete matching in this subgraph. ■

As an example, the deficiency of the graph in Fig. 11-6*a* is equal to 2 because for the subset $A = \{x_3, x_4, x_5\}$, $R(A) = \{y_2\}$. After the deletion of the vertices x_3 and x_4, together with the edges incident with them, there is a complete matching in the remaining subgraph shown in Fig. 11-6*b*.

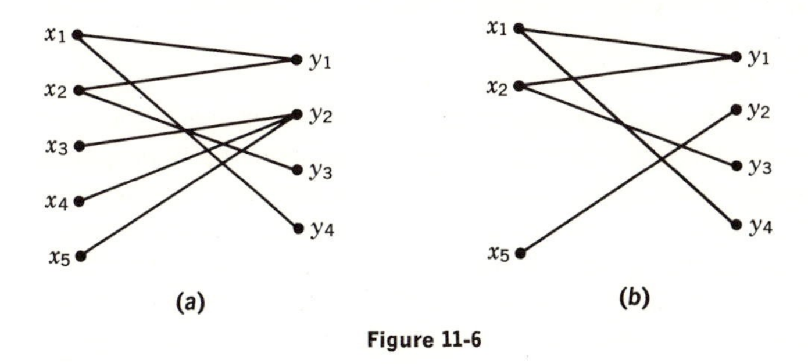

Figure 11-6

Example 11-5 A telephone switching network was built to route phone calls through incoming lines to outgoing trunks. There are 60 incoming lines that are divided into three groups I, II, and III.

There are 20 lines in each group. A group-I line is connected in such a way that it can be switched to one of eight outgoing trunks. A group-II line can be switched to one of four outgoing trunks, and a group-III line can be switched to one of two outgoing trunks. There are 48 outgoing trunks. Moreover, engineering considerations limit the number of incoming lines that can be switched to an outgoing trunk to six or less. Without any further knowledge about the switching network, we want to know at least how many calls will be routed to outgoing trunks when there are calls at all of the 60 lines. This is a problem of finding a maximal matching of the 60 incoming lines into the 48 outgoing trunks. Corresponding to the switching network let us define a bipartite graph in which the set X is the set of the 60 incoming lines and the set Y is the set of the 48 outgoing trunks. There is an edge between an incoming line and an outgoing trunk if the incoming line can be switched to the outgoing trunk. We wish to establish an upper bound on the value of the deficiency of the bipartite graph. The deficiency of a subset A of X containing p group-I lines, q group-II lines, and r group-III lines is

$$\delta(A) = |A| - |R(A)| = p + q + r - |R(A)|$$

Since

$$|R(A)| \geq \frac{8p + 4q + 2r}{6}$$

we have

$$\delta(A) \leq (p + q + r) - \frac{8p + 4q + 2r}{6} = -\frac{p}{3} + \frac{q}{3} + \frac{2r}{3}$$

It follows that the maximal value of $\delta(A)$ for all subsets A of X occurs at $p = 0$, $q = 20$, and $r = 20$. Therefore,

$$\delta(G) \leq 20$$

We conclude that no matter how the switching network was built, at least 40 of the 60 incoming lines will be connected to outgoing trunks. This result actually confirms our intuition that the most ineffective way to build the switching network is to try to accommodate all of the 40 group-II and group-III incoming lines with 20 outgoing trunks. In this case, it is clear that only 20 of these 40 lines will be routed when there are calls on all 40 incoming lines. ■

Example 11-6 A (0,1)-matrix is a matrix with 0's and 1's as its entries. A line of a matrix is either a row or a column of the matrix. We want to prove the following result, which is known as the König-

Egerváry theorem: The minimal number of lines containing all the 1's in a (0,1)-matrix is equal to the maximal number of 1's, no two of which are in one line of the matrix.

For a given $m \times n$ (0,1)-matrix W, let us construct a bipartite graph G with $x_1, x_2, \ldots, x_m$ in X corresponding to the m rows of W and with $y_1, y_2, \ldots, y_n$ in Y corresponding to the n columns of W. In the graph G, there is an edge joining x_i and y_j if and only if the entry in the ith row and the jth column of W is a 1. Since a matching of X into Y corresponds to a selection of the 1's in the matrix such that no two of the 1's are in one line, according to Theorem 11-2, the maximal number of 1's in the matrix, no two of which are in one line, is equal to $|X| - \delta(G)$.

Clearly, the minimal number of lines containing all the 1's in W is larger than or equal to the maximal number of 1's, no two of which are in one line. On the other hand, let A be a subset of X such that $\delta(A)$ is equal to $\delta(G)$. We observe that the rows corresponding to the vertices in $X - A$ together with the columns corresponding to the vertices in $R(A)$ contain all the 1's in W, as illustrated in Fig. 11-7.

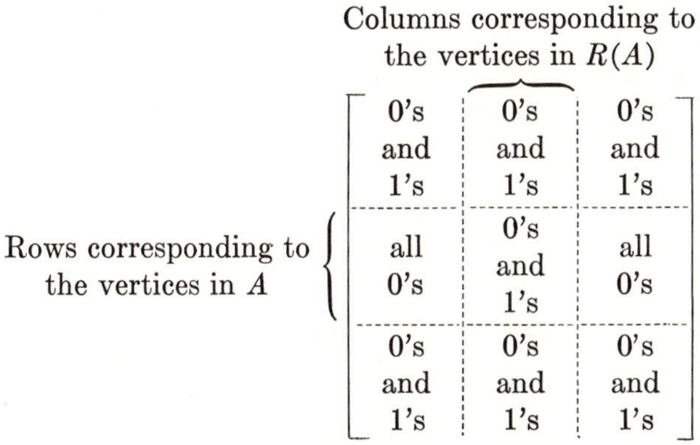

		Columns corresponding to the vertices in $R(A)$	
	0's and 1's	0's and 1's	0's and 1's
Rows corresponding to the vertices in A	all 0's	0's and 1's	all 0's
	0's and 1's	0's and 1's	0's and 1's

Figure 11-7

The total number of these rows and columns is equal to

$$|X - A| + |R(A)| = |X| - |A| + |R(A)| = |X| - \delta(G)$$

It follows that the minimal number of lines containing all the 1's in W must be less than or equal to $|X| - \delta(G)$. Therefore, we conclude that the minimal number of lines containing all the 1's in W is equal to the maximal number of 1's, no two of which are in one line, $|X| - \delta(G)$. ■

★11-4 AN ALTERNATIVE APPROACH

It is interesting to observe that Theorems 11-1 and 11-2 can be proved using the max-flow min-cut theorem for transport networks. Such an alternative point of view not only sheds more light on the problem but also suggests a procedure for finding a complete matching or a maximal matching in a bipartite graph.

An alternative proof of Theorem 11-1 In a bipartite graph G, let $x_1, x_2, \ldots, x_m$ be the vertices in X, and let $y_1, y_2, \ldots, y_n$ be the vertices in Y as shown in Fig. 11-8*a*. We construct a transport

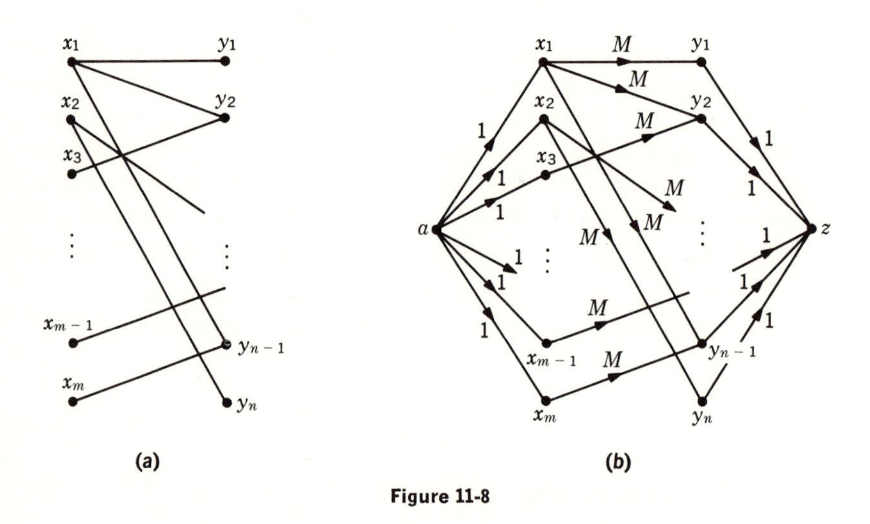

Figure 11-8

network by adding a source a and $|X|$ edges, one from a to each of the vertices in X, and by adding a sink z and $|Y|$ edges, one from each of the vertices in Y to z. Let the capacity of each of these edges be 1. Also let the edges between X and Y be directed from X to Y with their capacities equal to M where M is an integer larger than $|X|$. Such a transport network is shown in Fig. 11-8*b*. It is clear that there is a complete matching of X into Y if and only if there exists a maximal flow in the transport network, the value of which is equal to $|X|$. (Corollary 10-2.1 assures that if the value of the maximal flows in the network is $|X|$, there exists a maximal flow in which the flow in each edge is either 0 or 1.) We shall prove that there is a maximal flow of value $|X|$ in the transport network if and only if $|A| \leq |R(A)|$ for every subset A of X.

Suppose that there is a subset A of X which is such that $|A| > |R(A)|$. Consider the cut $(P,\bar{P})$ where the set P contains the source a, the vertices in A, and the vertices in $R(A)$ as shown in Fig. 11-9. Since there is no edge joining the vertices in A and the

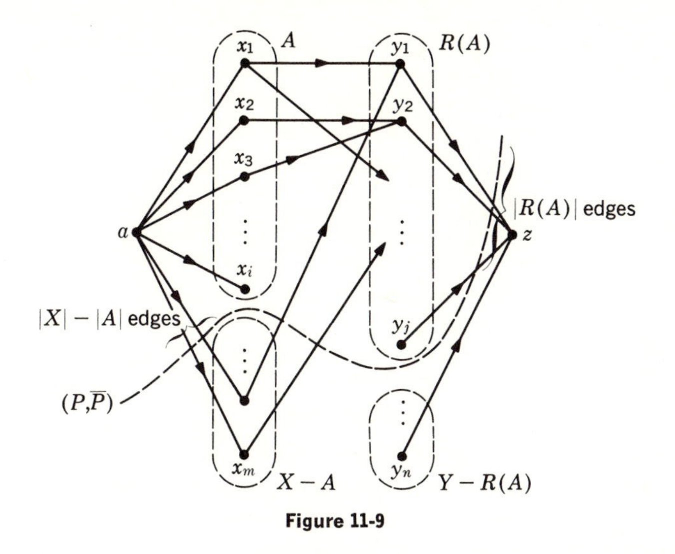

Figure 11-9

vertices in $Y - R(A)$, the capacity of the cut, $\alpha(P,\bar{P})$, is $|X| - |A| + |R(A)|$, which is less than $|X|$. According to the max-flow min-cut theorem (Theorem 10-2), there cannot exist a flow of value $|X|$ in the network. Therefore, there is no complete matching of X into Y in the bipartite graph G.

Suppose that the condition $|A| \leq |R(A)|$ is satisfied for every subset A of X. We want to show that the capacity of any cut $(P,\bar{P})$ in the transport network is equal to or larger than $|X|$. Suppose that the set P contains the source a, the vertices in a subset A of X, and the vertices in a subset B of Y as shown in Fig. 11-10. (The subsets A and B can be empty.) If there is any edge joining the vertices in A and the vertices in $Y - B$, then $\alpha(P,\bar{P}) \geq M$. If there is no edge joining the vertices in A and the vertices in $Y - B$, then $\alpha(P,\bar{P}) = |X| - |A| + |B|$. However, because $|B| \geq |R(A)|$,

$$\alpha(P,\bar{P}) \geq |X| - |A| + |R(A)| \geq |X|$$

Since the capacity of any cut $(P,\bar{P})$ is equal to or larger than $|X|$, it follows that there exists a complete matching of X into Y in the bipartite graph G. ■

We have not only obtained a necessary and sufficient condition on the existence of a complete matching in a bipartite graph, but also, in the

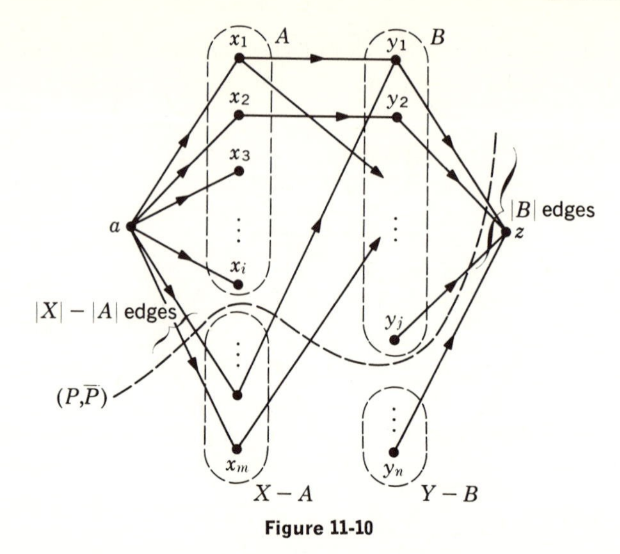

Figure 11-10

proof, we have shown that the problem of finding a complete matching can be reduced to that of constructing a maximal flow in a transport network. It follows that the labeling procedure presented in Chap. 10 can be applied to finding a complete matching in a bipartite graph.

An alternative proof of Theorem 11-2 We follow the alternative proof of Theorem 11-1 and construct a transport network corresponding to the bipartite graph G as illustrated in Fig. 11-8.

We show first that there is a cut in the transport network the capacity of which is equal to $|X| - \delta(G)$. Let A be a subset of X such that $\delta(A) = \delta(G)$. Let $(P,\bar{P})$ be a cut where the set P contains the source a, the vertices in A, and the vertices in $R(A)$ as shown in Fig. 11-9. Since there is no edge joining the vertices in A and the vertices in $Y - R(A)$,

$$\alpha(P,\bar{P}) = |X| - |A| + |R(A)| = |X| - \delta(G)$$

We show next that the capacity of any cut $(P,\bar{P})$ in the transport network is larger than or equal to $X - \delta(G)$. Suppose that the set P contains the source a, the vertices in a subset A of X, and the vertices in a subset B of Y as shown in Fig. 11-10. (The subsets A and B can be empty.) If there is any edge joining the vertices in A and the vertices in $Y - B$, then $\alpha(P,\bar{P}) \geq M$. If there is no edge joining the vertices in A and the vertices in $Y - B$, then $\alpha(P,\bar{P}) = |X| - |A| + |B|$. However, because $|B| \geq |R(A)|$,

$$\alpha(P,\bar{P}) \geq |X| - |A| + |R(A)| \geq |X| - \delta(G)$$

Therefore, we conclude that there exists a maximal flow in the transport network the value of which is equal to $|X| - \delta(G)$. ■

11-5 SUMMARY AND REFERENCES

In addition to its application to the solution of the personnel assignment problem, the theory of matching has some very interesting combinatorial implications and far-reaching consequences. There is an impressive literature on the subject. See, for example, the survey paper by Mirsky and Perfect [5].

The notion of matching in a bipartite graph can be extended to that in an arbitrary undirected graph. In an undirected graph G having no loops, a subset of the edges defines a matching in G if no two edges in the subset are incident with the same vertex. In other words, a matching is a pairing of a subset of the vertices of G. A matching is said to be perfect if all the vertices of G are paired.

On the topic of matching in a bipartite graph, see Chaps. 9 and 10 of Berge [1], Chap. 7 of Ore [6], and Chap. 5 of Ryser [7]. An algorithm for finding a maximal matching in a bipartite graph which is known as the Hungarian method can be found in Berge [1] as well as in the original papers by Egerváry [3] and by Kuhn [4] (see Prob. 11-11). For the topic of matching in an arbitrary graph, see Tutte [8], Edmonds [2], and Chap. 18 of Berge [1].

1. Berge, C.: "The Theory of Graphs and Its Applications," John Wiley & Sons, Inc., New York, 1962.
2. Edmonds, J.: Paths, Trees, and Flowers, *Nat. Bur. St. Rept.* (1964).
3. Egerváry, E.: On Combinatorial Properties of Matrices, translated by H. W. Kuhn, *Office Naval Res. Logist. Project Rept., Dept. of Math., Princeton University*, 1953.
4. Kuhn, H. W.: The Hungarian Method for the Assignment Problem, *Naval Res. Logist. Quart.*, **2**:83–97 (1955).
5. Mirsky, L., and H. Perfect: Systems of Representatives, *J. Math. Anal. Appl.*, **3**:520–568 (1966).
6. Ore, O.: "Theory of Graphs," American Mathematical Society, Providence, R.I., 1962.
7. Ryser, H. J.: "Combinatorial Mathematics," published by The Mathematical Association of America, distributed by John Wiley & Sons, Inc., New York, 1963.
8. Tutte, W. T.: The Factorization of Linear Graphs, *J. London Math. Soc.*, **22**:107–111 (1947).

PROBLEMS

11-1. Six senators A, B, C, D, E, and F are members of five committees. The memberships of the committees are $\{B,C,D\}$, $\{A,E,F\}$, $\{A,B,E,F\}$, $\{A,B,D,F\}$, and $\{A,B,C\}$.

The activities of each committee are to be reviewed by a senator who is not on the committee. Can five distinct senators be selected? How?

11-2. There are rs couples at a dance. The men are divided into r groups, with s men in each group, according to their ages. The women are also divided into r groups, with s women in each group, according to their heights. Show that r couples can be selected such that every age group and every height group will be represented.

11-3. (*a*) Let π_1 and π_2 be two partitions on a set of m elements, both of which contain exactly r disjoint subsets. State a necessary and sufficient condition for the possibility of selecting r of the m elements such that the r disjoint subsets in π_1 as well as the r disjoint subsets in π_2 are represented.

(*b*) The numbers 1, 2, . . . , 20 are partitioned into four disjoint subsets according to their remainders when they are divided by 4. They are also partitioned into four disjoint subsets according to the number of prime factors they contain, namely, $\{\{1,2,3,5,7,11,13,17,19\}, \{4,6,9,10,14,15\}, \{8,12,18,20\}, \{16\}\}$. Is it possible to select four numbers such that there is a representative for each possible remainder when divided by 4 and a representative for each possible number of prime factors?

11-4. A factor of an undirected graph (containing no loops) is defined as a circuit or an edge-disjoint union of circuits that contains all vertices of the graph. (If a factor is a circuit, it is a Hamiltonian circuit.) Show that a necessary and sufficient condition for a graph to possess a factor is that every subset of vertices A has the property $|A| \leq |R(A)|$ where $R(A)$ is the set of vertices that are adjacent to one or more vertices in A.

11-5. A group of p boys $x_1, x_2, \ldots, x_p$ are ranked according to their unpopularities; that is, x_1 knows the same number or fewer girls than x_2 does, x_2 knows the same number or fewer girls than x_3 does, and so on. A group of q girls $y_1, y_2, \ldots, y_q$ are ranked according to their popularities; that is, y_1 knows the same number or more boys than y_2 does, y_2 knows the same number or more boys than y_3 does, and so on. Show that the p boys can be matched with p of the q girls if the number of girls that the least popular k boys know exceeds the number of boys that the most popular $k - 1$ girls know for $1 \leq k \leq p$.

11-6. Let G be a bipartite graph with X and Y being the two disjoint sets of vertices. Show that $\beta(G) = |Y| + \delta(G)$ where $\beta(G)$ is the independence number and $\delta(G)$ is the deficiency of the graph G.

11-7. In a bipartite graph G with X and Y being the two disjoint set of vertices, let A be a subset of X such that $\delta(A) = \delta(G)$. Show that $\delta(X - A) \leq 0$.

11-8. Let G be a bipartite graph with X and Y being the two disjoint sets of vertices. If there are four or more edges incident from each vertex in X and there are five or less edges incident into each vertex in Y, show that $\delta(G) \leq 2$ for $|X| \leq 10$.

11-9. Let G be a directed graph with n vertices for which the incoming and outgoing degree of each vertex is k. Show that G can be decomposed into edge-disjoint circuits.

Hint: In the $n \times n$ (0,1)-matrix $A = [a_{ij}]$ defined as

$$a_{ij} = \begin{cases} 1 & \text{if } (v_i,v_j) \in G \\ 0 & \text{otherwise} \end{cases}$$

each row and column contains k 1's. Show that A can be decomposed as a sum of permutation matrices.

11-10. Let A be an $m \times n$ (0,1)-matrix ($m \leq n$). If each row in A has exactly k 1's and each column in A has k or less 1's, show that

$$A = P_1 + P_2 + \cdots + P_k$$

where the P's are $m \times n$ (0,1)-matrices that have a 1 in each row and no more than one 1 in each column.

11-11. In this problem, a method of constructing a maximal matching in a bipartite graph, known as the *Hungarian method,* is studied. Let G be a bipartite graph, and let W denote the set of edges in a matching from X to Y. A vertex (in X or Y) is said to be unsaturated if no edge incident with it is included in W. A path in G is said to be an alternate path if it contains edges in W and $E - W$ alternately, where E denotes the set of edges in G.

The Hungarian method is based on the following result: A matching is maximal if and only if there is no alternate path connecting two distinct unsaturated vertices.

(*a*) Show that a matching is not maximal if there exists such a path by showing that for each such path, the number of vertices in X matched into Y can be increased by 1.

(*b*) Label some of the vertices of G as follows:

1. Every unsaturated vertex in X is labeled.
2. Any vertex (in X or Y) connected to an unsaturated vertex in X by an alternate path is labeled.

What is the deficiency of the subset of X that contains all the labeled vertices in X? Show that a matching is maximal if no alternate path connecting two distinct unsaturated vertices exists.

(*c*) Suggest a construction procedure for finding a maximal matching in a bipartite graph, based on the above result.

Chapter 12
Linear Programming

12-1 INTRODUCTION

We shall first motivate the subject of linear programming with some examples illustrating the class of problems to which this technique is applicable. Consider a machine shop with three types of machines, A, B, and C. There are 20 type-A machines, 30 type-B machines, and 15 type-C machines, and these machines are used to manufacture four different kinds of products 1, 2, 3, and 4. To manufacture 1 lb of product 1, we would use a type-A machine for 2 hr, a type-B machine for 0.5 hr, and a type-C machine for 1.5 hr. To manufacture 1 lb of product 2, we would use a type-A machine for 2 hr, a type-B machine for 2 hr, and a type-C machine for 1 hr. To manufacture 1 lb of product 3, we would use a type-A machine for 0.5 hr, a type-B machine for 1 hr, and a type-C machine for 3 hr. To manufacture 1 lb of product 4, we would use a type-A machine for 1.5 hr, a type-B machine for 2 hr, and a type-C machine for 1.5 hr. The profit for manufacturing (and, of course, subsequently selling) 1 lb of each of the four products is \$3.5, \$4.2, \$6.5, and \$3.8, respectively. Suppose that each machine can be operated for no more than 60 hr in a week. How many pounds of these products should be manufactured every week for the largest total profit?

Let x_1, x_2, x_3, and x_4 denote the numbers of pounds of the four products that are to be manufactured. We want to choose the values of x_1, x_2, x_3, and x_4 so that the total profit

$$3.5x_1 + 4.2x_2 + 6.5x_3 + 3.8x_4$$

is maximized. Because of the availability of the machines, the total operating time for each type of machines cannot exceed the limits 60×20 (type-A machines), 60×30 (type-B machines), and 60×15 (type-C machines) hr, respectively. Since it takes a type-A machine 2 hr to manufacture each pound of product 1, and 2, 0.5, and 1.5 hr to manufacture each pound of products 2, 3, and 4, respectively, we have the constraint

$$2x_1 + 2x_2 + 0.5x_3 + 1.5x_4 \leq 1{,}200$$

Similarly, for the type-B and type-C machines, we have the constraints

$$0.5x_1 + 2x_2 + x_3 + 2x_4 \leq 1{,}800$$

and

$$1.5x_1 + x_2 + 3x_3 + 1.5x_4 \leq 900$$

Furthermore, since no negative amount of a product can be manufactured, we also have the conditions

$$x_1 \geq 0 \qquad x_2 \geq 0 \qquad x_3 \geq 0 \qquad x_4 \geq 0$$

Linear programming deals with problems in which linear functions† are to be optimized (maximized or minimized) subject to constraints specified by linear inequalities and linear equations‡ and to the condition that all the variables must assume nonnegative values. The linear function to be optimized is called the *objective function.* The linear inequalities and equations will be referred to as the *linear constraints.* The condition that the variables must assume nonnegative values will be referred to as the *nonnegative condition.* Thus, the general formulation of linear programming problems is as follows: To optimize the objective function

$$C = c_1x_1 + c_2x_2 + \cdots + c_rx_r$$

† A function of the form $c_1x_1 + c_2x_2 + \cdots + c_rx_r$ is a linear function of the r variables $x_1, x_2, \ldots, x_r$, where the c's are constants.

‡ Again, a linear inequality (or equation) is an inequality (or equation) of the form $c_1x_1 + c_2x_2 + \cdots + c_rx_r\{\leq, =, \geq\}b$, where the c's and b are constants.

subject to the linear constraints

$$a_{11}x_1 + a_{12}x_2 + \cdots + a_{1r}x_r \{\leq, =, \geq\} b_1$$
$$a_{21}x_1 + a_{22}x_2 + \cdots + a_{2r}x_r \{\leq, =, \geq\} b_2$$
$$\cdots\cdots\cdots\cdots\cdots\cdots$$
$$a_{i1}x_1 + a_{i2}x_2 + \cdots + a_{ir}x_r \{\leq, =, \geq\} b_i$$
$$\cdots\cdots\cdots\cdots\cdots\cdots$$
$$a_{m1}x_1 + a_{m2}x_2 + \cdots + a_{mr}x_r \{\leq, =, \geq\} b_m$$

and to the nonnegative condition

$$x_1 \geq 0, x_2 \geq 0, x_3 \geq 0, \ldots, x_r \geq 0$$

There are many problems in engineering, management, and social sciences that can be formulated as linear programming problems. For example, in the problem of finding a maximal flow in a transport network discussed in Sec. 10-2, we wish to maximize the linear function

$$\sum_{\text{all } i} \phi(i,z)$$

subject to the linear constraints

$$\alpha(i,j) - \phi(i,j) \geq 0 \qquad \text{for all } i \text{ and } j$$
$$\sum_{\text{all } i} \phi(i,j) - \sum_{\text{all } k} \phi(j,k) = 0 \qquad \text{for all } j \text{ except the source } a \text{ and the sink } z$$

and the nonnegative condition

$$\phi(i,j) \geq 0 \qquad \text{for all } i \text{ and } j$$

As another example, consider a racetrack betting problem. Suppose that r horses have been entered in a race and that we are offered the odds a_i on horse i for $1 \leq i \leq r$; that is, we shall receive a_i dollars for every dollar we bet on horse i if it should win. Ignoring probabilistic considerations, we want to distribute the total amount of money we have (say a dollar) over the wagers so that our return will be as large as possible, assuming that the outcome of the race will be the most unfavorable one. Let $x_1, x_2, \ldots, x_r$ be the stakes we place on horses $1, 2, \ldots, r$, respectively. Clearly, there are the constraints

$$x_1 + x_2 + \cdots + x_r = 1$$

and

$$x_1 \geq 0, x_2 \geq 0, \ldots, x_r \geq 0$$

Our return will be

$$a_1x_1 - x_2 - x_3 - \cdots - x_r$$

or

$$-x_1 + a_2x_2 - x_3 - \cdots - x_r$$
$$\cdots\cdots\cdots\cdots\cdots$$

or

$$-x_1 - x_2 - x_3 - \cdots + a_rx_r$$

depending on the outcome of the race. Therefore, what we are to maximize is the smallest of these r quantities.

Let v denote the smallest of the r possible returns. We can write

$$\begin{aligned}
&a_1x_1 - x_2 - x_3 - \cdots - x_r \geq v \\
&-x_1 + a_2x_2 - x_3 - \cdots - x_r \geq v \\
&\cdots\cdots\cdots\cdots\cdots\cdots \\
&-x_1 - x_2 - x_3 - \cdots + a_rx_r \geq v
\end{aligned}$$

Let $y_1, y_2, \ldots, y_r$ denote the differences between v and the quantities on the left-hand sides of these r inequalities. Then

$$\begin{aligned}
&a_1x_1 - x_2 - x_3 - \cdots - x_r - y_1 = v \\
&-x_1 + a_2x_2 - x_3 - \cdots - x_r - y_2 = v \\
&\cdots\cdots\cdots\cdots\cdots\cdots\cdots \\
&-x_1 - x_2 - x_3 - \cdots + a_rx_r - y_r = v
\end{aligned}$$

It is clear that all the y's are nonnegative, that is,

$$y_1 \geq 0, y_2 \geq 0, \ldots, y_r \geq 0$$

We have now a typical linear programming problem, in which the function

$$C = a_1x_1 - x_2 - x_3 - \cdots - x_r - y_1$$

is to be maximized, subject to the linear constraints

$$\begin{aligned}
&a_1x_1 - x_2 - x_3 - \cdots - x_r - y_1 = -x_1 + a_2x_2 - x_3 - \cdots \\
&\qquad\qquad\qquad\qquad\qquad\qquad - x_r - y_2 \\
&a_1x_1 - x_2 - x_3 - \cdots - x_r - y_1 = -x_1 - x_2 + a_3x_3 - \cdots \\
&\qquad\qquad\qquad\qquad\qquad\qquad - x_r - y_3 \\
&\cdots\cdots\cdots\cdots\cdots\cdots\cdots\cdots\cdots\cdots \\
&a_1x_1 - x_2 - x_3 - \cdots - x_r - y_1 = -x_1 - x_2 - x_3 - \cdots \\
&\qquad\qquad\qquad\qquad\qquad\qquad + a_rx_r - y_r \\
&x_1 + x_2 + x_3 + \cdots + x_r = 1
\end{aligned}$$

together with the nonnegative condition

$$\begin{aligned}
&x_1 \geq 0, x_2 \geq 0, \ldots, x_r \geq 0 \\
&y_1 \geq 0, y_2 \geq 0, \ldots, y_r \geq 0
\end{aligned}$$

12-2 OPTIMAL FEASIBLE SOLUTIONS

In a linear programming problem, a set of the values of the variables $x_1, x_2, \ldots, x_r$ that satisfy the linear constraints and the nonnegative condition is called a *feasible solution.* A feasible solution that optimizes the objective function is called an *optimal feasible solution.*

Consider the example of maximizing the function

$$C = x_1 + x_2$$

subject to the linear constraints

$$\begin{aligned} 4x_1 + 5x_2 &\leq 10 \\ 5x_1 + 2x_2 &\leq 10 \\ 3x_1 + 8x_2 &\leq 12 \end{aligned} \tag{12-1}$$

and to the nonnegative condition

$$\begin{aligned} x_1 &\geq 0 \\ x_2 &\geq 0 \end{aligned} \tag{12-2}$$

In Fig. 12-1*a*, we observe that the straight line $4x_1 + 5x_2 = 10$ divides the x_1-x_2 plane into two half-planes. The points in the shaded half-plane, such as $x_1 = 1$ and $x_2 = 1$, satisfy the inequality $4x_1 + 5x_2 < 10$. The points in the unshaded half-plane, such as $x_1 = 2$ and $x_2 = 1$, satisfy the inequality $4x_1 + 5x_2 > 10$. Finally, the points on the line, such as $x_1 = 5/4$ and $x_2 = 1$, satisfy the equation $4x_1 + 5x_2 = 10$. Similarly, the straight line $5x_1 + 2x_2 = 10$ also divides the x_1-x_2 plane into two half-planes, one containing points that satisfy the inequality $5x_1 + 2x_2 < 10$ and the other one containing points that satisfy the inequality $5x_1 + 2x_2 > 10$. Therefore, the points in the shaded area in Fig. 12-1*b*, which is the intersection of the half-plane containing points that satisfy the inequality $4x_1 + 5x_2 < 10$ and the half-plane containing points that satisfy the inequality $5x_1 + 2x_2 < 10$, satisfy both the inequalities $4x_1 + 5x_2 < 10$ and $5x_1 + 2x_2 < 10$. It follows that the points in the shaded region (including points on the boundary) in Fig. 12-1*c*, which is the intersection of five half-planes and is bounded by the five straight lines

$$\begin{aligned} 4x_1 + 5x_2 &= 10 \\ 5x_1 + 2x_2 &= 10 \\ 3x_1 + 8x_2 &= 12 \\ x_1 &= 0 \\ x_2 &= 0 \end{aligned} \tag{12-3}$$

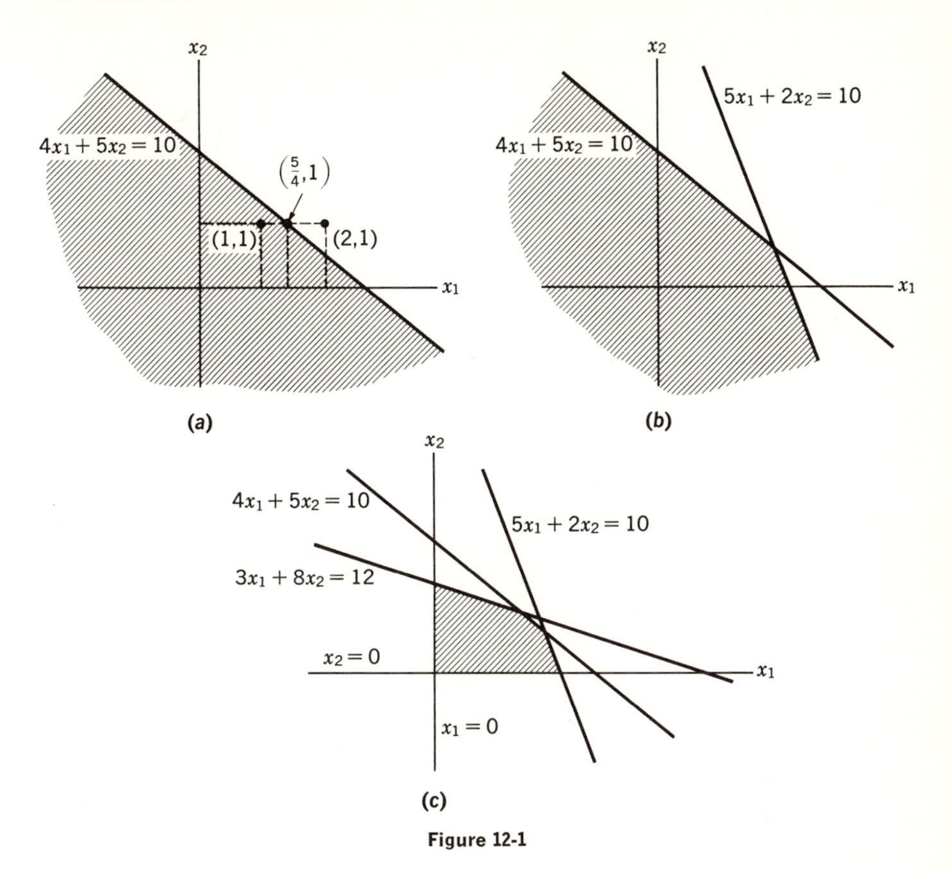

Figure 12-1

satisfy all the constraints in (12-1) and (12-2). In other words, the points in the shaded region are feasible solutions to the linear programming problem. The region that contains all the feasible solutions is called the *feasible region.*

The problem of maximizing the objective function $C = x_1 + x_2$ is that of picking a point in the feasible region at which the value of $x_1 + x_2$ is the largest. This can be done graphically by drawing a family of parallel straight lines of the form $x_1 + x_2 = k$, as shown in Fig. 12-2. Among all the straight lines in the family that intersect the feasible region, there is one for which the value of k is the largest. The intersections of this straight line with the feasible region are then the optimal solutions to the problem. As shown in Fig. 12-2, the straight line $x_1 + x_2 = 40/17$ is tangent to the feasible region at $x_1 = 30/17$ and $x_2 = 10/17$. Therefore, for $x_1 = 30/17$ and $x_2 = 10/17$, the objective function attains its maximal value of $40/17$.

Similarly, we see in Fig. 12-2 that the optimal solution minimizing

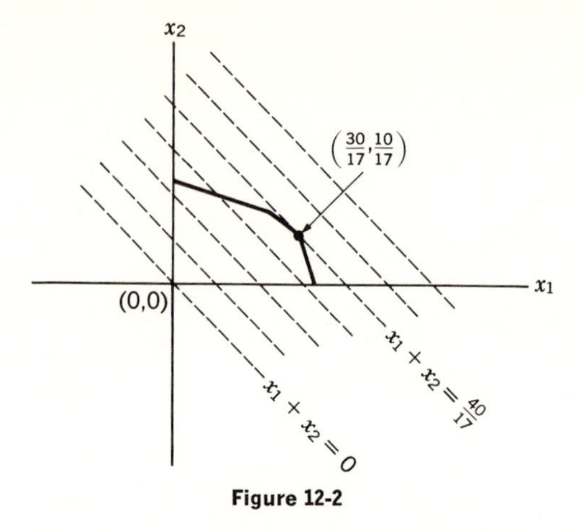

Figure 12-2

the objective function $C = x_1 + x_2$ is $x_1 = 0$ and $x_2 = 0$. When the function $C = x_1 - x_2$ is to be optimized subject to the constraints in (12-1) and (12-2), we draw a family of parallel straight lines of the form $x_1 - x_2 = k$ as shown in Fig. 12-3. At $x_1 = 2$ and $x_2 = 0$, the function

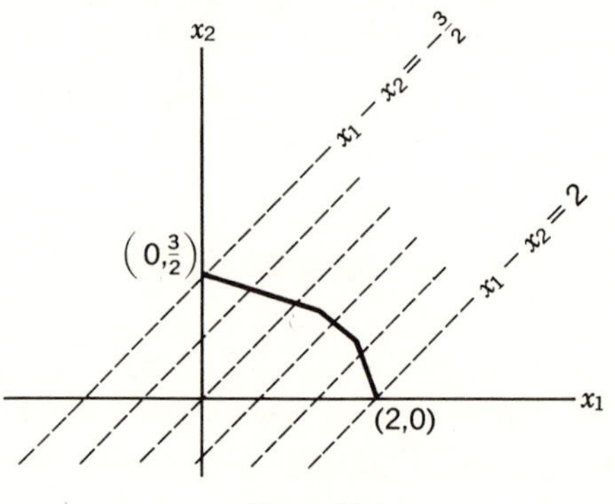

Figure 12-3

$x_1 - x_2$ attains its maximal value of 2; at $x_1 = 0$ and $x_2 = 3/2$, the function $x_1 - x_2$ attains its minimal value of $-3/2$.

It is no sheer coincidence that all the optimal feasible solutions are at the corners of the feasible region in the previous example. Since, for the optimization of an objective function of two variables, each of the linear constraints corresponds to a straight line in the x_1-x_2 plane, it follows that the feasible region is always a convex polygon.† (The convex

† As defined in elementary plane geometry, a convex polygon is a polygon with no internal angle being larger than 180°. An alternative definition is that a polygon is convex if the straight-line segment joining any two interior or boundary points lies either within or along the boundary of the polygon.

polygon can be unbounded, such as the shaded area in Fig. 12-4, or it can be a straight-line segment in the case in which one of the linear constraints is an equality.) Since an optimal feasible solution is at a point where a

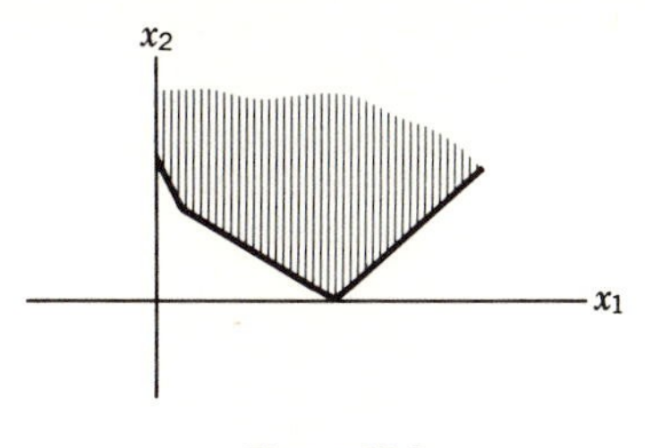

Figure 12-4

member of the family of parallel straight lines corresponding to the objective function is tangent to the feasible region, we conclude that one of the corners of the feasible region will be an optimal feasible solution. We should note that a linear programming problem may have many optimal feasible solutions when the family of straight lines corresponding to the objective function is parallel to one side of the feasible region. However, our statement that one of the corners of the feasible region is an optimal feasible solution is still valid.

For the optimization of linear functions of two variables, the geometric argument is very descriptive and clear. It can also be extended to the general case of optimizing a linear function of r variables. We shall do so now for the purpose of providing a geometric interpretation to the algebraic results that will be proved rigorously in Sec. 12-3 (Theorem 12-2).

In r-dimensional space, a point is identified by its r coordinates; and a linear equation of the form

$$a_{i1}x_1 + a_{i2}x_2 + \cdots + a_{ir}x_r = b_i$$

represents a hyperplane. (A hyperplane is a plane in a higher-dimensional space.) A hyperplane divides the r-dimensional space into two half-spaces. The coordinates of the points in one half-space satisfy the linear inequality

$$a_{i1}x_1 + a_{i2}x_2 + \cdots + a_{ir}x_r > b_i$$

whereas the coordinates of the points in the other half-space satisfy the linear inequality

$$a_{i1}x_1 + a_{i2}x_2 + \cdots + a_{ir}x_r < b_i$$

The coordinates of the points on the hyperplane satisfy the equation

$$a_{i1}x_1 + a_{i2}x_2 + \cdots + a_{ir}x_r = b_i$$

Therefore, in a linear programming problem with r variables and m linear constraints, the feasible region is a convex polyhedron† in r-dimensional space bounded by the m hyperplanes corresponding to the linear constraints and the r hyperplanes corresponding to the nonnegative condition

$$
\begin{aligned}
a_{11}x_1 + a_{12}x_2 + \cdots + a_{1r}x_r &= b_1 \\
a_{21}x_1 + a_{22}x_2 + \cdots + a_{2r}x_r &= b_2 \\
\cdots\cdots\cdots\cdots\cdots\cdots & \\
a_{i1}x_1 + a_{i2}x_2 + \cdots + a_{ir}x_r &= b_i \\
\cdots\cdots\cdots\cdots\cdots\cdots & \\
a_{m1}x_1 + a_{m2}x_2 + \cdots + a_{mr}x_r &= b_m \\
x_1 &= 0 \\
x_2 &= 0 \\
\cdots & \\
x_r &= 0
\end{aligned}
\tag{12-4}
$$

It follows that an optimal feasible solution is at a corner of the polyhedron where a member of the family of hyperplanes of the form

$$c_1x_1 + c_2x_2 + \cdots + c_rx_r = k$$

is tangent to the polyhedron.‡

Similar to the case in which the intersection of two straight lines is a point in two-dimensional space, the intersection of r hyperplanes is a point in r-dimensional space. Given the linear equations of r hyperplanes, we can solve them as a set of simultaneous equations to find the coordinates of their intersection. (Obviously, if this set of equations is not solvable, then the hyperplanes do not intersect at a point. Also, if this set of equations does not have a unique solution, then the hyperplanes intersect at more than one point.) Therefore, we can find the coordinates of the intersections of the boundary hyperplanes by solving sets of r equations from the $m + r$ equations in (12-4). Clearly, there are $\binom{m+r}{r}$ sets of equations corresponding to $\binom{m+r}{r}$ possible intersections.

It should now be clear why our discussion is restricted to a limited

† Again, a polyhedron is convex if the straight-line segment joining any two points in the polyhedron or on the boundary hyperplanes lies entirely in the polyhedron or on the boundary hyperplanes.

‡ This is just an informal way of motivating the algebraic proof to be presented in Sec. 12-3. The reader need not try very hard to visualize the geometric picture.

class of problems, namely, the optimization of linear functions subject to linear constraints. For this class of problems, since an optimal feasible solution is always at a corner of the feasible region, we need at most to examine the value of the objective function at each corner of the feasible region. When the objective function is not a linear function and the feasible region is not a convex polyhedron, an optimal feasible solution, in general, can be at any point within the feasible region. The problem of finding an optimal feasible solution is then considerably more complicated.

For those who are familiar with the method of Lagrange multipliers in differential calculus, the reason that such a method is not useful in the solution of linear programming problems should also be clear now; namely, an optimal feasible solution at a corner of the feasible region is an absolute maximum or minimum, whereas the method of Lagrange multipliers is useful only in the determination of relative maxima and minima. The distinction between a relative maximum and an absolute maximum is illustrated graphically in Fig. 12-5.

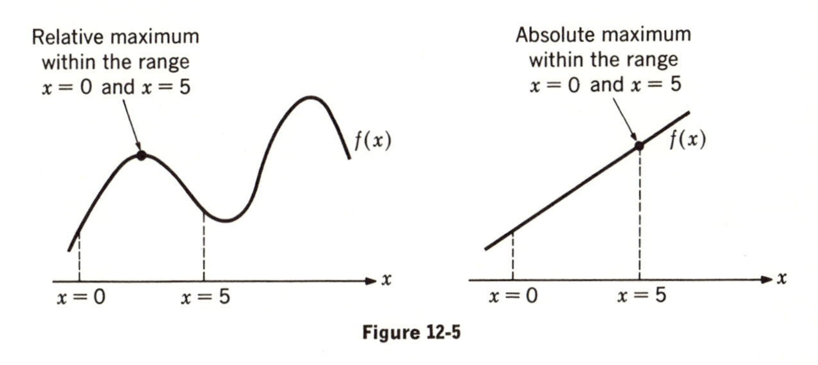

Figure 12-5

12-3 SLACK VARIABLES

Since an optimal solution to a linear programming problem is at one of the corners of the feasible region, it seems that we can always solve the problem by exhaustively examining the values of the objective function at the corners of the feasible region. Unfortunately, the number of corners of the feasible region increases quite rapidly when the number of variables and the number of linear constraints increase. For example, in optimizing an objective function of four variables subject to six linear constraints, since the intersection of every four hyperplanes is possibly a corner of the feasible region, there are $\binom{10}{4} = 210$ intersections to be examined (the feasible region is bounded by 10 hyperplanes, six corresponding to the linear constraints and four corresponding to the nonnegative condition). An exhaustive search would amount to the solution of 210 sets of simultaneous equations. The exhaustive method, therefore, is basically impractical, and we should look for different methods of solution. One such method will be discussed in Sec. 12-4. Moreover, because not every intersection of the hyperplanes is a corner of the feasible region, even if we are willing to exhaustively solve the 210 sets of simultaneous equations, we still have the problem of weeding out those intersections that are not corners of the feasible region. Let us recall the example presented in Sec. 12-2 which we repeat here for convenience. The problem is to maximize the objective function $C = x_1 + x_2$, subject to the linear constraints

$$\begin{aligned} 4x_1 + 5x_2 &\leq 10 \\ 5x_1 + 2x_2 &\leq 10 \\ 3x_1 + 8x_2 &\leq 12 \end{aligned} \tag{12-1}$$

and to the nonnegative condition

$$x_1 \geq 0 \qquad x_2 \geq 0 \tag{12-2}$$

In Fig. 12-1c, we see that the intersection of the two straight lines $5x_1 + 2x_2 = 10$ and $3x_1 + 8x_2 = 12$ is not a corner of the feasible region, whereas the intersection of the two straight lines $5x_1 + 2x_2 = 10$ and $4x_1 + 5x_2 = 10$ is such a corner. This problem of identifying the corners of the feasible region is handled by the addition of "slack variables," which we now discuss.

Consider the linear constraint $4x_1 + 5x_2 \leq 10$. By adding an extra variable x_3, we can write the inequality as the equation

$$4x_1 + 5x_2 + x_3 = 10$$

At a point in the x_1-x_2 plane, the variable x_3 assumes a positive value if $4x_1 + 5x_2 < 10$, assumes a negative value if $4x_1 + 5x_2 > 10$, and is equal to zero if $4x_1 + 5x_2 = 10$. In other words, in the shaded half-plane in Fig. 12-1a, $x_3 > 0$; in the unshaded half-plane, $x_3 < 0$; on the straight line $4x_1 + 5x_2 = 10$, $x_3 = 0$. This suggests that instead of talking about the shaded half-plane and the unshaded half-plane, we can use x_3 as a third coordinate to distinguish the two half-planes into which the x_1-x_2 plane is divided by the straight line $4x_1 + 5x_2 = 10$. Therefore, for a point in the plane, besides the two coordinates x_1 and x_2, there is the third coordinate x_3, as shown in Fig. 12-6. These coordinates are not independent coordinates in that the values of any two of the three coordinates determine a point in the plane and thus determine the value of the remaining coordinate. It should also be pointed out that the

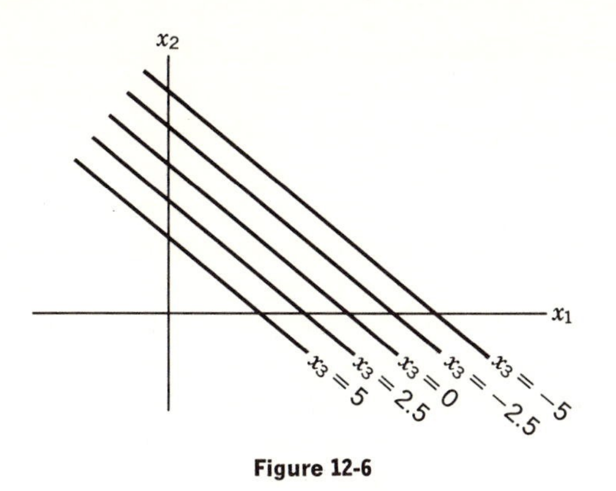

Figure 12-6

introduction of the third coordinate is just an extension of a familiar concept. Just as the sign of the coordinate x_1 of a point indicates the side of the vertical axis in which the point lies, and the sign of the coordinate x_2 of a point indicates the side of the horizontal axis in which the point lies, the sign of the coordinate x_3 of a point indicates the side of the straight line $4x_1 + 5x_2 = 10$ in which the point lies.

Extending this idea, we rewrite the constraint $5x_1 + 2x_2 \leq 10$ as $5x_1 + 2x_2 + x_4 = 10$ and rewrite the constraint $3x_1 + 8x_2 \leq 12$ as $3x_1 + 8x_2 + x_5 = 12$. Thus, a point in the x_1-x_2 plane will have five coordinates. For example, the coordinates of the origin of the x_1-x_2 plane are $x_1 = 0$, $x_2 = 0$, $x_3 = 10$, $x_4 = 10$, and $x_5 = 12$. Notice that a point in the feasible region must have nonnegative values for all five of its coordinates, whereas one or more of the coordinates of a point outside of the feasible region is negative.

Thus, by introducing the slack variables, we have reformulated the linear programming problem. The problem now is to maximize the objective function $C = x_1 + x_2$ subject to the linear constraints

$$\begin{aligned} 4x_1 + 5x_2 + x_3 &= 10 \\ 5x_1 + 2x_2 + x_4 &= 10 \\ 3x_1 + 8x_2 + x_5 &= 12 \end{aligned} \tag{12-5}$$

and to the nonnegative condition

$$x_1 \geq 0 \qquad x_2 \geq 0 \qquad x_3 \geq 0 \qquad x_4 \geq 0 \qquad x_5 \geq 0$$

To find the intersection of two straight lines, clearly, we can determine its x_1 and x_2 coordinates by solving the two corresponding equations among the five equations in (12-3) and then determine its x_3, x_4, and x_5 coordinates from the three equations in (12-5). For example, we solve the simultaneous equations

$$5x_1 + 2x_2 = 10$$
$$x_2 = 0$$

to obtain $x_1 = 2$ and $x_2 = 0$ and then use the equations in (12-5) to obtain $x_3 = 2$, $x_4 = 0$, and $x_5 = 6$, which gives us the five coordinates of the intersection of the straight line $5x_1 + 2x_2 = 10$ and the horizontal axis. However, a slightly different point of view can be taken. Because a zero coordinate identifies one of the five boundary straight lines, two zero coordinates identify the intersection of two of the five boundary straight lines. For example, that $x_2 = 0$ identifies the horizontal axis, and that $x_4 = 0$ identifies the straight line $5x_1 + 2x_2 = 10$. To find the intersection of these two straight lines, we shall solve for the coordinates x_1, x_3, and x_5 from the equations in (12-5) after setting $x_2 = 0$ and $x_4 = 0$. Thus,

$$4x_1 + x_3 = 10$$
$$5x_1 = 10$$
$$3x_1 + x_5 = 12$$

Therefore, we see that the coordinates of the intersection of two straight lines can be found by setting the corresponding coordinates to zero and solving for the remaining coordinates from the equations in (12-5). Clearly, if the intersection is a corner of the feasible region, it has two zero coordinates and three positive coordinates.

The notion of the introducing of new coordinates carries over to the general case. Thus, by introducing the new variable x_{r+i}, we change a linear inequality of the form

$$a_{i1}x_1 + a_{i2}x_2 + \cdots + a_{ir}x_r \leq b_i$$

into an equation of the form

$$a_{i1}x_1 + a_{i2}x_2 + \cdots + a_{ir}x_r + x_{r+i} = b_i$$

together with the constraint

$$x_{r+i} \geq 0$$

Also, by introducing the new variable x_{r+j}, we change a linear inequality of the form

$$a_{j1}x_1 + a_{j2}x_2 + \cdots + a_{jr}x_r \geq b_j$$

into an equation of the form

$$a_{j1}x_1 + a_{j2}x_2 + \cdots + a_{jr}x_r - x_{r+j} = b_j$$

together with the constraint

$$x_{r+j} \geq 0$$

A new variable introduced to change an inequality into an equation is called a *slack variable* (because it "takes up the slack"). Since the feasible region will be on one side or the other of the hyperplane corresponding to a linear inequality, the positiveness of the slack variable assures that a point is in the "proper side" of the hyperplane. In this way, we reformulate a linear programming problem as one of optimizing a linear function

$$C = c_1x_1 + c_2x_2 + \cdots + c_rx_r$$

subject to the linear constraints

$$\begin{array}{l} a_{11}x_1 + a_{12}x_2 + \cdots + a_{1r}x_r \pm x_{r+1} = b_1 \\ a_{21}x_1 + a_{22}x_2 + \cdots + a_{2r}x_r \pm x_{r+2} = b_2 \\ \cdots\cdots\cdots\cdots\cdots\cdots\cdots\cdots \\ a_{i1}x_1 + a_{i2}x_2 + \cdots + a_{ir}x_r \pm x_{r+i} = b_i \\ \cdots\cdots\cdots\cdots\cdots\cdots\cdots\cdots \\ a_{m1}x_1 + a_{m2}x_2 + \cdots + a_{mr}x_r \pm x_{r+m} = b_m \end{array} \tag{12-6}$$

and to the nonnegative condition

$$x_1 \geq 0, x_2 \geq 0, \ldots, x_r \geq 0$$
$$x_{r+1} \geq 0, x_{r+2} \geq 0, \ldots, x_{r+m} \geq 0$$

Recall that the feasible region is a convex polyhedron in r-dimensional space bounded by the $r + m$ hyperplanes corresponding to the $r + m$ equations in (12-4). It follows that a point in r-dimensional space is in the feasible region if and only if all its $r + m$ coordinates (including, now, the values of the m slack variables) are nonnegative, while a corner of the feasible region has r zero coordinates, corresponding to the r hyperplanes meeting at that point, and m positive coordinates.† To find the coordinates of a possible corner of the feasible region, we can solve the set of r of the $r + m$ equations in (12-4) corresponding to the r intersect-

† When more than r of the $r + m$ coordinates of a corner are equal to zero, it is a degenerate case. This case corresponds to more than r hyperplanes meeting at the same point.

ing hyperplanes. Alternatively, by setting r of the $r + m$ variables in the m equations in (12-6) to zero, we can solve for the remaining m variables as the coordinates of a corner of the feasible region.† If the solution yields nonnegative values for all the remaining m variables, we will have obtained the coordinates of a corner of the feasible region.

As defined earlier, a point in the feasible region is called a feasible solution to the problem. A feasible solution is said to be a *basic feasible solution* if r of its $r + m$ coordinates are zeros. (A feasible solution is said to be a *degenerate basic feasible solution* if more than r of its $r + m$ coordinates are zeros.) Clearly, a basic feasible solution corresponds to a corner of the feasible region. In a basic feasible solution, the variables that assume nonzero values are called *basic variables,* whereas the others are called *nonbasic variables.*

Now, we proceed to prove the statement made in Sec. 12-2 that one of the corners of the feasible region is an optimal solution to a linear programming problem. To prepare the way, we first prove the following theorem.

Theorem 12-1 *If a linear programming problem has a feasible solution, it also has a basic feasible solution.*

Proof Suppose that $\hat{x}_1, \hat{x}_2, \ldots, \hat{x}_{r+m}$ is a feasible solution. If all of $\hat{x}_1, \hat{x}_2, \ldots, \hat{x}_{r+m}$ are zero, they constitute a basic feasible solution. (In fact, this is a degenerate case in which the feasible region consists only of this one point.) If not all of them are zero, we can assume, without loss of generality, that

$$\hat{x}_1 > 0, \hat{x}_2 > 0, \ldots, \hat{x}_k > 0, \hat{x}_{k+1} = \hat{x}_{k+2} = \cdots = \hat{x}_{r+m} = 0$$

If $k \leq m$, again, we have a basic solution. Thus, the only case we have to examine is that when k is greater than m.

† When one of the linear constraints is an equality of the form

$$a_{i1}x_1 + a_{i2}x_2 + \cdots + a_{ir}x_r = b_i$$

we can replace it by the two inequalities

$$a_{i1}x_1 + a_{i2}x_2 + \cdots + a_{ir}x_r \geq b_i$$
$$a_{i1}x_1 + a_{i2}x_2 + \cdots + a_{ir}x_r \leq b_i$$

and then introduce the corresponding slack variables. Alternatively, we can leave the equality as it is. Then, since we need not introduce a slack variable for this constraint, the total number of variables will be $r + m - 1$. To find the coordinates of the intersection of r hyperplanes, we shall set $r - 1$ of the variables to zero and solve the m linear constraints for the m remaining variables.

Let us rewrite the m linear constraints in (12-6) as

$$\begin{aligned} a_{11}x_1 + a_{12}x_2 + \cdots + a_{1k}x_k &= -a_{1,k+1}x_{k+1} - a_{1,k+2}x_{k+2} \\ &\quad - \cdots - a_{1r}x_r \mp x_{r+1} \\ &\quad + b_1 \\ a_{21}x_1 + a_{22}x_2 + \cdots + a_{2k}x_k &= -a_{2,k+1}x_{k+1} - a_{2,k+2}x_{k+2} \\ &\quad - \cdots - a_{2r}x_r \mp x_{r+2} \\ &\quad + b_2 \\ \cdots\cdots\cdots\cdots\cdots\cdots\cdots\cdots \\ a_{m1}x_1 + a_{m2}x_2 + \cdots + a_{mk}x_k &= -a_{m,k+1}x_{k+1} - a_{m,k+2}x_{k+2} \\ &\quad - \cdots - a_{mr}x_r \mp x_{r+m} \\ &\quad + b_m \end{aligned} \tag{12-7}$$

In other words, we rewrite the constraints as m equations in k unknowns $x_1, x_2, \ldots, x_k$, by treating the variables $x_{k+1}, x_{k+2}, \ldots, x_{r+m}$ as constants. Suppose that q of these m equations are independent equations in the k unknowns ($q \leq m < k$). We can solve these q independent equations and express q of the k variables $x_1, x_2, \ldots, x_k$ in terms of the remaining variables. Without loss of generality, assume that we solve the first q equations in (12-7) for $x_1, x_2, \ldots, x_q$ and obtain

$$\begin{aligned} x_1 &= w_1 - v_{1,q+1}x_{q+1} - v_{1,q+2}x_{q+2} - \cdots - v_{1,r+m}x_{r+m} \\ x_2 &= w_2 - v_{2,q+1}x_{q+1} - v_{2,q+2}x_{q+2} - \cdots - v_{2,r+m}x_{r+m} \\ \cdots\cdots\cdots\cdots\cdots\cdots\cdots\cdots \\ x_q &= w_q - v_{q,q+1}x_{q+1} - v_{q,q+2}x_{q+2} - \cdots - v_{q,r+m}x_{r+m} \end{aligned} \tag{12-8}$$

where the v's and w's are constants.

Moreover, we rewrite the remaining $m - q$ equations in (12-7) as follows. [For the case $q = m$, all m equations in (12-7) have been rewritten as those in (12-8).] Since the left-hand side of the $(q + 1)$st equation in (12-7) is a linear combination of the left-hand sides of the first q equations, we can eliminate the variables $x_1, x_2, \ldots, x_k$ in the $(q + 1)$st equation and rewrite it as one that contains only the variables $x_{k+1}, x_{k+2}, \ldots, x_{r+m}$. Using the same argument for the $(q + 2)$nd, $\ldots$, mth equations, we can rewrite the last $m - q$ equations in (12-7) as

$$\begin{aligned} 0 &= w_{q+1} - v_{q+1,k+1}x_{k+1} - v_{q+1,k+2}x_{k+2} - \cdots - v_{q+1,r+m}x_{r+m} \\ 0 &= w_{q+2} - v_{q+2,k+1}x_{k+1} - v_{q+2,k+2}x_{k+2} - \cdots - v_{q+2,r+m}x_{r+m} \\ \cdots\cdots\cdots\cdots\cdots\cdots\cdots\cdots \\ 0 &= w_m - v_{m,k+1}x_{k+1} - v_{m,k+2}x_{k+2} - \cdots - v_{m,r+m}x_{r+m} \end{aligned} \tag{12-9}$$

In other words, the m linear constraints in (12-7) are rewritten as those in (12-8) and in (12-9).† Since the variable x_{q+1} does not appear in any of the constraints in (12-9), these constraints will not be violated when we decrease the value of $\hat{x}_{q+1}$ in the feasible solution $\hat{x}_1, \hat{x}_2, \ldots, \hat{x}_{r+m}$. However, to satisfy the constraints in (12-8), the values of $\hat{x}_1, \hat{x}_2, \ldots, \hat{x}_q$ should be changed accordingly. We can decrease the value of $\hat{x}_{q+1}$ either until it becomes zero or the value of one of $\hat{x}_1, \hat{x}_2, \ldots, \hat{x}_q$ becomes zero. In any case, we shall have a new feasible solution in which only $k - 1$ of the variables have nonzero values.

The argument can be repeated until a basic feasible solution is obtained. ∎

We prove now the following desired result.

Theorem 12-2 *If a linear programming problem has an optimal feasible solution, it also has an optimal basic feasible solution.*

Proof In a fashion similar to that in the proof of Theorem 12-1, let us suppose that $\hat{x}_1, \hat{x}_2, \ldots, \hat{x}_{r+m}$ is an optimal feasible solution in which

$$\hat{x}_1 > 0, \hat{x}_2 > 0, \ldots, \hat{x}_k > 0, \hat{x}_{k+1} = \hat{x}_{k+2} = \cdots = \hat{x}_{r+m} = 0$$

with $k > m$. We proceed as in the proof of Theorem 12-1 to express the variables $x_1, x_2, \ldots, x_q$ in terms of the variables $x_{q+1}, x_{q+2}, \ldots, x_{r+m}$ to obtain (12-8). Substituting the equations in (12-8) into the objective function, we can express the objective function in terms of the variables $x_{q+1}, x_{q+2}, \ldots, x_{r+m}$ as follows:

$$C = w_0 - v_{0,q+1}x_{q+1} - v_{0,q+2}x_{q+2} - \cdots - v_{0,r+m}x_{r+m}$$

Since $\hat{x}_{q+1}$ has a nonzero positive value, we claim that $v_{0,q+1}$ must be equal to zero. If $v_{0,q+1}$ is positive, we can decrease the value of $\hat{x}_{q+1}$ until it becomes zero or the value of one of $\hat{x}_1, \hat{x}_2, \ldots, \hat{x}_q$ becomes zero according to the constraints in (12-8). The value of C is then increased. If $v_{0,q+1}$ is negative, we can increase the value of $\hat{x}_{q+1}$ until the value of one of $\hat{x}_1, \hat{x}_2, \ldots, \hat{x}_q$ becomes zero according to the constraints in (12-8). The value of C is then

† Let us point out that the linear constraints in (12-8) and (12-9) are equivalent to the linear constraints in (12-7) in the sense that a set of the values of the variables $x_1, x_2, \ldots, x_{r+m}$ satisfies the constraints in (12-7) if and only if it satisfies the constraints in (12-8) and (12-9). The validity of this statement is due to the fact that not only are the constraints in (12-8) and (12-9) linear combinations of the constraints in (12-7), but the constraints in (12-7) can also be expressed as linear combinations of the constraints in (12-8) and (12-9).

increased. If C is already a maximum, neither of these two cases is possible.† We can, therefore, decrease the value of $\hat{x}_{q+1}$ until it becomes zero or until the value of one of $\hat{x}_1, \hat{x}_2, \ldots, \hat{x}_q$ becomes zero according to the constraints in (12-8) without changing the value of the objective function.

Such an argument can be repeated until we have an optimal basic feasible solution. ■

Theorems 12-1 and 12-2 affirm our geometric argument that when searching for an optimal feasible solution to a linear programming problem, we need only examine values of the objective function at the corners of the feasible region.

12-4 THE SIMPLEX METHOD

With the results in Secs. 12-2 and 12-3, the problem of optimizing a linear objective function is greatly simplified, since we need only compute the values of the objective function at the corners of the feasible region. However, as was also pointed out in Sec. 12-2, when the number of variables and the number of linear constraints increase, to compute the value of the objective function at every corner becomes a very tedious job. The *simplex method* is one of the most useful methods for finding an optimal feasible solution without examining exhaustively the values of the objective function at all the corners. The method starts at one corner of the feasible region and searches for another corner at which the value of the objective function will be improved. Such a search is repeated iteratively until an optimal basic feasible solution is found. In this way, one never examines corners at which there is no improvement in the value of the objective function over those values examined previously.

Since the problem of minimizing the objective function

$$C = c_1x_1 + c_2x_2 + c_3x_3 + \cdots + c_rx_r$$

under a set of constraints is equivalent to the problem of maximizing the objective function

$$C = -c_1x_1 - c_2x_2 - c_3x_3 - \cdots - c_rx_r$$

under the same set of constraints, without loss of generality, we shall conduct our discussion in terms of the maximization of linear objective functions.

As pointed out earlier, not every basic feasible solution is examined in the simplex method. Therefore, we must know the answers to these questions: After examining the value of the objective function at a corner, how do we find another corner at which the value of the objective

† If the problem is a minimization problem, a similar argument can be made.

function will be larger? How do we know that an optimal solution has been found when the corresponding corner of the feasible region is reached?

The answers to these two questions hinge on the observation that the linear constraints can be rewritten to express the basic variables in terms of the nonbasic variables. Recall that a basic feasible solution can be found by setting r of the $r + m$ variables (the nonbasic variables) to zero and solving the m linear constraints in (12-6) for the m remaining variables (the basic variables). Instead of setting the r nonbasic variables to zero, we can treat them as constants and solve the m linear constraints to express the m basic variables in terms of the r nonbasic variables.† It follows that we can substitute these expressions of the basic variables in terms of the nonbasic variables into the objective function so that the objective function becomes a function of the nonbasic variables only.

Let us illustrate this by an example in which we are to maximize the function $C = x_1 + 4x_2$ subject to the constraints

$$4x_1 + 5x_2 \leq 10$$
$$5x_1 + 2x_2 \leq 10$$
$$-7x_1 + 4x_2 \leq 4$$
$$x_1 \geq 0 \qquad x_2 \geq 0$$

With the addition of the slack variables x_3, x_4, and x_5, these constraints become

$$4x_1 + 5x_2 + x_3 = 10$$
$$5x_1 + 2x_2 + x_4 = 10$$
$$-7x_1 + 4x_2 + x_5 = 4$$
$$x_1 \geq 0 \qquad x_2 \geq 0 \qquad x_3 \geq 0 \qquad x_4 \geq 0 \qquad x_5 \geq 0$$

As shown in Fig. 12-7, there are five basic feasible solutions corresponding to the five corners of the feasible region (each corner is identified by the

† The linear constraints obtained this way are equivalent to the constraints in (12-6). See the footnote in the proof of Theorem 12-1.

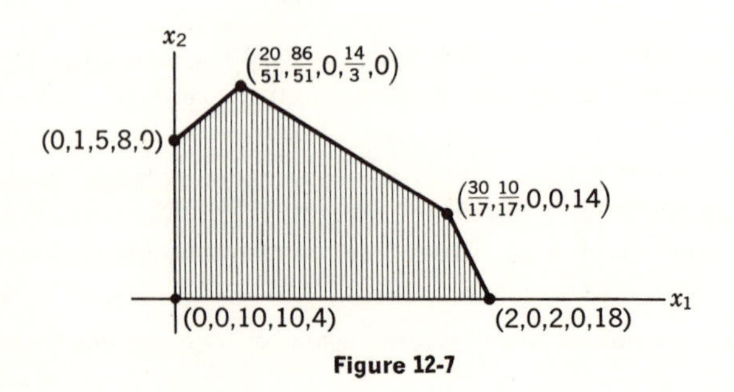

Figure 12-7

values of its five coordinates). The table below lists the different expressions for the objective function in terms of the nonbasic variables corresponding to the basic feasible solutions:

Basic feasible solution	*Nonbasic variables*	*Objective function*
(0,0,10,10,4)	x_1, x_2	$C - x_1 - 4x_2 = 0$
(0,1,5,8,0)	x_1, x_5	$C - 8x_1 + x_5 = 4$
($^{20}/_{51}$,$^{86}/_{51}$,0,$^{14}/_{3}$,0)	x_3, x_5	$C + {}^{32}/_{51}x_3 + {}^{11}/_{51}x_5 = {}^{364}/_{51}$
($^{30}/_{17}$,$^{10}/_{17}$,0,0,14)	x_3, x_4	$C + {}^{18}/_{17}x_3 - {}^{11}/_{17}x_4 = {}^{70}/_{17}$
(2,0,2,0,18)	x_2, x_4	$C - {}^{18}/_{5}x_2 + {}^{1}/_{5}x_4 = 2$

From these five different expressions for the objective function, we can tell that at the corner ($^{20}/_{51}$,$^{86}/_{51}$,0,$^{14}/_{3}$,0) the objective function attains its maximal value of $^{364}/_{51}$. This conclusion comes from the fact that the coefficients of both x_3 and x_5 in the expression

$$C + {}^{32}/_{51}x_3 + {}^{11}/_{51}x_5 = {}^{364}/_{51}$$

are positive. When the values of both x_3 and x_5 are zero, the value of C is $^{364}/_{51}$, whereas at any other point in the feasible region where one or both of x_3 and x_5 assume positive values, the value of C is less than $^{364}/_{51}$.

In the general case, when the objective function is expressed in terms of the nonbasic variables $x_{i_1}, x_{i_2}, \ldots, x_{i_r}$; that is, when

$$C + v_{0,i_1}x_{i_1} + v_{0,i_2}x_{i_2} + \cdots + v_{0,i_r}x_{i_r} = w_0$$

we can conclude that the objective function attains its maximal value w_0 at the corresponding basic feasible solution if all the coefficients $v_{0,i_1}, v_{0,i_2}, \ldots, v_{0,i_r}$ in the expression are positive.

Returning to the illustrative example, suppose that in searching for an optimal basic feasible solution we start at the corner (0,1,5,8,0). In terms of the nonbasic variables x_1 and x_5, the expression for the objective function is $C - 8x_1 + x_5 = 4$. Because the coefficient of x_1 is negative, we see that the value of C will be increased if the value of x_1 is increased from zero to a positive value; that is, the nonbasic variable x_1 is made a basic variable. Moreover, since we wish to move from one corner to another corner of the feasible region, we shall at the same time make one of the variables x_2, x_3, and x_4 a nonbasic variable when x_1 is made a basic variable. On the other hand, it would not be profitable to make x_5 a basic variable because its coefficient in the expression for the objective function is positive. Suppose that we start, instead, at the corner (0,0,10,10,4). In terms of the nonbasic variables x_1 and x_2 the expression for the objective function is

$$C - x_1 - 4x_2 = 0$$

It is clear that either making x_1 a basic variable or making x_2 a basic variable will increase the value of the objective function. In the simplex method, we shall change only one nonbasic variable into a basic variable at a time. (In other words, we move only between adjacent corners of the feasible region.) In this case, the choice between making x_1 or making x_2 a basic variable is quite arbitrary.

In the general case, when the objective function is expressed in terms of the nonbasic variables $x_{i_1}, x_{i_2}, \ldots, x_{i_r}$ as

$$C + v_{0,i_1}x_{i_1} + v_{0,i_2}x_{i_2} + \cdots + v_{0,i_r}x_{i_r} = w_0$$

we can increase the value of C by making the nonbasic variable x_{i_j} a basic variable if its coefficient v_{0,i_j} is negative.†

This brings up the next question: When a nonbasic variable is to be changed into a basic variable to increase the value of the objective function, how do we decide which basic feasible solution should be examined next? By increasing the value of the nonbasic variable from zero until the value of one of the basic variables becomes zero, we determine the next basic feasible solution. We shall illustrate the procedure by solving the preceding example, starting with the basic feasible solution (0,0,10, 10,4). The basic variables and the objective function can be expressed in terms of the nonbasic variables x_1 and x_2. Thus,

$$\begin{aligned} x_3 &= -4x_1 - 5x_2 + 10 \\ x_4 &= -5x_1 - 2x_2 + 10 \\ x_5 &= 7x_1 - 4x_2 + 4 \\ C &= x_1 + 4x_2 \end{aligned}$$

However, anticipating the *tableau format* to be presented in Sec. 12-5, these equations are rewritten so that only constants appear on the right-hand sides of the equations. Thus,

$$x_3 + 4x_1 + 5x_2 = 10 \tag{12-10}$$

$$x_4 + 5x_1 + 2x_2 = 10 \tag{12-11}$$

$$x_5 - 7x_1 + 4x_2 = 4 \tag{12-12}$$

$$C - x_1 - 4x_2 = 0 \tag{12-13}$$

† As mentioned above, if two or more of the v's are negative, one of the corresponding nonbasic variables can be chosen arbitrarily to become a basic variable. Although any choice leads eventually to an optimal solution, the number of iterations to reach the optimal solution may be different for different choices. There is no known selection rule that will enable us to make a choice that requires the least number of iterations. A rule of thumb usually followed is to choose the variable with a negative coefficient that is largest in magnitude.

As pointed out earlier in this section, either x_1 or x_2 can be made a basic variable. Suppose that we choose to increase the value of x_2 while keeping the value of x_1 at zero. According to Eq. (12-10), the value of x_2 cannot be increased beyond $10/5$, since any further increment will make the value of x_3 negative. Similarly, Eq. (12-11) limits the increment to $10/2$, and Eq. (12-12) limits the increment to $4/4$. Thus, the value of x_2 can be increased to $4/4$ $(= 1)$, which is the smallest of the three limits. We håve now another basic feasible solution $x_1 = 0$, $x_2 = 1$, $x_3 = 5$, $x_4 = 8$, and $x_5 = 0$ in which x_1 and x_5 are nonbasic variables. Expressing the basic variables and the objective function in terms of the nonbasic variables, we obtain

$$x_3 + 51/4x_1 - 5/4x_5 = 5 \tag{12-14}$$

$$x_4 + 17/2x_1 - 1/2x_5 = 8 \tag{12-15}$$

$$x_2 - 7/4x_1 + 1/4x_5 = 1 \tag{12-16}$$

$$C - 8x_1 + x_5 = 4 \tag{12-17}$$

Repeating the same argument, we note from Eq. (12-17) that x_1 should be made a basic variable. Because the coefficient of x_1 is negative in Eq. (12-16), increasing the value of x_1 will cause a corresponding increment in the value of x_2. Therefore, Eq. (12-16) does not place any limit on the increment of the value of x_1. In the constraints (12-14) and (12-15), we observe that the increment of the value of x_1 is limited by (12-14), in which the value of x_3 becomes 0 when the value of x_1 becomes $20/51$. Thus, the next basic feasible solution to be examined is $x_1 = 20/51$, $x_2 = 86/51$, $x_3 = 0$, $x_4 = 14/3$, and $x_5 = 0$. Expressing the basic variables and the objective function in terms of x_3 and x_5, we obtain

$$x_1 + 4/51x_3 - 5/51x_5 = 20/51 \tag{12-18}$$

$$x_4 - 2/3x_3 + 1/3x_5 = 14/3 \tag{12-19}$$

$$x_2 + 7/51x_3 + 4/51x_5 = 86/51 \tag{12-20}$$

$$C + 32/51x_3 + 11/51x_5 = 364/51 \tag{12-21}$$

Since the coefficients of both x_3 and x_5 in the objective function are positive, we know that the objective function attained its maximal value which is $364/51$.

This completes the illustration of the simplex method for solving linear programming problems. We now summarize the procedure as follows:

1. Slack variables are added to transform the linear constraints into linear equations.

2. An initial basic feasible solution is picked arbitrarily (see Sec. 12-6 on how an initial solution can be found).
3. The basic variables and the objective function are expressed in terms of the nonbasic variables in the basic feasible solution.
4. According to the signs of the coefficients of the nonbasic variables in the expression for the objective function, a nonbasic variable that has a negative coefficient is picked to become a basic variable. The value of this nonbasic variable is increased until the value of one of the basic variables becomes zero.
5. Steps (3) and (4) are repeated until the coefficients of the nonbasic variables in the expression for the objective function are all positive.

12-5 THE TABLEAU FORMAT

A tableau format that can be used for systematically carrying out the simplex method is introduced in this section. In an alternative representation, Eqs. (12-10) to (12-13) are rewritten in tabular form as

	x_1	x_2	
x_3	4	5	10
x_4	5	2	10
x_5	-7	4	4
C	-1	-4	0

The nonbasic variables are written in the top row of the table. The remaining rows are the expressions for the basic variables and the objective function in terms of the nonbasic variables. Specifically, the numbers in the two columns under x_1 and x_2 are their coefficients in the expressions for x_3, x_4, x_5, and the objective function C. The rightmost column contains the constants on the right-hand sides of Eqs. (12-10) to (12-13).

To determine which of the variables x_3, x_4, and x_5 will become a nonbasic variable when x_2 is made a basic variable, we compute the ratio of the number in the rightmost column and the number under the x_2 column for each of the rows x_3, x_4, and x_5. (In this case, the ratios are 10/5, 10/2, 4/4.) The variable in the row that has the smallest ratio is the variable that will become a nonbasic variable when the value of x_2 is made as large as possible. In this case, it is the variable x_5 that will become zero when the value of x_2 is increased from zero to $4/4$. (The reader should recall the reasoning given in Sec. 12-4, which we shall not repeat here.)

To express the basic variables and the objective function in terms of the nonbasic variables x_1 and x_5, instead of solving a set of simultaneous equations, we first rewrite Eq. (12-12) as

$$x_2 - \tfrac{7}{4}x_1 + \tfrac{1}{4}x_5 = 1$$

that is, we rewrite the equation such that the coefficient of x_2 is unity. We can then rewrite Eqs. (12-10), (12-11), and (12-13) by substituting this expression for x_2 into them. The resultant tableau as well as the computation involved is shown in the following:

	x_1	x_5	
x_3	$4 - 5 \times (-\tfrac{7}{4}) = {}^{51}\!/_{4}$	$-\tfrac{5}{4}$	$10 - 5 \times \tfrac{4}{4} = 5$
x_4	$5 - 2 \times (-\tfrac{7}{4}) = {}^{17}\!/_{2}$	$-\tfrac{2}{4} = -\tfrac{1}{2}$	$10 - 2 \times \tfrac{4}{4} = 8$
x_2	$-\tfrac{7}{4}$	$\tfrac{1}{4}$	$\tfrac{4}{4} = 1$
C	$-1 - (-4) \times (-\tfrac{7}{4}) = -8$	$\tfrac{4}{4} = 1$	$0 - (-4) \times \tfrac{4}{4} = 4$

Again, this tableau is just an alternative representation for Eqs. (12-14) to (12-17).

The steps can now be repeated, and we have the following tableau corresponding to Eqs. (12-18) to (12-21):

	x_3	x_5	
x_1	${}^{4}\!/_{51}$	$-{}^{5}\!/_{51}$	${}^{20}\!/_{51}$
x_4	$\dfrac{-{}^{17}\!/_{2}}{{}^{51}\!/_{4}} = -\tfrac{2}{3}$	$-\tfrac{1}{2} - ({}^{17}\!/_{2}) \times (-{}^{5}\!/_{51}) = \tfrac{1}{3}$	$8 - ({}^{17}\!/_{2}) \times ({}^{20}\!/_{51}) = {}^{14}\!/_{3}$
x_2	$\dfrac{-(-\tfrac{7}{4})}{{}^{51}\!/_{4}} = {}^{7}\!/_{51}$	$\tfrac{1}{4} - (-\tfrac{7}{4}) \times (-{}^{5}\!/_{51}) = {}^{4}\!/_{51}$	$1 - (-\tfrac{7}{4}) \times ({}^{20}\!/_{51}) = {}^{86}\!/_{51}$
C	$\dfrac{-(-8)}{{}^{51}\!/_{4}} = {}^{32}\!/_{51}$	$1 - (-8) \times (-{}^{5}\!/_{51}) = {}^{11}\!/_{51}$	$4 - (-8) \times ({}^{20}\!/_{51}) = {}^{364}\!/_{51}$

We use this example to introduce the tableau format, which is actually just an alternative representation for the linear constraints and the objective function. The reason for using such a representation is that there is a simple set of rules by which one tableau can be transformed into another tableau when interchanging a nonbasic variable and a basic variable.

We now present the tableau format in general terms. Without loss of generality, suppose that we start the simplex method at the basic solution in which $x_1, x_2, \ldots, x_m$ are basic variables and $x_{m+1}, x_{m+2}, \ldots, x_{r+m}$ are nonbasic variables.† According to our previous discus-

† There is no loss in generality since we can always rename the variables.

sion, we can rewrite the linear constraints and the objective function as

$$x_1 + v_{1,m+1}x_{m+1} + v_{1,m+2}x_{m+2} + \cdots + v_{1,m+j}x_{m+j} + \cdots + v_{1,r+m}x_{r+m} = w_1$$
$$x_2 + v_{2,m+1}x_{m+1} + v_{2,m+2}x_{m+2} + \cdots + v_{2,m+j}x_{m+j} + \cdots + v_{2,r+m}x_{r+m} = w_2$$
$$\cdots\cdots\cdots\cdots\cdots\cdots\cdots\cdots\cdots$$
$$x_k + v_{k,m+1}x_{m+1} + v_{k,m+2}x_{m+2} + \cdots + v_{k,m+j}x_{m+j} + \cdots + v_{k,r+m}x_{r+m} = w_k$$
$$\cdots\cdots\cdots\cdots\cdots\cdots\cdots\cdots\cdots$$
$$x_m + v_{m,m+1}x_{m+1} + v_{m,m+2}x_{m+2} + \cdots + v_{m,m+j}x_{m+j} + \cdots + v_{m,r+m}x_{r+m} = w_m$$
$$C + v_{0,m+1}x_{m+1} + v_{0,m+2}x_{m+2} + \cdots + v_{0,m+j}x_{m+j} + \cdots + v_{0,r+m}x_{r+m} = w_0$$

The corresponding tableau is

	x_{m+1}	x_{m+2}	$\cdots$	x_{m+j}	$\cdots$	x_{r+m}	
x_1	$v_{1,m+1}$	$v_{1,m+2}$	$\cdots$	$v_{1,m+j}$	$\cdots$	$v_{1,r+m}$	w_1
x_2	$v_{2,m+1}$	$v_{2,m+2}$	$\cdots$	$v_{2,m+j}$	$\cdots$	$v_{2,r+m}$	w_2
$\vdots$	$\cdots$	$\cdots$	$\cdots$	$\cdots$	$\cdots$	$\cdots$	$\vdots$
x_k	$v_{k,m+1}$	$v_{k,m+2}$	$\cdots$	$v_{k,m+j}$	$\cdots$	$v_{k,r+m}$	w_k
$\vdots$	$\cdots$	$\cdots$	$\cdots$	$\cdots$	$\cdots$	$\cdots$	$\vdots$
x_m	$v_{m,m+1}$	$v_{m,m+2}$	$\cdots$	$v_{m,m+j}$	$\cdots$	$v_{m,r+m}$	w_m
C	$v_{0,m+1}$	$v_{0,m+2}$	$\cdots$	$v_{0,m+j}$	$\cdots$	$v_{0,r+m}$	w_0

To determine which of the variables $x_1, x_2, \ldots, x_m$ will become a nonbasic variable when the value of a nonbasic variable x_{m+j} is increased from zero to as large a positive value as possible, the ratios $w_1/v_{1,m+j}$, $w_2/v_{2,m+j}$, $\ldots$, $w_m/v_{m,m+j}$ are computed. Suppose that the ratio $w_k/v_{k,m+j}$ is the smallest of all the *positive* quantities. Then x_k will become a nonbasic variable when the value of x_{m+j} is increased, making it a basic variable. The coefficient $v_{k,m+j}$ is called the *pivot.* The row that contains the pivot is called the *pivotal row,* and the column that contains the pivot is called the *pivotal column.* To compute the tableau for the basic variables $x_1, x_2, \ldots, x_{k-1}, x_{m+j}, x_{k+1}, \ldots, x_m$ (with rows arranged in this order) and the nonbasic variables $x_{m+1}, x_{m+2}, \ldots, x_{m+j-1}, x_k, x_{m+j+1}, \ldots, x_{r+m}$ (with columns arranged in this order),† we can transform the previous tableau according to the following rules:

† In other words, the variables x_k and x_{m+j} interchange their positions in the tableau.

1. The pivot is replaced by its reciprocal.
2. The other entries in the pivotal row are divided by the pivot.
3. The other entries in the pivotal column are divided by the pivot with their signs reversed.
4. For the other entries in the tableau, $v_{p,m+q}$ $(p \neq k;\, q \neq j)$ is replaced by $v_{p,m+q} - \left(v_{p,m+j}\dfrac{v_{k,m+q}}{v_{k,m+j}}\right)$, w_p $(p \neq k)$ is replaced by $w_p - \left(v_{p,m+j}\dfrac{w_k}{v_{k,m+j}}\right)$.

Instead of giving a proof of the validity of these rules of transformation, we remind the reader that such a transformation is just a systematic way of carrying out the substitution steps illustrated earlier in this section.

12-6 COMPLICATIONS AND THEIR RESOLUTIONS

When the simplex method is applied to the solution of linear programming problems, some complications may arise. We shall discuss these complications and their resolutions in this section.

Complication 1 The value of the objective function is unbounded. Consider the maximization of the objective function

$$C = x_1 + x_2$$

subject to the constraints

$$-x_1 + 2x_2 \leq 2$$
$$x_1 - x_2 \geq 2$$
$$x_1 \geq 0 \qquad x_2 \geq 0$$

Adding the slack variables x_3 and x_4, we obtain

$$-x_1 + 2x_2 + x_3 = 2$$
$$x_1 - x_2 - x_4 = 2$$
$$x_1 \geq 0 \qquad x_2 \geq 0 \qquad x_3 \geq 0 \qquad x_4 \geq 0$$

Graphically, we see in Fig. 12-8 that the feasible region is an unbounded one.

To solve the problem by the simplex method, suppose that we start with the basic feasible solution $x_1 = 2$, $x_2 = 0$, $x_3 = 4$, and $x_4 = 0$. The

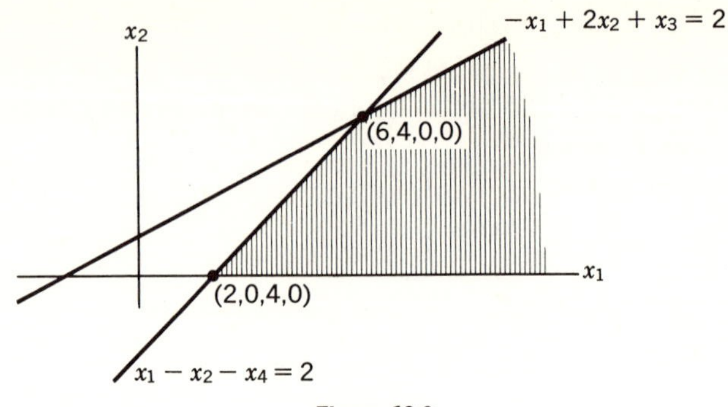

Figure 12-8

corresponding tableau is

	x_2	x_4	
x_1	-1	-1	2
x_3	1	-1	4
C	-2	-1	2

Since the coefficients of both x_2 and x_4 are negative, either the value of x_2 or the value of x_4 can be increased to increase the value of the objective function. Suppose that we choose to increase the value of x_4. It can be seen that increasing the value of x_4 will increase the value of x_1, the value of x_3, and the value of the objective function. Therefore, there is no limit to the increment of the value of x_4 and, consequently, no bound to the value of the objective function.

Suppose that we happen to choose to increase the value of x_2 and make it a basic variable. In that case, the tableau becomes

	x_3	x_4	
x_1	1	-2	6
x_2	1	-1	4
C	2	-3	10

To further increase the value of the objective function, we shall increase the value of x_4, and, again, we conclude that the value of the objective function is unbounded.

As can be observed in this example, when there is a column in the tableau that contains all negative entries, the value of the objective function is unbounded. (By convention, a maximization problem is said to be *not* possessing an optimal feasible solution if the value of the objective function is unbounded.)

Complication 2 There is more than one optimal feasible solution. Suppose the function $C = 2x_1 + 2x_2$ is to be maximized subject to the constraints

$$-x_1 + x_2 \leq 1$$
$$x_1 + x_2 \leq 3$$
$$x_1 \geq 0 \qquad x_2 \geq 0$$

Adding the slack variables x_3 and x_4, we obtain

$$-x_1 + x_2 + x_3 = 1$$
$$x_1 + x_2 + x_4 = 3$$
$$x_1 \geq 0 \qquad x_2 \geq 0 \qquad x_3 \geq 0 \qquad x_4 \geq 0$$

Graphically, the feasible region is shown in Fig. 12-9. It can be seen that any point on the heavy segment of the line $x_1 + x_2 + x_4 = 3$ is an optimal feasible solution to the problem.

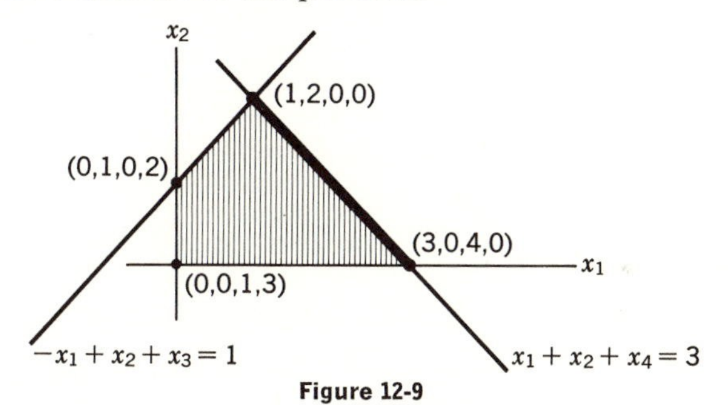

Figure 12-9

When the simplex method is employed, suppose that we start with the feasible solution $x_1 = 0$, $x_2 = 1$, $x_3 = 0$, and $x_4 = 2$. The corresponding tableau is

	x_1	x_3	
x_2	-1	1	1
x_4	2	-1	2
C	-4	2	2

Making x_1 a basic variable and x_4 a nonbasic variable, we obtain

	x_4	x_3	
x_2	$1/2$	$1/2$	2
x_1	$1/2$	$-1/2$	1
C	2	0	6

In the objective function, the coefficient of x_3 is 0. This means that changing the value of x_3 will not change the value of the objective function.

Therefore, we conclude that in the expression for the objective function in terms of the nonbasic variables, if all the coefficients are nonnegative and one or more of the coefficients are zero, then there is more than one optimal solution to the problem.

Complication 3 There is no obvious basic feasible solution. To carry out the simplex method, we need, initially, a basic feasible solution. Sometimes an initial basic feasible solution can be found by inspection. However, there are also cases in which the existence of a basic feasible solution is not obvious. Let us consider the example of maximizing the function $C = 5x_1 + 3x_2$ subject to the constraints

$$
\begin{aligned}
-4x_1 - 5x_2 &\leq -10 \\
5x_1 + 2x_2 &\leq 10 \\
3x_1 + 8x_2 &\leq 12 \\
x_1 \geq 0 \qquad x_2 &\geq 0
\end{aligned}
$$

Adding the slack variables x_3, x_4, and x_5, we obtain

$$
\begin{aligned}
-4x_1 - 5x_2 + x_3 &= -10 \\
5x_1 + 2x_2 + x_4 &= 10 \\
3x_1 + 8x_2 + x_5 &= 12
\end{aligned}
$$

$$x_1 \geq 0 \qquad x_2 \geq 0 \qquad x_3 \geq 0 \qquad x_4 \geq 0 \qquad x_5 \geq 0$$

Since the origin of the x_1-x_2 plane is not in the feasible region, $x_1 = 0$ and $x_2 = 0$ (together with $x_3 = -10$, $x_4 = 10$, and $x_5 = 12$) is not a feasible solution. It seems that we would, therefore, have difficulty in starting the simplex method.

Suppose that we have a different problem, that is, maximizing the function $C = 5x_1 + 3x_2 - Mx_0$, where M is a very large constant, subject to the constraints

$$
\begin{aligned}
-4x_1 - 5x_2 - x_0 &\leq -10 \\
5x_1 + 2x_2 &\leq 10 \\
3x_1 + 8x_2 &\leq 12
\end{aligned}
$$

$$x_0 \geq 0 \qquad x_1 \geq 0 \qquad x_2 \geq 0$$

Adding the slack variables x_3, x_4, and x_5, we obtain

$$
\begin{aligned}
-4x_1 - 5x_2 - x_0 + x_3 &= -10 \\
5x_1 + 2x_2 + x_4 &= 10 \\
3x_1 + 8x_2 + x_5 &= 12
\end{aligned}
$$

$$x_0 \geq 0 \qquad x_1 \geq 0 \qquad x_2 \geq 0 \qquad x_3 \geq 0 \qquad x_4 \geq 0 \qquad x_5 \geq 0$$

Observe that $x_1 = 0$, $x_2 = 0$, $x_3 = 0$, $x_0 = 10$, $x_4 = 10$, and $x_5 = 12$ is a feasible solution. Therefore, the simplex method can be used to find the maximal value of the function $5x_1 + 3x_2 - Mx_0$. However, since M is a very large number an optimal solution to the first problem together with $x_0 = 0$ is an optimal solution to the second problem. On the other hand, an optimal solution to the second problem in which x_0 is a nonbasic variable is also an optimal solution to the first problem. When the simplex method is applied to solve the second problem, the sequence of tableaux is

	x_1	x_2	x_3	
x_0	4	5	-1	10
x_4	5	2	0	10
x_5	3	8	0	12
C	$-(5 + 4M)$	$-(3 + 5M)$	M	$-10M$

	x_1	x_5	x_3	
x_0	$17/8$	$-5/8$	-1	$5/2$
x_4	$17/4$	$-1/4$	0	7
x_2	$3/8$	$1/8$	0	$3/2$
C	$-31/8 - 17/8 M$	$3/8 + 5/8 M$	M	$9/2 - 5/2 M$

	x_0	x_5	x_3	
x_1	$8/17$	$-5/17$	$-8/17$	$20/17$
x_4	-2	1	2	2
x_2	$-3/17$	$4/17$	$3/17$	$18/17$
C	$31/17 + M$	$-13/17$	$-31/17$	$154/17$

	x_0	x_5	x_4	
x_1	0	$-1/17$	$4/17$	$28/17$
x_3	-1	$1/2$	$1/2$	1
x_2	0	$5/34$	$-3/34$	$15/17$
C	M	$5/34$	$31/34$	$185/17$

Therefore, we conclude that for the first problem, an optimal solution is $x_1 = 28/17$, $x_2 = 15/17$, $x_3 = 1$, $x_4 = 0$, and $x_5 = 0$, at which the value of the objective function is $185/17$.

In general, a linear programming problem is said to be *augmented* when new variables are added to or subtracted from the linear constraints and when terms containing these new variables are added to or subtracted from the objective function. These new variables are called *artificial variables.* It is clear that an initial basic feasible solution for the augmented problem will become evident when artificial variables are added to or subtracted from the linear constraints appropriately. Moreover, in a maximization problem,† when the term

$$-M(x_0 + x_0' + x_0'' + \cdots)$$

is added to the objective function, where x_0, x_0', x_0'', . . . are artificial variables and M is a very large number, an optimal solution to the augmented problem in which all the artificial variables are nonbasic variables will also be an optimal solution to the original problem.

Complication 4 There is no feasible solution. When a linear programming problem has no obvious initial basic feasible solution, it is possible that the problem has no feasible solution at all. However, the situation will become evident when the augmentation procedure in (3) is applied. Consider the problem of maximizing the function

$$C = 5x_1 + 3x_2$$

subject to the constraints

$$4x_1 + 5x_2 \le 10$$
$$5x_1 + 2x_2 \ge 10$$
$$3x_1 + 8x_2 \ge 12$$
$$x_1 \ge 0 \qquad x_2 \ge 0$$

Adding the slack variables x_3, x_4, and x_5, we obtain

$$4x_1 + 5x_2 + x_3 = 10$$
$$5x_1 + 2x_2 - x_4 = 10$$
$$3x_1 + 8x_2 - x_5 = 12$$
$$x_1 \ge 0 \qquad x_2 \ge 0 \qquad x_3 \ge 0 \qquad x_4 \ge 0 \qquad x_5 \ge 0$$

Since there is no obvious feasible solution to the problem, let us consider the augmented problem of maximizing the objective function

$$C = 5x_1 + 3x_2 - M(x_0 + x_0')$$

subject to the constraints

$$4x_1 + 5x_2 \le 10$$
$$5x_1 + 2x_2 + x_0 \ge 10$$
$$3x_1 + 8x_2 + x_0' \ge 12$$
$$x_1 \ge 0 \qquad x_2 \ge 0 \qquad x_0 \ge 0 \qquad x_0' \ge 0$$

† In a minimization problem, the term $M(x_0 + x_0' + x_0'' + \cdots)$ should be added.

where x_0 and x_0' are artificial variables. Introducing the slack variables to the constraints, we obtain

$$4x_1 + 5x_2 + x_3 = 10$$
$$5x_1 + 2x_2 + x_0 - x_4 = 10$$
$$3x_1 + 8x_2 + x_0' - x_5 = 12$$
$$x_1 \geq 0 \quad x_2 \geq 0 \quad x_0 \geq 0 \quad x_0' \geq 0$$
$$x_3 \geq 0 \quad x_4 \geq 0 \quad x_5 \geq 0$$

Since $x_1 = 0$, $x_2 = 0$, $x_0 = 10$, $x_0' = 12$, $x_3 = 10$, $x_4 = 0$, and $x_5 = 0$ is a basic feasible solution to the augmented problem, we can solve the augmented problem by the simplex method. The sequence of tableaux is

	x_1	x_2	x_4	x_5	
x_3	4	5	0	0	10
x_0	5	2	-1	0	10
x_0'	3	8	0	-1	12
C	$-5 - 8M$	$-3 - 10M$	M	M	$-22M$

	x_1	x_0'	x_4	x_5	
x_3	$17/8$	$-5/8$	0	$5/8$	$5/2$
x_0	$17/4$	$-1/4$	-1	$1/4$	7
x_2	$3/8$	$1/8$	0	$-1/8$	$3/2$
C	$-31/8 - 17/4M$	$3/8 + 5/4M$	M	$-3/8 - 1/4M$	$9/2 - 7M$

	x_3	x_0'	x_4	x_5	
x_1	$8/17$	$-5/17$	0	$5/17$	$20/17$
x_0	-2	1	-1	-1	2
x_2	$-3/17$	$4/17$	0	$-4/17$	$18/17$
C	$31/17 + 2M$	$-13/17$	M	$13/17 + M$	$154/17 - 2M$

	x_3	x_0	x_4	x_5	
x_1	$-2/17$	$5/17$	$-5/17$	0	$30/17$
x_0'	-2	1	-1	-1	2
x_2	$5/17$	$-4/17$	$4/17$	0	$10/17$
C	$5/17 + 2M$	$13/17$	$-13/17 + M$	M	$180/17 - 2M$

Therefore, the maximal value for the objective function $5x_1 + 3x_2 - M(x_0 + x_0')$ is $180/17 - 2M$, which is a very large negative quantity. This indicates that there is no feasible solution to the original problem, because if there is, such a feasible solution together with $x_0 = 0$ and $x_0' = 0$ will be a feasible solution to the augmented problem, which will give the objective function $5x_1 + 3x_2 - M(x_0 + x_0')$ a value larger than $180/17 - 2M$.

In general, when not all the artificial variables are nonbasic variables in an optimal solution to the augmented problem, we conclude that there is no feasible solution to the original problem.

★12-7 DUALITY

Consider the linear programming problem of maximizing the objective function

$$C = c_1x_1 + c_2x_2 + \cdots + c_rx_r$$

subject to the constraints

$$\begin{aligned} a_{11}x_1 + a_{12}x_2 + \cdots + a_{1r}x_r &\leq b_1 \\ a_{21}x_1 + a_{22}x_2 + \cdots + a_{2r}x_r &\leq b_2 \\ \cdots\cdots\cdots\cdots\cdots\cdots \\ a_{m1}x_1 + a_{m2}x_2 + \cdots + a_{mr}x_r &\leq b_m \\ x_1 \geq 0, x_2 \geq 0, \ldots, x_r &\geq 0 \end{aligned} \tag{12-22}$$

Also consider the linear programming problem of minimizing the objective function

$$B = b_1y_1 + b_2y_2 + \cdots + b_my_m$$

subject to the constraints

$$\begin{aligned} a_{11}y_1 + a_{21}y_2 + \cdots + a_{m1}y_m &\geq c_1 \\ a_{12}y_1 + a_{22}y_2 + \cdots + a_{m2}y_m &\geq c_2 \\ \cdots\cdots\cdots\cdots\cdots\cdots \\ a_{1r}y_1 + a_{2r}y_2 + \cdots + a_{mr}y_m &\geq c_r \\ y_1 \geq 0, y_2 \geq 0, \ldots, y_m &\geq 0 \end{aligned} \tag{12-23}$$

As can be seen, there is a close relationship between these two problems. Specifically, we observe the following properties:

1. The first problem is a *maximization* problem, and the second problem is a *minimization* problem.
2. In the linear constraints in the maximization problem, the combinations of the values of the variables are all constrained to be *less than*

or equal to the constants $b_1, b_2, \ldots, b_m$. In the linear constraints in the minimization problem, the combinations of the values of the variables are all constrained to be *larger than or equal to* the constants $c_1, c_2, \ldots, c_r$.

3. The constants on the right-hand sides of the linear constraints in the maximization problem, $b_1, b_2, \ldots, b_m$, become the coefficients of the objective function in the minimization problem. The constants on the right-hand sides of the linear constraints in the minimization problem, $c_1, c_2, \ldots, c_r$, become the coefficients of the objective function in the maximization problem.
4. The r coefficients $a_{i1}, a_{i2}, \ldots, a_{ir}$ in the ith linear constraint in the maximization problem become the coefficients of y_i in the r linear constraints in the minimization problem for $1 \le i \le m$.

For two linear programming problems related in this way, the second problem is called the *dual problem* of the first problem. Correspondingly, the first problem is called the *primal problem.*

Since the dual problem of a maximization problem is a minimization problem, we would expect that the definition of duality can be extended such that the dual problem of a minimization problem is a maximization problem. The first problem above can be reformulated as a problem of minimizing the objective function

$$C = -c_1x_1 - c_2x_2 - \cdots - c_rx_r$$

subject to the constraints

$$\begin{aligned}
-a_{11}x_1 - a_{12}x_2 - \cdots - a_{1r}x_r &\ge -b_1 \\
-a_{21}x_1 - a_{22}x_2 - \cdots - a_{2r}x_r &\ge -b_2 \\
\cdots\cdots\cdots\cdots\cdots\cdots\cdots\cdots \\
-a_{m1}x_1 - a_{m2}x_2 - \cdots - a_{mr}x_r &\ge -b_m \\
x_1 \ge 0, x_2 \ge 0, \ldots, x_r \ge 0
\end{aligned}$$

Also, the second problem above can be reformulated as a problem of maximizing the objective function

$$B = -b_1y_1 - b_2y_2 - \cdots - b_my_m$$

subject to the constraints

$$\begin{aligned}
-a_{11}y_1 - a_{21}y_2 - \cdots - a_{m1}y_m &\le -c_1 \\
-a_{12}y_1 - a_{22}y_2 - \cdots - a_{m2}y_m &\le -c_2 \\
\cdots\cdots\cdots\cdots\cdots\cdots\cdots\cdots \\
-a_{1r}y_1 - a_{2r}y_2 - \cdots - a_{mr}y_m &\le -c_r \\
y_1 \ge 0, y_2 \ge 0, \ldots, y_m \ge 0
\end{aligned}$$

Notice now that the primal problem is reformulated as a minimization problem, and the dual problem is reformulated as a maximization problem. It is immediately clear that the dual of the dual of a linear programming problem is the problem itself. We can, therefore, speak of two problems being the dual of each other, since either one can be the primal problem with the other being its dual problem. (The reader is reminded of the discussion concerning dual graphs in Chap. 8.)

The following theorems show the implication of the concept of duality.

Theorem 12-3 *Let $\hat{x}_1, \hat{x}_2, \ldots, \hat{x}_r$ be a feasible solution to the primal problem, and let $\hat{y}_1, \hat{y}_2, \ldots, \hat{y}_m$ be a feasible solution to its dual problem. Then,*

$$c_1\hat{x}_1 + c_2\hat{x}_2 + \cdots + c_r\hat{x}_r \le b_1\hat{y}_1 + b_2\hat{y}_2 + \cdots + b_m\hat{y}_m$$

Proof From the linear constraints in (12-22), we have

$$\begin{aligned}
a_{11}\hat{x}_1 + a_{12}\hat{x}_2 + \cdots + a_{1r}\hat{x}_r &\le b_1 \\
a_{21}\hat{x}_1 + a_{22}\hat{x}_2 + \cdots + a_{2r}\hat{x}_r &\le b_2 \\
\cdots\cdots\cdots\cdots\cdots\cdots\cdots\cdots \\
a_{m1}\hat{x}_1 + a_{m2}\hat{x}_2 + \cdots + a_{mr}\hat{x}_r &\le b_m
\end{aligned}$$

Multiplying both sides of the ith inequality by $\hat{y}_i$ for $i = 1, 2, \ldots, m$ and summing both sides of the inequalities, we obtain

$$\sum_{i=1}^{m} \Big(\sum_{j=1}^{r} a_{ij}\hat{x}_j \Big) \hat{y}_i \le \sum_{i=1}^{m} b_i\hat{y}_i \tag{12-24}$$

Similarly, from the linear constraints in (12-23), we have

$$\begin{aligned}
a_{11}\hat{y}_1 + a_{21}\hat{y}_2 + \cdots + a_{m1}\hat{y}_m &\ge c_1 \\
a_{12}\hat{y}_1 + a_{22}\hat{y}_2 + \cdots + a_{m2}\hat{y}_m &\ge c_2 \\
\cdots\cdots\cdots\cdots\cdots\cdots\cdots\cdots \\
a_{1r}\hat{y}_1 + a_{2r}\hat{y}_2 + \cdots + a_{mr}\hat{y}_m &\ge c_r
\end{aligned}$$

Multiplying both sides of the jth inequality by $\hat{x}_j$ for $j = 1, 2, \ldots, r$ and summing both sides of the inequalities, we obtain

$$\sum_{j=1}^{r} \Big(\sum_{i=1}^{m} a_{ij}\hat{y}_i \Big) \hat{x}_j \ge \sum_{j=1}^{r} c_j\hat{x}_j \tag{12-25}$$

Combining (12-24) and (12-25), we obtain

$$\sum_{j=1}^{r} c_j\hat{x}_j \le \sum_{i=1}^{m} b_i\hat{y}_i \quad \blacksquare$$

Corollary 12-3.1 *Let $\hat{x}_1, \hat{x}_2, \ldots, \hat{x}_r$ and $\hat{y}_1, \hat{y}_2, \ldots, \hat{y}_m$ be feasible solutions to the primal and the dual problems, respectively. Both of the solutions are optimal feasible solutions to their respective problems if*

$$\sum_{j=1}^{r} c_j\hat{x}_j = \sum_{i=1}^{m} b_i\hat{y}_i$$

The next theorem shows that the existence of an optimal feasible solution to the primal problem guarantees the existence of an optimal feasible solution to the dual problem.

Theorem 12-4 *If the primal problem has an optimal feasible solution, then the dual problem also has an optimal feasible solution.*

Proof After adding the slack variables, the m linear constraints in the primal problem are rewritten as m linear equations. Thus,

$$\begin{aligned}
a_{11}x_1 + a_{12}x_2 + \cdots + a_{1r}x_r + x_{r+1} &= b_1 \\
a_{21}x_1 + a_{22}x_2 + \cdots + a_{2r}x_r + x_{r+2} &= b_2 \\
\cdots\cdots\cdots\cdots\cdots\cdots\cdots\cdots &\\
a_{m1}x_1 + a_{m2}x_2 + \cdots + a_{mr}x_r + x_{r+m} &= b_m
\end{aligned}$$

Also, the objective function is written as

$$C - c_1x_1 - c_2x_2 - \cdots - c_rx_r = 0$$

To simplify the notation later on, we rewrite the m linear equations as

$$\begin{aligned}
a_{11}x_1 + a_{12}x_2 + \cdots + a_{1r}x_r + a_{1,r+1}x_{r+1} + \cdots &\\
+ a_{1,r+m}x_{r+m} &= b_1 \\
a_{21}x_1 + a_{22}x_2 + \cdots + a_{2r}x_r + a_{2,r+1}x_{r+1} + \cdots &\\
+ a_{2,r+m}x_{r+m} &= b_2 \qquad (12\text{-}26) \\
\cdots\cdots\cdots\cdots\cdots\cdots\cdots\cdots\cdots\cdots &\\
a_{m1}x_1 + a_{m2}x_2 + \cdots + a_{mr}x_r + a_{m,r+1}x_{r+1} + \cdots &\\
+ a_{m,r+m}x_{r+m} &= b_m
\end{aligned}$$

with

$$\begin{aligned}
&a_{1,r+1} = 1,\ a_{1,r+2} = a_{1,r+3} = \cdots = a_{1,r+m} = 0 \\
&a_{2,r+1} = 0,\ a_{2,r+2} = 1,\ a_{2,r+3} = a_{2,r+4} = \cdots = a_{2,r+m} = 0 \\
&\cdots\cdots\cdots\cdots\cdots\cdots\cdots\cdots\cdots\cdots \\
&a_{m,r+1} = a_{m,r+2} = \cdots = a_{m,r+m-1} = 0,\ a_{m,r+m} = 1
\end{aligned}$$

Also, we rewrite the objective function as

$$C - c_1x_1 - c_2x_2 - \cdots - c_rx_r - c_{r+1}x_{r+1} - \cdots - c_{r+m}x_{r+m} = 0 \tag{12-27}$$

with

$$c_{r+1} = c_{r+2} = \cdots = c_{r+m} = 0$$

Suppose that we solve the problem by the simplex method and obtain the following tableau as the final one. (As we can always relabel the variables, there is no loss in generality in assuming that $x_1, x_2, \ldots, x_m$ are basic variables in the basic feasible optimal solution.)

	x_{m+1}	x_{m+2}	$\cdots$	x_{r+m}	
x_1	$v_{1,m+1}$	$v_{1,m+2}$	$\cdots$	$v_{1,r+m}$	w_1
x_2	$v_{2,m+1}$	$v_{2,m+2}$	$\cdots$	$v_{2,r+m}$	w_2
$\cdot$					$\cdot$
$\cdot$	$\cdots$	$\cdots$	$\cdots$	$\cdots$	$\cdot$
$\cdot$					$\cdot$
x_m	$v_{m,m+1}$	$v_{m,m+2}$	$\cdots$	$v_{m,r+m}$	w_m
C	$v_{0,m+1}$	$v_{0,m+2}$	$\cdots$	$v_{0,r+m}$	w_0

We recall that the tableau is just a shorthand representation of the m equations

$$\begin{aligned} x_1 + v_{1,m+1}x_{m+1} + v_{1,m+2}x_{m+2} + \cdots + v_{1,r+m}x_{r+m} &= w_1 \\ x_2 + v_{2,m+1}x_{m+1} + v_{2,m+2}x_{m+2} + \cdots + v_{2,r+m}x_{r+m} &= w_2 \\ \cdots\cdots\cdots\cdots\cdots\cdots\cdots\cdots\cdots\cdots\cdots\cdots \\ x_m + v_{m,m+1}x_{m+1} + v_{m,m+2}x_{m+2} + \cdots + v_{m,r+m}x_{r+m} &= w_m \end{aligned} \tag{12-28}$$

and the objective function

$$C + v_{0,m+1}x_{m+1} + v_{0,m+2}x_{m+2} + \cdots + v_{0,r+m}x_{r+m} = w_0 \tag{12-29}$$

Moreover, the m equations in (12-28) are obtained by solving the m equations in (12-26). It follows that each of the equations in (12-28) can be expressed as a linear combination of the m equations in (12-26). Suppose that the equations in (12-28) can be expressed as

$$\pi_{11}(12\text{-}26.1) + \pi_{21}(12\text{-}26.2) + \cdots + \pi_{m1}(12\text{-}26.m)$$
$$\pi_{12}(12\text{-}26.1) + \pi_{22}(12\text{-}26.2) + \cdots + \pi_{m2}(12\text{-}26.m)$$
$$\cdots\cdots\cdots\cdots\cdots\cdots\cdots\cdots$$
$$\pi_{1m}(12\text{-}26.1) + \pi_{2m}(12\text{-}26.2) + \cdots + \pi_{mm}(12\text{-}26.m)$$

where the π's are constants and we use (12-26.1), (12-26.2), . . . , (12-26.m) to denote the first, second, . . . , mth equation in (12-26). Thus the m equations in (12-28) can be written as

$$\begin{aligned}
&\sum_{i=1}^{m} \pi_{i1}a_{i1}x_1 + \sum_{i=1}^{m} \pi_{i1}a_{i2}x_2 + \cdots + \sum_{i=1}^{m} \pi_{i1}a_{i,r+m}x_{r+m} \\
&\qquad\qquad = \sum_{i=1}^{m} \pi_{i1}b_i \\
&\sum_{i=1}^{m} \pi_{i2}a_{i1}x_1 + \sum_{i=1}^{m} \pi_{i2}a_{i2}x_2 + \cdots + \sum_{i=1}^{m} \pi_{i2}a_{i,r+m}x_{r+m} \\
&\qquad\qquad = \sum_{i=1}^{m} \pi_{i2}b_i \qquad (12\text{-}30) \\
&\cdots\cdots\cdots\cdots\cdots\cdots\cdots\cdots \\
&\sum_{i=1}^{m} \pi_{im}a_{i1}x_1 + \sum_{i=1}^{m} \pi_{im}a_{i2}x_2 + \cdots + \sum_{i=1}^{m} \pi_{im}a_{i,r+m}x_{r+m} \\
&\qquad\qquad = \sum_{i=1}^{m} \pi_{im}b_i
\end{aligned}$$

Recall that the expression for the objective function in (12-29) is obtained from the expression in (12-27) by eliminating the variables $x_1, x_2, \ldots, x_m$. Therefore, Eq. (12-29) can be expressed as

$$(12\text{-}27) + c_1(12\text{-}28.1) + c_2(12\text{-}28.2) + \cdots + c_m(12\text{-}28.m)$$

Since the equations in (12-28) can be rewritten as those in (12-30), Eq. (12-29) can be rewritten as

$$\begin{aligned}
C &+ \left(-c_1 + \sum_{j=1}^{m} c_j \sum_{i=1}^{m} \pi_{ij}a_{i1}\right) x_1 + \left(-c_2 + \sum_{j=1}^{m} c_j \sum_{i=1}^{m} \pi_{ij}a_{i2}\right) x_2 \\
&+ \cdots + \left(-c_{r+m} + \sum_{j=1}^{m} c_j \sum_{i=1}^{m} \pi_{ij}a_{i,r+m}\right) x_{r+m} \\
&\qquad = \sum_{j=1}^{m} c_j \sum_{i=1}^{m} \pi_{ij}b_i \qquad (12\text{-}31)
\end{aligned}$$

Note that

$$\sum_{j=1}^{m} c_j \sum_{i=1}^{m} \pi_{ij}a_{ik} = \sum_{i=1}^{m} a_{ik} \sum_{j=1}^{m} c_j\pi_{ij} \qquad k = 1, 2, \ldots, r + m$$

If we let

$$\delta_i = \sum_{j=1}^{m} c_j \pi_{ij} \qquad i = 1, 2, \ldots, m$$

then Eq. (12-31) can be written as

$$C + \Big(-c_1 + \sum_{i=1}^{m} a_{i1}\delta_i\Big) x_1 + \Big(-c_2 + \sum_{i=1}^{m} a_{i2}\delta_i\Big) x_2 + \cdots + \Big(-c_{r+m} + \sum_{i=1}^{m} a_{i,r+m}\delta_i\Big) x_{r+m} = \sum_{i=1}^{m} b_i\delta_i \quad (12\text{-}32)$$

Comparing Eq. (12-32) with Eq. (12-29), we observe that

$$-c_k + \sum_{i=1}^{m} a_{ik}\delta_i \geq 0 \qquad k = 1, 2, \ldots, r + m \qquad (12\text{-}33)$$

because all the v's in (12-29) are nonnegative. For the k's corresponding to the m slack variables, (12-33) is reduced to

$$\delta_i \geq 0 \qquad i = 1, 2, \ldots, m$$

For the remaining k's, (12-33) can be rewritten as

$$\sum_{i=1}^{m} a_{ik}\delta_i \geq c_k \qquad k = 1, 2, \ldots, r$$

It follows that $\delta_1, \delta_2, \ldots, \delta_m$ satisfies the constraints in (12-23) and is a feasible solution to the dual problem. Moreover, according to (12-32), the optimal value of the objective function for the primal problem is $\sum_{i=1}^{m} b_i\delta_i$. Therefore, according to Corollary 12-3.1, $\delta_1, \delta_2, \ldots, \delta_m$ is also an optimal feasible solution to the dual problem. ∎

Example 12-1 Find the minimal value of the function

$$C = x_1 + x_2$$

subject to the constraints

$$4x_1 + 5x_2 \leq 10$$
$$5x_1 + 2x_2 \leq 10$$
$$3x_1 + 8x_2 \leq 12$$
$$-7x_1 + 4x_2 \leq 7$$
$$8x_1 + 5x_2 \geq 1$$
$$x_1 \geq 0 \qquad x_2 \geq 0$$

Because there is no obvious feasible solution to the problem, we need to augment the problem before the application of the simplex method. However, we can also solve the dual problem. The problem is to maximize the objective function

$$B = -10y_1 - 10y_2 - 12y_3 - 7y_4 + y_5$$

subject to the constraints

$$-4y_1 - 5y_2 - 3y_3 + 7y_4 + 8y_5 \leq 1$$
$$-5y_1 - 2y_2 - 8y_3 - 4y_4 + 5y_5 \leq 1$$
$$y_1 \geq 0 \qquad y_2 \geq 0 \qquad y_3 \geq 0 \qquad y_4 \geq 0 \qquad y_5 \geq 0$$

Adding the slack variables to the linear constraints, we obtain

$$-4y_1 - 5y_2 - 3y_3 + 7y_4 + 8y_5 + y_6 = 1$$
$$-5y_1 - 2y_2 - 8y_3 - 4y_4 + 5y_5 + y_7 = 1$$
$$y_1 \geq 0 \qquad y_2 \geq 0 \qquad y_3 \geq 0 \qquad y_4 \geq 0 \qquad y_5 \geq 0 \qquad y_6 \geq 0 \qquad y_7 \geq 0$$

The sequence of tableaux is

	y_1	y_2	y_3	y_4	y_5	
y_6	-4	-5	-3	7	8	1
y_7	-5	-2	-8	-4	5	1
B	10	10	12	7	-1	0

	y_1	y_2	y_3	y_4	y_6	
y_5	$-1/2$	$-5/8$	$-3/8$	$7/8$	$1/8$	$1/8$
y_7	$-5/2$	$9/8$	$-49/8$	$-67/8$	$-5/8$	$3/8$
B	$19/2$	$75/8$	$93/8$	$63/8$	$1/8$	$1/8$

Therefore, we conclude that the minimal value of the objective function $C = x_1 + x_2$ of the primal problem is equal to $1/8$. ■

12-8 SUMMARY AND REFERENCES

The application of linear programming to assignment problems, transportation problems, industrial problems, economic theory, and many other fields can be found in the many books and papers in the literature. See Dantzig [2] for an extensive list of references.

The simplex method offers an effective method of solution to linear programming problems. Moreover, the systematic way in which tableaux can be transformed makes the method most suitable for digital computation. As a matter of fact, linear programming problems with hundreds of variables and constraints have been solved on digital computers. In most computation centers, there are library routines solving linear programming problems that a programmer can look up and use. For a discussion of the computational considerations, see Chap. 25 of Ralston and Wilf [6].

In applying the simplex method, complication may arise when a degenerate basic feasible solution is present. In this case, the value of the objective function may not increase when we move from one basic feasible solution to another. Fortunately, such a case is rare and can be handled by a perturbation method due to Charnes [1]. See also Chap. 14 of Garvin [3] and Chap. 6 of Hadley [4].

The implication of duality is more extensive than the presentation in Sec. 12-7. See, for example, Probs. 12-10 and 12-13.

When the objective function is not a linear function or the constraints are not linear inequalities or equations, the optimization problem is no longer a linear programming problem, and the method of solution developed in this chapter cannot be applied. Mathematically, the techniques of nonlinear programming and dynamic programming, which are useful in solving problems like these, are more complicated than the technique of linear programming. There is an extensive literature on the subject of nonlinear programming. See, for example, Hadley [5]. In Chap. 13, we shall have an introductory discussion on the subject of dynamic programming.

Two introductory books whose level of presentation is the same as ours are Garvin [3] and Vajda [7]. For more complete coverage, see Dantzig [2], Hadley [4], and Vajda [8].

1. Charnes, A.: Optimality and Degeneracy in Linear Programming, *Econometrica*, **20**:160–170 (1952).
2. Dantzig, G. B.: "Linear Programming and Extensions," Princeton University Press, Princeton, N.J., 1963.
3. Garvin, W. W.: "Introduction to Linear Programming," McGraw-Hill Book Company, New York, 1960.
4. Hadley, G.: "Linear Programming," Addison-Wesley Publishing Company, Inc., Reading, Mass., 1962.
5. Hadley, G.: "Nonlinear and Dynamic Programming," Addison-Wesley Publishing Company, Inc., Reading, Mass., 1964.
6. Ralston, A., and H. S. Wilf (eds.): "Mathematical Methods for Digital Computers," vol. I, John Wiley & Sons, Inc., New York, 1960.
7. Vajda, S.: "The Theory of Games and Linear Programming," John Wiley & Sons, Inc., New York, 1956.
8. Vajda, S.: "Mathematical Programming," Addison-Wesley Publishing Company, Inc., Reading, Mass., 1961.

PROBLEMS

12-1. Maximize the function $C = 3x_1 + 6x_2 + 2x_3$ subject to the constraints

$$3x_1 + 4x_2 + x_3 \leq 2$$
$$x_1 + 3x_2 + 2x_3 \leq 1$$
$$x_1 \geq 0 \qquad x_2 \geq 0 \qquad x_3 \geq 0$$

12-2. Use the simplex method to show that the value of the objective function in the following linear programming problem is unbounded. Maximize the function $C = 3x_1 + 4x_2$ subject to the constraints

$$x_1 + 2x_2 \geq 6$$
$$2x_1 + x_2 \geq 6$$
$$x_1 \geq 0 \qquad x_2 \geq 0$$

(*Note:* $x_1 = 2$ and $x_2 = 2$ is a basic feasible solution.)

12-3. Minimize the function $C = x_1 - x_2$ subject to the constraints

$$4x_1 + 5x_2 \leq 10$$
$$5x_1 + 2x_2 \leq 1$$
$$3x_1 + 8x_2 \leq 12$$
$$x_1 \geq 0 \qquad x_2 \geq 0$$

Use the simplex method and start with the basic feasible solution $x_1 = 0$ and $x_2 = 0$.

12-4. Maximize the function $C = 3x_1 + 4x_2 - 2x_3$ subject to the constraints

$$3x_1 + 4x_2 + x_3 \leq 2$$
$$2x_1 - 3x_2 + x_3 \geq -4$$
$$x_1 \leq 0 \qquad x_2 \geq 0 \qquad x_3 \leq 0$$

12-5. Maximize the function $C = 5x_1 + 2x_2$ subject to the constraints

$$-x_1 + x_2 \leq 5$$
$$10x_1 + x_2 \leq 10$$
$$x_2 \geq 0$$

by

(*a*) Using a geometric method.

(*b*) Formulating the problem as a linear programming problem and using the simplex method to solve it.

12-6. Solve the dual of the following linear programming problem: Maximize the function $C = -3x_1 + 4x_2 - x_3$ subject to the constraints

$$-x_1 + x_2 + x_3 \leq 5$$
$$x_1 - x_2 - 2x_3 \geq -6$$
$$5x_1 + x_2 - 3x_3 = 4$$
$$x_1 \geq 0 \qquad x_2 \geq 0 \qquad x_3 \geq 0$$

12-7. An automobile manufacturer produces two basic car models A and B. They are sold to car dealers at a profit of \$200 per car A and \$100 per car B. A requires, on the average, 150 man-hours for assembly, 50 man-hours for painting and finishing, and 10 man-hours for checking out and testing. B averages 60 man-hours for assembly, 40 man-hours for painting and finishing, and 20 man-hours for checkout and testing. During each production run, there are 30,000 man-hours available in the assembly shops, 13,000 man-hours in the painting and finishing shops, and 5,000 man-hours in the checking and testing division. Use the simplex method to determine the number of each model the manufacturer should plan to produce so as to realize the greatest possible profit from each production run.

12-8. The values of a continuous function $f(x)$ at $x = 0, 1, 2$, and 3 are

x	0	1	2	3
$f(x)$	2	-2	1	4

$f(x)$ may be approximated by a second-degree polynomial

$$g(x) = a_0 + a_1x + a_2x^2$$

such that the maximal value of the approximation error $|g(x) - f(x)|$ at $x = 0, 1, 2$, and 3 is minimized.

(*a*) Formulate this minimization problem as a linear programming problem.

(*b*) Formulate the dual problem.

12-9. A pound of liver contains 0.1 unit of protein and 0.04 unit of iron. A pound of beef contains 0.3 unit of protein and 0.016 unit of iron. Both liver and beef cost \$1/lb. The minimal daily requirement for a student is 0.5 unit of protein and 0.048 unit of iron.

(*a*) What quantities of liver and beef should a student consume daily to meet his minimal requirement at the smallest total cost?

(*b*) Formulate the dual of the linear programming problem formulated in (*a*).

12-10. Prove that if the kth constraint in a linear programming problem is an equality, then the corresponding variable y_k in the dual problem is unrestricted in sign, and conversely.

12-11. Describe how the simplex method may be applied directly to a linear programming problem in which the objective function is to be minimized, without reformulation as a maximization problem.

12-12. When there is no obvious basic feasible solution to a linear programming problem, one can be found by introducing artificial variables, as discussed in Sec. 12-6. Another method for finding a basic feasible solution is presented in this problem. Consider the example used in Sec. 12-6 and repeated below.

Maximize the function $C = 5x_1 + 3x_2$ subject to the constraints

$$-4x_1 - 5x_2 \leq -10$$

$$5x_1 + 2x_2 \leq 10$$

$$3x_1 + 8x_2 \leq 12$$

$$x_1 \geq 0 \qquad x_2 \geq 0$$

(*a*) Use the simplex method to minimize the function $C' = x_0$ subject to the constraints

$$-4x_1 - 5x_2 - x_0 \leq -10$$
$$5x_1 + 2x_2 \leq 10$$
$$3x_1 + 8x_2 \leq 12$$
$$x_1 \geq 0 \qquad x_2 \geq 0 \qquad x_0 \geq 0$$

What is the significance of the optimal solution to this problem?

(*b*) Use the optimal solution obtained in part (*a*) to solve the original maximization problem.

(*c*) State a general procedure for finding a basic feasible solution to a linear programming problem. Under what condition will a linear programming problem possess no feasible solution?

12-13. In this problem another method of solving linear programming problems, known as the *dual simplex method,* is studied. In the dual simplex method, a sequence of basic solutions is generated, each having the property that the coefficients in the bottom row of the corresponding tableau are all positive. The method terminates when a basic feasible solution is found. Consider the problem of maximizing the function $C = -x_1 - 2x_2 - 5x_3$ subject to the constraints

$$3x_1 - 5x_2 + 2x_3 \leq -3$$
$$x_1 + x_2 - 3x_3 \leq 1$$
$$2x_1 - x_2 - 3x_3 \leq -2$$
$$x_1 \geq 0 \qquad x_2 \geq 0 \qquad x_3 \geq 0$$

Adding slack variables, these constraints become

$$3x_1 - 5x_2 + 2x_3 + x_4 = -3$$
$$x_1 + x_2 - 3x_3 + x_5 = 1$$
$$2x_1 - x_2 - 3x_3 + x_6 = -2$$

It is clear that $x_1 = 0$, $x_2 = 0$, $x_3 = 0$, $x_4 = -3$, $x_5 = 1$, and $x_6 = -2$ constitutes a basic solution, although it is not a feasible one.

(*a*) Set up the tableau corresponding to this basic solution.

(*b*) Since both x_4 and x_6 assume negative values, one should be made a nonbasic variable. If x_4 is made a nonbasic variable, which variable should be made a basic variable? Set up the new tableau.

(*c*) At this point x_6 should be made a nonbasic variable. Which variable should be made a basic variable? Why? Complete the problem.

(*d*) State in general terms the steps necessary to carry out the dual simplex method.

12-14. To start the dual simplex method, a basic solution is needed for which the coefficients in the bottom row of the corresponding tableau are all positive. Consider the problem of maximizing the function $C = x_1 + 4x_2 + 2x_3$ subject to the constraints

$$3x_1 + x_2 + 4x_3 \leq 2$$
$$x_1 + 2x_2 + x_3 \leq 5$$
$$x_1 \geq 0 \qquad x_2 \geq 0 \qquad x_3 \geq 0$$

Since there is no obvious basic solution that satisfies the condition, the additional constraint

$$x_1 + x_2 + x_3 \leq M$$

is introduced, where M is a very large number. Clearly, this does not change the problem. Adding slack variables, the constraints become

$$\begin{aligned} 3x_1 + x_2 + 4x_3 + x_4 &= 2 \\ x_1 + 2x_2 + x_3 + x_5 &= 5 \\ x_1 + x_2 + x_3 + x_6 &= M \end{aligned}$$

A basic solution in which x_6 and two of the three variables x_1, x_2, and x_3 are nonbasic variables will have all positive coefficients in the bottom row of the corresponding tableau.

(*a*) Find this basic solution.

(*b*) Solve the problem using the dual simplex method.

(*c*) Maximize the objective function $C' = x_1 - 4x_2 + 2x_3$ under the same set of constraints using the dual simplex method.

(*d*) State in general terms the steps necessary to find a basic solution for starting the dual simplex method.

Chapter 13
Dynamic Programming

13-1 INTRODUCTION

Consider the operation of a machine shop in which there are 100 machines. These machines can be used to manufacture two different kinds of products, A and B. The total amount of product A manufactured by one machine in a week will be sold for a profit of 300 dollars and the total amount of product B manufactured by one machine in a week will be sold for a profit of 500 dollars. However, after each week's operation, 30 percent of the machines making product A will become worn out beyond repair, whereas 60 percent of the machines making product B will become worn out beyond repair. Suppose that machines are assigned to manufacture one of the two kinds of products on a weekly basis. Our question is as follows: Over a three-week period, how should the machines be allocated to the two products A and B such that the total profit is maximized?

This problem can be formulated and solved as a linear programming problem. Let x_1, x_2, and x_3 denote the numbers of machines allocated to product A in the first, second, and third week, respectively. Also, let

y_1, y_2, and y_3 denote the numbers of machines allocated to product B in the first, second, and third week, respectively. Thus, the problem is that of maximizing the objective function

$$C = 300(x_1 + x_2 + x_3) + 500(y_1 + y_2 + y_3)$$

subject to the constraints

$$x_1 + y_1 = 100$$
$$x_2 + y_2 = 0.7x_1 + 0.4y_1$$
$$x_3 + y_3 = 0.7x_2 + 0.4y_2$$
$$x_1 \geq 0 \qquad x_2 \geq 0 \qquad x_3 \geq 0 \qquad y_1 \geq 0 \qquad y_2 \geq 0 \qquad y_3 \geq 0$$

However, this problem can also be solved in an alternative way. Let us consider the operation in the last week of the three-week period. Suppose that n machines are still in working order at that time. If x of the n machines are assigned to product A and $n - x$ of the n machines are assigned to product B, the profit from the week's operation is

$$300x + 500(n - x) = 500n - 200x$$

Since in the last week there is no need to consider the number of machines that will remain usable later on, we should simply maximize the profit, $500n - 200x$. Clearly, by setting x to 0, that is, using all the machines for product B, we shall obtain the largest possible profit, $500n$ dollars. This, of course, is an expected result. Since the profit from manufacturing product B is higher than that from manufacturing product A, we should assign all the machines to product B. Also notice that since the allocation of machines in the last week of the three-week period cannot change the profit from the operation in the preceding weeks, the optimal decision for the last week is simply the one that makes the best use of the n machines that are in working order.

Anticipating the functional equation formulation to be presented later on, we introduce some symbolism. Let $f_1(n)$ denote the maximal profit from the operation of n machines for the period of one week. Then we can write

$$f_1(n) = \max_{0 \leq x \leq n} [300x + 500(n - x)]\dagger = 500n$$

Now, consider the allocation problem in the second week. Suppose that n machines are still in working order at that time. If x of the n

† The notation means the maximal value of the expression within the brackets subject to the constraint $0 \leq x \leq n$.

machines are assigned to product A and $n - x$ of the n machines are assigned to product B, then the profit from the second week's operation is

$$300x + 500(n - x) = 500n - 200x$$

Moreover, the number of machines that will still be usable in the third week is

$$0.7x + 0.4(n - x) = 0.4n + 0.3x$$

Therefore, the total profit from the two weeks' operation is equal to the sum of the profit from the second week's operation, $500n - 200x$, and the profit from the operation in the last week with $0.4n + 0.3x$ usable machines. According to our earlier discussion, with $0.4n + 0.3x$ usable machines the largest profit from the last week's operation will be

$$f_1(0.4n + 0.3x) = 500(0.4n + 0.3x)$$

Thus, the largest total profit from the two weeks' operation is

$$(500n - 200x) + 500(0.4n + 0.3x)$$

for an appropriately chosen value of x.

Again, in a more formal way, let $f_2(n)$ denote the largest total profit from the operation in two weeks starting with n usable machines. Then,

$$\begin{aligned} f_2(n) &= \max_{0 \le x \le n} [300x + 500(n - x) + f_1(0.4n + 0.3x)] \\ &= \max_{0 \le x \le n} [300x + 500(n - x) + 500(0.4n + 0.3x)] \\ &= \max_{0 \le x \le n} [700n - 50x] \\ &= 700n \end{aligned}$$

where the value of x is chosen to be zero; that is, in the second week, all the machines should be assigned to product B. As pointed out above, in the third week all the machines should also be assigned to product B. It follows that by assigning all the usable machines to product B in both the second and third weeks, we shall bring in the largest profit, which is $700n$ dollars. Of course, the allocation of machines in the second week does not affect the profit from the first week's operation. Therefore, the optimal decision at the second week is the one that makes the best use of the n usable machines in the two-week period consisting of the second and third weeks.

Similarly, consider the operation in the whole three-week period. Let $f_3(n)$ denote the largest total profit with n initially usable machines. Let x denote the number of machines assigned to product A in the first week. Then,

$$\begin{aligned} f_3(n) &= \max_{0 \le x \le n} [300x + 500(n - x) + f_2(0.4n + 0.3x)] \\ &= \max_{0 \le x \le n} [300x + 500(n - x) + 700(0.4n + 0.3x)] \\ &= \max_{0 \le x \le n} [780n + 10x] \\ &= 790n \end{aligned}$$

where the value of x is chosen to be n. This means that during the first week's operation we should use all the machines to manufacture product A, although we should assign all the usable machines to product B in the second and third weeks. Such a decision confirms our intuition that although the immediate profit is less, assigning the machines to product A is still a right move on a long-term basis, as the attrition of machines is lower. With $n = 100$, we have

$$f_3(100) = 79{,}000$$

which is the maximal profit from the operation in a three-week period.

Comparing the linear programming formulation with the method of solution that is outlined above, we realize that in the former approach the assignments of machines for all three weeks are determined simultaneously, whereas in the latter approach the assignments are determined in a step-by-step manner. In the linear programming approach, one solves an optimization problem involving three variables x_1, x_2, and x_3 (the other three variables y_1, y_2, and y_3 can be eliminated by using the equality constraints); on the other hand, in the new approach, one successively solves three optimization problems involving a single variable. The difference will become even more significant when we consider the operation of the machine shop for a period of 10 or 100 weeks, as the reader can see.

Let us modify the machine-shop problem by assuming that the profit from the sale of the total amount of product A manufactured by m machines in a week is $3m^2$ dollars and that the profit from the sale of the total amount of product B manufactured by m machines in a week is $4m^2$ dollars. Notice that the problem is no longer a linear programming problem. However, the step-by-step computation procedure employed above can still be applied. Let $f_k(n)$ denote the maximal profit from the operation in a k-week period starting with n machines.

Let x denote the number of machines assigned to product A. Then,

$$\begin{aligned} f_1(n) &= \max_{0 \leq x \leq n} [3x^2 + 4(n - x)^2] \\ &= \max_{0 \leq x \leq n} [4n^2 - 8nx + 7x^2] \\ &= 4n^2 \end{aligned}$$

where x is chosen to be equal to 0. Similarly,

$$\begin{aligned} f_2(n) &= \max_{0 \leq x \leq n} [3x^2 + 4(n - x)^2 + f_1(0.4n + 0.3x)] \\ &= \max_{0 \leq x \leq n} [3x^2 + 4(n - x)^2 + 4(0.4n + 0.3x)^2] \\ &= \max_{0 \leq x \leq n} [4.64n^2 - 7.04nx + 7.36x^2] \\ &= 4.96n^2 \end{aligned}$$

where x is chosen to be equal to n; and

$$\begin{aligned} f_3(n) &= \max_{0 \leq x \leq n} [3x^2 + 4(n - x)^2 + f_2(0.4n + 0.3x)] \\ &= \max_{0 \leq x \leq n} [3x^2 + 4(n - x)^2 + 4.96(0.4n + 0.3x)^2] \\ &= \max_{0 \leq x \leq n} [4.7936n^2 - 6.8096nx + 7.4464x^2] \\ &= 5.4304n^2 \end{aligned}$$

where x is chosen to be equal to n.

Notice now that in both the first and second weeks we should assign all the machines to product A. Only in the third week should we use the machines for product B. Following such a policy, the total profit is

$$f_3(100) = 5.4304 \times (100)^2 = 54{,}304$$

The example in this section illustrates the solution of a class of problems by a recursive computational procedure which is referred to as *dynamic programming*. In this chapter, we shall present the basic idea of dynamic programming together with a few illustrative examples. The details of this subject are beyond our scope of discussion.†

13-2 THE PRINCIPLE OF OPTIMALITY

The technique of dynamic programming is quite powerful in solving the decision problems of multistage processes. A *multistage process* is one

† For further details, see Bellman [1] and Nemhauser [4].

that has a certain number of successive stages of operation. For example, in the machine-shop problem in Sec. 13-1 the operation can be divided into three stages, namely, the first, second, and third week. A multistage process is characterized by a certain number of parameters, which are called the *state variables.* In other words, the values of the state variables at each stage fully describe the status of the process at that stage. A combination of the values of the state variables is called a *state* of the process. For example, in the machine-shop problem, the state variable is the number of machines that are in working order. Also, consider the problem of making daily assignment of men and women workers to perform a certain number of jobs, some of which are open only to men and some of which are open only to women. The state variables are then the numbers of men and women who report to work on each day.

A multistage process is said to be *Markovian,* if at any stage the behavior of the process depends solely on the current values of the state variables. For example, in the machine-shop problem, the operation of the shop depends solely on the number of machines that are in working order in each week. Throughout our discussion, a multistage process shall be understood to be Markovian, unless otherwise specified. At each stage of a multistage process, a *decision* is a choice among a certain number of possible actions. The number of alternative actions can be either finite or infinite. In the machine-shop problem, the decision we have to make each week is the number of machines to be assigned to product A. The quantity to be optimized in a decision problem is called the *objective function.* Clearly, this function in the machine-shop problem is the total profit in the three-week period. A decision at each stage not only affects the value of the objective function, but also determines the values of the state variables at the subsequent stage. In the machine-shop problem, the decision on the number of machines to be assigned to each of the two products not only affects the total profit, but also determines the number of usable machines for the subsequent week. Since the process is Markovian, a decision at each stage can be made solely on the basis of the values of the state variables. In a multistage decision problem, a *policy* is any rule for making a decision at each of the stages that yields an allowable sequence of decisions. For example, in the machine-shop problem, a policy consists of the numbers of machines to be assigned to product A in each of the first, second, and third weeks. An *optimal policy* is one which optimizes the objective function.

The formulation and the solution of multistage decision problems are based on the *principle of optimality:*

An optimal policy is one where, whatever the initial state of the process and the initial decision, the remaining decisions must constitute an optimal policy with regard to the new state of the process resulting from the first decision.

The principle of optimality is simply a formalization of an intuitive notion which can be justified by the following argument. Suppose that there is an optimal policy that does not satisfy the principle of optimality; that is, for a certain initial state and a certain initial decision, the subsequent decisions based on the policy do not constitute an optimal policy with regard to the new state resulting from the initial decision. This means that if the process *starts initially* at this new state, the policy will not yield a sequence of optimal decisions, which is a contradiction to the assumption that the policy is an optimal policy.

The recursive computational procedure in solving multistage decision problems is a direct consequence of the principle of optimality. In searching for an optimal decision at any stage of the process, one needs only to look for a decision that will optimize the activities in the subsequent stages. Therefore, as illustrated by the machine-shop problem in Sec. 13-1, one starts the computation by searching for an optimal decision for the last stage of the process and then works "backward" to determine an optimal policy.

Let us consider a multistage process with r state variables. Let $f_k(x_1,x_2, \ldots ,x_r)$ denote the optimal value of the objective function for a k-stage operation that starts with the initial values of the state variables being $x_1, x_2, \ldots , x_r$. For $i = 1, 2, \ldots$, let $g_i(x_1,x_2, \ldots ,x_r)$ denote the return (the contribution to the value of the objective function) in one stage of operation when the values of the state variables are $x_1, x_2, \ldots , x_r$ and the ith decision is made. Also, for $i = 1, 2, \ldots$, let $h_i(x_1,x_2, \ldots ,x_r)$† denote the new values of the state variables when the values of the state variables are $x_1, x_2, \ldots , x_r$ and the ith decision is made. Let $G(u,v)$ denote the value of the objective function for a k-stage operation, where u is the return in one stage of operation (for a certain decision) and v is the return in the subsequent $k - 1$ stages of operation (for a corresponding sequence of decisions). Formally, the recursive computational step based on the principle of optimality can be written as

$$f_k(x_1,x_2, \ldots ,x_r) = \underset{i}{\text{optimize}} \; [G(g_i(x_1,x_2, \ldots ,x_r),f_{k-1}(h_i(x_1,x_2, \ldots ,x_r)))]\ddagger \qquad (13\text{-}1)$$

Using Eq. (13-1), we can compute $f_k(x_1,x_2, \ldots ,x_r)$ when the value of

† The value of the function h_i is an ordered r-tuple.

‡ The notation means to optimize the value of the expression in the brackets by choosing an optimal value for i.

$f_{k-1}(x_1,x_2, \ldots ,x_r)$ is known. It follows that a step-by-step computation can be carried out recursively. Before studying the next example, we suggest that the reader review the machine-shop problem in Sec. 13-1.

Example 13-1 Consider the problem of distributing $6,000 among three methods A, B, and C of marketing a new product. The projected profit from each method as a function of the amount of money spent is shown in the following table (the functions are denoted by g_A, g_B, and g_C, respectively):

	Amount spent						
Profit function	0	1,000	2,000	3,000	4,000	5,000	6,000
g_A	10,000	12,000	15,000	18,500	24,000	30,000	33,000
g_B	12,000	18,000	20,000	22,500	24,000	25,000	26,000
g_C	6,000	12,000	20,000	22,000	24,000	26,000	28,000

Let $f_1(a)$ denote the total profit when a dollars is allocated to method A. Let $f_2(a)$ denote the total profit when a dollars is distributed between methods A and B following an optimal policy. Let $f_3(a)$ denote the total profit when a dollars is distributed among methods A, B, and C following an optimal policy. Clearly,

$$f_1(a) = g_A(a) \tag{13-2}$$

$$f_2(a) = \max_{0 \leq x \leq a} [g_B(x) + f_1(a - x)] \tag{13-3}$$

$$f_3(a) = \max_{0 \leq x \leq a} [g_C(x) + f_2(a - x)] \tag{13-4}$$

According to Eq. (13-2), we have

a	$f_1(a)$
0	10,000
1,000	12,000
2,000	15,000
3,000	18,500
4,000	24,000
5,000	30,000
6,000	33,000

In a step-by-step computation based on Eq. (13-3), we obtain

a	$g_B(x) + f_1(a - x)$							$f_2(a)$
	$a - x$							
	0	1,000	2,000	3,000	4,000	5,000	6,000	
0	22,000							22,000
1,000	28,000	24,000						28,000
2,000	30,000	30,000	27,000					30,000
3,000	32,500	32,000	33,000	30,500				33,000
4,000	34,000	34,500	35,000	36,500	36,000			36,500
5,000	35,000	36,000	37,500	38,500	42,000	42,000		42,000
6,000	36,000	37,000	39,000	41,000	44,000	48,000	45,000	48,000

Similarly, based on Eq. (13-4), we obtain

a	$g_C(x) + f_2(a - x)$							$f_3(a)$
	$a - x$							
	0	1,000	2,000	3,000	4,000	5,000	6,000	
0	28,000							28,000
1,000	34,000	34,000						34,000
2,000	42,000	40,000	36,000					42,000
3,000	44,000	48,000	42,000	39,000				48,000
4,000	46,000	50,000	50,000	45,000	42,500			50,000
5,000	48,000	52,000	52,000	53,000	48,500	48,000		53,000
6,000†	50,000	54,000	54,000	55,000	56,500	54,000	54,000	56,500

† In obtaining the answer to this problem, only the entries in this row need to be computed. The entries in other rows are included for completeness.

Since $f_3(6{,}000) = 56{,}500$, we conclude that the maximal total profit will be \$56,500. Moreover, because

$$f_3(6{,}000) = \max_{0 \le x \le 6{,}000} [g_C(x) + f_2(6{,}000 - x)] = 56{,}500$$

at $x = 2{,}000$ and because

$$f_2(4{,}000) = \max_{0 \le x \le 4{,}000} [g_B(x) + f_1(4{,}000 - x)] = 36{,}500$$

at $x = 1{,}000$, the optimal way of distributing the money is to allocate \$3,000 to method A, \$1,000 to method B, and \$2,000 to method C. ■

13-3 FUNCTIONAL EQUATIONS

According to our discussion in Sec. 13-2, Eq. (13-1) is used for a step-by-step computation of the value of $f_k(x_1,x_2, \ldots ,x_r)$ for $k = 1, 2, \ldots$. Actually, Eq. (13-1) can also be viewed as a *functional equation* relating the two functions $f_k(x_1,x_2, \ldots ,x_r)$ and $f_{k-1}(x_1,x_2, \ldots ,x_r)$. Therefore, instead of using a step-by-step computation, we can also solve the functional equation to obtain directly the function $f_k(x_1,x_2, \ldots ,x_r)$ in closed form. By solving a functional equation of the form in Eq. (13-1) we mean finding $f_t(x_1,x_2, \ldots ,x_r)$ for $t = 1, 2, \ldots$ such that when the expression for $f_{k-1}(x_1,x_2, \ldots ,x_r)$ is substituted into the expression

$$[G(g_i(x_1,x_2, \ldots ,x_r),f_{k-1}(h_i(x_1,x_2, \ldots ,x_r)))]$$

and a value of i that optimizes the expression is chosen, the expression

$$\underset{i}{\text{Optimize}}\ [G(g_i(x_1,x_2, \ldots ,x_r),f_{k-1}(h_i(x_1,x_2, \ldots ,x_r)))]$$

becomes $f_k(x_1,x_2, \ldots ,x_r)$. Unfortunately, not all functional equations can be solved in closed form. The existence and uniqueness problems of the solution of a functional equation are mathematically too involved to be covered here.† We shall limit ourselves to some illustrative examples.

Example 13-2 Find the values of $p_1, p_2, \ldots , p_n$ such that the quantity

$$p_1 \log p_1 + p_2 \log p_2 + \cdots + p_n \log p_n$$

is minimized subject to the constraints

$$p_1 \geq 0, p_2 \geq 0, \ldots , p_n \geq 0$$

$$p_1 + p_2 + \cdots + p_n = 1$$

This problem can be viewed as a multistage decision problem in which the total amount of 1 is to be divided into n parts in n steps. If we let $f_k(a)$ denote the minimal value of $\sum_{i=1}^{k} p_i \log p_i$ when the quantity a is to be divided into k nonnegative parts $p_1, p_2, \ldots , p_k$, we have the functional equation

$$f_k(a) = \min_{0 \leq x \leq a} [x \log x + f_{k-1}(a - x)]$$

† See Bellman [1].

With this functional equation and the initial condition

$$f_1(a) = a \log a$$

we can compute $f_k(a)$ and thus eventually $f_n(1)$ in a step-by-step manner. However, it is observed that the solution to the functional equation is

$$f_t(y) = y \log \frac{y}{t}\dagger \qquad t = 1, 2, \ldots, n$$

This can be checked as follows: We have

$$\begin{aligned} x \log x + f_{k-1}(a - x) &= x \log x + (a - x) \log \frac{a - x}{k - 1} \\ &= x \log x + (a - x) \log (a - x) \\ &\qquad - (a - x) \log (k - 1) \end{aligned}$$

To find the value of x $(0 \leq x \leq a)$ that minimizes the value of the expression, we set the derivative of the expression to zero. Thus,

$$\begin{aligned} &\frac{d}{dx} [x \log x + (a - x) \log (a - x) - (a - x) \log (k - 1)] \\ &\qquad = \log x + 1 - \log (a - x) - 1 + \log (k - 1) = 0 \end{aligned}$$

that is,

$$\frac{x}{a - x} = \frac{1}{k - 1}$$

or

$$x = \frac{a}{k}$$

It follows that

$$\begin{aligned} \min_{0 \leq x \leq a} [x \log x + f_{k-1}(a - x)] &= \frac{a}{k} \log \frac{a}{k} + f_{k-1}\left[\frac{(k - 1)a}{k}\right] \\ &= \frac{a}{k} \log \frac{a}{k} + \frac{(k - 1)a}{k} \log \frac{a}{k} \\ &= a \log \frac{a}{k} = f_k(a) \end{aligned}$$

Therefore, we have

$$f_n(1) = \log \frac{1}{n}$$

† Note that we use y as the independent variable in the function f_t for the sole reason of avoiding confusion in notations.

as the minimal value of the quantity $\sum_{i=1}^{n} p_i \log p_i$. Moreover, we observe that

$$p_1 = p_2 = \cdots = p_n = \frac{1}{n} \quad \blacksquare$$

A multistage process consisting of a large number of stages can be approximated as an infinite-stage process. In that case, the functions $f_t(x_1,x_2, \ldots ,x_r)$ for all t can be approximated by one function

$$f(x_1,x_2, \ldots ,x_r)$$

That is, at each stage, we have the functional equation

$$f(x_1,x_2, \ldots ,x_r) = \underset{i}{\text{optimize}}\ [G(g_i(x_1,x_2, \ldots ,x_r),f(h_i(x_1,x_2, \ldots ,x_r)))]$$

Let us consider the following example.

Example 13-3 Consider the operation of a cab company that has a fleet of n cars. Each month, a car in operation can bring in a profit of a dollars. After each month's operation, only a fraction of the cars remain in good running condition, and the others will be sent to the junk yard. Let b denote such a fraction. A car can be sold at any time for c dollars (we assume that there is no depreciation, since that can be accounted for in the fraction b). What is the optimal policy for the owner of the company in terms of selling or keeping his cars so that the total profit is maximized?

The operation can be approximated as a process with an infinite number of stages. Letting $f(u)$ denote the total profit which results from starting with u cabs and following an optimal policy, we have the functional equation

$$f(u) = \max_{0 \le x \le u} [ax + c(u - x) + f(bx)]$$

where x denotes the number of cabs that are kept in operation for one month and $(u - x)$ denotes the number of cabs that are sold at the beginning of the operation.

For $[a/(1 - b)] - c \ge 0$, the functional equation is satisfied by

$$f(y) = \left(\frac{a}{1 - b}\right) y$$

This can be checked by observing that

$$\max_{0\le x\le u}\left[ax + c(u - x) + \left(\frac{a}{1-b}\right)bx\right]$$
$$= \max_{0\le x\le u}\left[\left(a - c + \frac{ab}{1-b}\right)x + cu\right]$$
$$= \max_{0\le x\le u}\left[\left(\frac{a}{1-b} - c\right)x + cu\right]$$
$$= \frac{a}{1-b}\,u$$

at $x = u$.

For $[a/(1 - b)] - c \le 0$, the functional equation is satisfied by

$$f(y) = cy$$

which can also be checked by observing that

$$\max_{0\le x\le u}[ax + c(u - x) + bcx] = \max_{0\le x\le u}[(a + bc - c)x + cu] = cu$$

at $x = 0$.

In summary, we have the following results:

	Optimal policy	*Total profit of the operation*
$\frac{a}{1-b} - c \ge 0$	Continue the operation indefinitely	$\frac{an}{1-b}$
$\frac{a}{1-b} - c \le 0$	Sell all cars at the beginning	cn

It should be noted that the results check out quite well with our intuition. Suppose that we can run a car for an infinitely long period; that is, imagine that the car is depreciated and becomes b car (a fraction of a car), and then b^2 car, and so on. The total return from the car is then

$$a + ab + ab^2 + \cdots + ab^r + \cdots = \frac{a}{1-b}$$

If $[a/(1 - b)] - c \ge 0$, clearly, we should keep the car running instead of selling it at the beginning. Moreover, from the relation

$$c \le a + bc \le a + ab + cb^2 \le \cdots$$
$$\le a + ab + ab^2 + \cdots + ab^{k-1} + cb^k \le \cdots \le \frac{a}{1-b}$$

we conclude that the longer we run the car before we sell it, the larger the total profit will be, since the expression $a + ab + ab^2 + \cdots + ab^{k-1} + cb^k$ is the total profit when we sell the car after running it for k months.

On the other hand, if $[a/(1 - b)] - c \leq 0$, we have the relation

$$c \geq a + bc \geq a + ab + cb^2 \geq \cdots$$
$$\geq a + ab + ab^2 + \cdots + ab^{k-1} + cb^k \geq \cdots \geq \frac{a}{1-b}$$

We conclude, therefore, that an outright sale of the cars is the best policy, whereas the longer we keep the cars, the smaller the total profit will be. ∎

13-4 SUMMARY AND REFERENCES

Although the presentation in this chapter is brief and introductory, we have covered the fundamental concepts in the solution of multistage decision problems by the technique of dynamic programming. The two major steps in problem solving are the formulation of the functional equation and the solution of the functional equation. With the functional equation, a step-by-step computation for the optimal decisions is possible, as illustrated by the machine-shop problem in Sec. 13-1. On the other hand, there is no general method of finding the solutions to functional equations in closed form. As a matter of fact, the determination of the existence and the uniqueness of the solutions to functional equations is no simple mathematical problem, as the reader can find out in Bellman's book [1]. At this point, it is instructive to recall our discussion on recurrence relations in Chap. 3 and to see the strong resemblance between the technique of recurrence relations in solving enumeration problems and the technique of dynamic programming in solving optimization problems.

Bellman, together with his disciples, has made significant contributions to the field. The book by Bellman [1] was the first book written on the topic and is still a most complete reference. See Chaps. 1, 2, and 3 of this book for the material presented here. The remainder of this book consists of the formulation of dynamic programming problems in many areas of application. Nemhauser's book [4] is more compatible with our level of presentation. See also Chaps. 10 and 11 of Hadley [3] and Chaps. 1, 2, 3, and 4 of Bellman and Dreyfus [2].

1. Bellman, R.: "Dynamic Programming," Princeton University Press, Princeton, N.J., 1957.
2. Bellman, R., and S. Dreyfus: "Applied Dynamic Programming," Princeton University Press, Princeton, N.J., 1962.
3. Hadley, G.: "Nonlinear and Dynamic Programming," Addison-Wesley Publishing Company, Inc., Reading, Mass., 1964.

4. Nemhauser, G. L.: "Introduction to Dynamic Programming," John Wiley & Sons, Inc., New York, 1966.

PROBLEMS

13-1. Find the maximal value of the product $x_1x_2 \cdots x_n$ subject to the constraints

$$x_1 + x_2 + \cdots + x_n = 1$$
$$x_1 \geq 0, x_2 \geq 0, \ldots, x_n \geq 0$$

13-2. Minimize the quantity $x_1^2 + x_2^2 + \cdots + x_n^2$ subject to the constraints

$$x_1 + x_2 + \cdots + x_n = 1$$
$$x_1 \geq 0, x_2 \geq 0, \ldots, x_n \geq 0$$

13-3. Maximize the quantity $x_1y_1 + x_2y_2 + \cdots + x_ny_n$ subject to the constraints

$$x_1 + x_2 + \cdots + x_n = a$$
$$y_1 + y_2 + \cdots + y_n = b$$
$$x_1 \geq 0, x_2 \geq 0, \ldots, x_n \geq 0$$
$$y_1 \geq 0, y_2 \geq 0, \ldots, y_n \geq 0$$

13-4. A particle is traveling on a plane from a point (x_1,y_1) to another point (x_2,y_2) at a constant speed c. Find the minimal traveling time by formulating the problem as a dynamic programming problem.

13-5. A chemical processing plant must purify 20,000 lb of a certain chemical compound in a period of four weeks. There are two purification methods that can be used. The first completely purifies x lb of the compound in one week at a cost of $0.60x^2$ dollars. The second, given x lb of the compound to be processed in one week, completely purifies 70 percent of this material and leaves 30 percent to be reprocessed during a subsequent week. It does this at a cost of $0.15x^2$ dollars. Determine an optimal policy for the operation.

13-6. Problem 13-5 is to be approximated as an infinite-stage process.

(*a*) Set up the functional equation.

(*b*) It is known that the solution to this functional equation is of the form $f(y) = my^2$. Determine the constant m.

(*c*) Compare the result in part (*b*) with the result in Prob. 13-5.

13-7. A furniture manufacturer has an order for 100 chairs to be delivered in three months. The costs of producing x chairs in the first, second, and third months are $120x$, $1.2x^2$, and $1.5x^2$ dollars, respectively. Determine an optimal policy for the number of chairs to be produced in each month.

13-8. Two types of machines, I and II, are available to make two kinds of products, A and B, in a five-week period. Machines are assigned to the products on a weekly basis. The weekly attrition rates for the machines, according to the product assigned, are

	Product A	*Product B*
Type I machines	30%	50%
Type II machines	50%	80%

A machine producing product A yields a profit of \$300 in a week, and a machine producing product B yields a profit of \$400 in a week. Determine an optimal policy for the operation.

13-9. Nineteen yards of material are available for making three kinds of dresses, which require 2, 3, and 5 yd of material, respectively. The profit from selling a dress of each kind is \$1.5, \$2.5, and \$4, respectively.

(*a*) Find *all* optimal ways of allocating the material to the three kinds of dresses so that the total profit is maximized.

(*b*) Find a way of allocating the material so that the total profit is the largest integer number of dollars.

13-10. An electronics firm has a contract to deliver the following numbers of radios in a four-month period:

	1*st month*	2*d month*	3*d month*	4*th month*
Number of radios	300	400	300	300

The labor costs are

	1*st month*	2*d month*	3*d month*	4*th month*
Labor cost per radio	\$10	\$10	\$12	\$13

The inventory cost for each month is \$1.5 for each radio in stock at the beginning of the month. The cost of setting up production each month is \$250. The number of radios produced must be an integer multiple of 100. Determine an optimal policy.

13-11. To advertise a new product, a company has the choice of purchasing commercial time from a radio station or from a television station. The subscription is to be made on a weekly basis. Determine an optimal policy for the operation during a three-week period, given the following data:

1. The estimated profit from sales during the dth consecutive week of radio (television) advertisement is $4{,}000/d$ $(6{,}000/d)$ dollars.
2. There is a fee of \$2,500 each time the company switches its advertising medium.

13-12. (*a*) Consider the operation of a machine shop in which there are 200 machines. A machine can be used to manufacture product A for a profit of \$300 per week or product B for a profit of \$400 per week. After each week's operation, 30 percent of the machines making product A are inoperable, and 50 percent of the machines making product B are inoperable. Find an optimal policy of operation for a four-week period.

(*b*) Several mechanics are available during the weekend preceding the third week to repair all inoperable machines resulting only from the second week's operation. When repaired machines are used to manufacture product A, 50 percent of them become inoperable after one week. For product B, 80 percent become inoperable after one week. The mechanics can be hired for \$36,000. Should they be hired?

(*c*) Repeat part (*b*) if the mechanics can repair all inoperable machines from previous weeks.

Chapter 14
Block Designs

14-1 INTRODUCTION

The effects of six different drugs, numbered 1, 2, 3, 4, 5, and 6, on human bodies are to be tested. There are 10 human subjects, $A, B, \ldots, I$, and J, who will use the drugs within a six-day period, one drug on each day. We want to set up a schedule for the subjects to take the drugs during the time the experiment is conducted. A simple schedule is shown in Table 14-1; namely, all the subjects take the same drug on each day of the experiment.

It is quite possible that the change in the physical condition of the subjects within the six-day period will affect the outcome of the experiment. For example, because of the aftereffect of the drugs taken in the earlier part of the week, the drugs taken in the latter part of the week might be found to be more powerful than they actually are. Therefore, for the schedule in Table 14-1, the effect of drug 6 might be enhanced in the experimental data because it is taken on the last day by every subject. The situation is remedied in the schedule in Table 14-2, where

Table 14-1

	Day					
Subject	*Mon.*	*Tues.*	*Wed.*	*Thurs.*	*Fri.*	*Sat.*
A	1	2	3	4	5	6
B	1	2	3	4	5	6
C	1	2	3	4	5	6
D	1	2	3	4	5	6
E	1	2	3	4	5	6
F	1	2	3	4	5	6
G	1	2	3	4	5	6
H	1	2	3	4	5	6
I	1	2	3	4	5	6
J	1	2	3	4	5	6

Table 14-2

	Day					
Subject	*Mon.*	*Tues.*	*Wed.*	*Thurs.*	*Fri.*	*Sat.*
A	3	1	6	5	4	2
B	6	1	5	2	4	3
C	4	3	2	6	5	1
D	1	5	6	2	3	4
E	2	4	3	5	1	6
F	3	6	4	1	2	5
G	6	5	3	1	2	4
H	2	4	1	3	6	5
I	4	2	3	5	1	6
J	5	3	6	2	4	1

the subjects take the drugs in arbitrary order, and there is no regularity in the days the subjects take the drugs. Therefore, when the experimental data on the effect of a drug on the 10 subjects are averaged, the effect of each of the drugs is measured more accurately.

Suppose now that the subjects are available only for a period of three days. To obtain equal amounts of data on the effect of the different drugs, we want to make up the schedule in such a way that each drug will be used by the same number of subjects. A simple calculation, $(10 \times 3)/6 = 5$, shows that each drug should be tested on five subjects. Tables 14-3, 14-4, and 14-5 show three different schedules that can be used.

The schedule in Table 14-3 is similar to the schedule in Table 14-1 in that there is no consideration given to the order in which the drugs are taken by the subjects. The schedule in Table 14-4 is a slight modification of the one in Table 14-3 so that there will be no regularity in the order the drugs are taken. However, since there are differences in the physical conditions of the subjects, it is also desirable to have information about the effect of two different drugs on the same subject. For the schedules in Table 14-1 and Table 14-2, since every subject will take all six kinds of drugs, such information is readily available. On the other hand, in the schedules in Table 14-3 and Table 14-4 some pairs of drugs (for example, drug 1 and drug 4) are not to be tested by the same subject.

Table 14-3

	Day		
Subject	*Mon.*	*Tues.*	*Wed.*
A	1	2	3
B	1	2	3
C	1	2	3
D	1	2	3
E	1	2	3
F	4	5	6
G	4	5	6
H	4	5	6
I	4	5	6
J	4	5	6

Table 14-4

	Day		
Subject	*Mon.*	*Tues.*	*Wed.*
A	3	2	1
B	2	3	1
C	3	1	2
D	1	3	2
E	1	2	3
F	4	5	6
G	6	5	4
H	4	6	5
I	5	4	6
J	5	6	4

Table 14-5

	Day		
Subject	*Mon.*	*Tues.*	*Wed.*
A	1	2	3
B	1	4	2
C	3	1	5
D	4	6	1
E	5	1	6
F	3	6	2
G	2	5	4
H	6	5	2
I	5	4	3
J	6	3	4

The schedule in Table 14-5 not only has the feature that each drug is tested by five subjects (as in Table 14-3) and the feature that the subjects will take the drugs in some random order (as in Table 14-4), but also has the feature that every two drugs are tested by exactly two subjects. (For example, drugs 1 and 2 are taken by subjects A and B, drugs 1 and 3 are taken by subjects A and C, and so on.)

This example illustrates the class of problems discussed under the heading *block designs* that is the topic of this chapter. By a block design we mean a selection of the subsets of a given set such that some prescribed conditions are satisfied. In some designs, the elements in each of the subsets are also to be ordered in a certain way. In the example we have just seen, each schedule is a selection of 10 subsets of the set of six different drugs. The schedules in Tables 14-1 and 14-2 are selections containing 10 subsets, each of which is simply the set of six drugs. The schedules in Tables 14-3, 14-4, and 14-5 are selections containing ten 3-subsets of the set of six drugs. Moreover, in the schedules in Tables 14-2, 14-4, and 14-5, the elements in the subsets are ordered in such a way that there is no regularity in the days the drugs are taken by the subjects.

14-2 COMPLETE BLOCK DESIGNS

Let $X = \{x_1, x_2, \ldots, x_v\}$ be a set of v objects. A *complete block design* of X is a certain number of replications of the set X with the objects in the replications arranged according to certain specifications. In other words, all the subsets in the selection are the set X itself. Both the schedules in Tables 14-1 and 14-2 are examples of complete block designs.

One class of block designs is the so-called *randomized complete block design*, as illustrated by the schedule in Table 14-2. In such designs, we want to randomly order the objects in each replication so that there will be no regularity in the ordering of the objects in the replications. Imagine a lottery containing $v!$ tickets with each of the $v!$ permutations of the v objects in $X = \{x_1, x_2, \ldots, x_v\}$ on a ticket. A certain number of random orderings of the objects can be obtained by drawing the corresponding number of tickets from the lottery (with each ticket replaced after it is drawn). There are tables containing randomly arranged numbers that can be looked up for such random orderings. These tables are called *random-number tables.*† In most computer installations there are also computer programs that can generate a sequence of randomly ordered numbers. Such programs are usually called *random-number generating routines.*‡

† See, for example, Kendall and Babington-Smith [9], or RAND [13].

‡ See, for example, Hamming [8] and Ralston and Wilf [12].

Another class of complete block designs can be grouped under the heading *Latin squares.* We have defined an $r \times n$ Latin rectangle in Chap. 11 as an $r \times n$ arrangement of the numbers $1, 2, 3, \ldots, n$ in such a way that no number appears twice in the same row or in the same column. An $n \times n$ Latin rectangle is called an $n \times n$ *Latin square,* or a Latin square of order n.

Let us continue with the example in Sec. 14-1. Suppose that only six subjects are available when the experiment is conducted. In addition to testing the effect of different drugs on the same subject, we also want to have some measurement of the effect of the drugs when taken on different days of the six-day period. Therefore, we want to have a schedule in which the six different drugs will be taken by the six subjects on each day. Such a schedule is shown in Table 14-6. Observe that the entries in this table form a 6×6 Latin square.

Table 14-6

	Day					
Subject	*Mon.*	*Tues.*	*Wed.*	*Thurs.*	*Fri.*	*Sat.*
A	1	2	3	4	5	6
B	2	3	4	5	6	1
C	3	4	5	6	1	2
D	4	5	6	1	2	3
E	5	6	1	2	3	4
F	6	1	2	3	4	5

If, instead of six subjects, only four subjects are available for the experiment, although we cannot have all the six different drugs tested on each day, we can have a schedule in which four different drugs will be tested on each day (no two subjects will take the same drug on the same day). In this case, the schedule is a 4×6 Latin rectangle.

The construction of Latin squares is a very simple matter. For example, we can construct a Latin square of order n by arranging the numbers $1, 2, 3, \ldots, n$ in that order in the first row; the numbers $2, 3, \ldots, n, 1$ in that order in the second row; . . . ; and the numbers $n, 1, 2, 3, \ldots, n-1$ in that order in the nth row. Also, according to the result in Example 11-4, we can always expand a given $r \times n$ Latin rectangle into an $(r+1) \times n$ Latin rectangle $(r < n)$. Starting with any permutation of the n numbers, which is a $1 \times n$ Latin rectangle, we can then construct an $n \times n$ Latin square in a step-by-step manner.

14-3 ORTHOGONAL LATIN SQUARES

Let A_1 and A_2 be two Latin squares of order n. Let $a_{ij}^{(1)}$ and $a_{ij}^{(2)}$ ($i = 1, 2, \ldots, n$; $j = 1, 2, \ldots, n$) denote the entries in the ith row and the jth column in A_1 and A_2, respectively. The two Latin squares A_1 and A_2 are said to be *orthogonal* if the n^2 ordered pairs $(a_{ij}^{(1)}, a_{ij}^{(2)})$ ($i = 1, 2, \ldots, n$; $j = 1, 2, \ldots, n$) are all distinct. In other words, if we superimpose the two squares to form an $n \times n$ square with ordered pairs as entries, the orthogonality condition means that the entries of the resultant square are all distinct. For example, Fig. 14-1a shows two orthogonal Latin squares, and Fig. 14-1b shows the resultant square when they are superimposed.

$$A_1 = \begin{bmatrix} 1 & 2 & 3 \\ 2 & 3 & 1 \\ 3 & 1 & 2 \end{bmatrix} \quad A_2 = \begin{bmatrix} 1 & 2 & 3 \\ 3 & 1 & 2 \\ 2 & 3 & 1 \end{bmatrix} \qquad \begin{bmatrix} (1,1) & (2,2) & (3,3) \\ (2,3) & (3,1) & (1,2) \\ (3,2) & (1,3) & (2,1) \end{bmatrix}$$

(a) (b)

Figure 14-1

Suppose that three different drugs for fever and three different drugs for headache are to be tested by three subjects in a three-day period. So that the effect of different drugs for fever on the same subject and their effect when they are used on the same day of the experimental period by different subjects can be compared, we shall use a schedule for the subjects to take the fever drug that is a 3×3 Latin square as shown in Table 14-7, where the fever drugs are numbered 1, 2, and 3. Similarly, we shall use a schedule which is also a 3×3 Latin square for the headache drugs as shown in Table 14-8, where the headache drugs are also numbered 1, 2, and 3. Since each subject takes a fever drug and a headache drug on each day, we shall have the opportunity of observing their combined effect. Given a nine-day experimental period, we can let each of the subjects try all the nine possible combinations

Table 14-7

	Day		
Subject	*Mon.*	*Tues.*	*Wed.*
A	1	2	3
B	2	3	1
C	3	1	2

Table 14-8

	Day		
Subject	*Mon.*	*Tues.*	*Wed.*
A	1	2	3
B	3	1	2
C	2	3	1

of the fever drugs and headache drugs. However, with only three days for the experiment we want to have two schedules such that each combination of fever drug and headache drug will be tested by a subject once. In this case, we use a pair of orthogonal Latin squares of order 3 to schedule the fever drugs and the headache drugs, as shown in Tables 14-7 and 14-8.

Let $A_1, A_2, \ldots, A_r$ be a set of Latin squares of order n. They are said to be a set of orthogonal Latin squares if every two of them are orthogonal. We shall now present some results concerning the existence of sets of orthogonal Latin squares. It should be pointed out that there are no orthogonal Latin squares of orders 1 and 2. Trivially, there is only one Latin square of order 1. There are two Latin squares of order 2 as shown in Fig. 14-2. However, they are not orthogonal. Therefore, when we talk about orthogonal Latin squares of order n, n is understood to be larger than or equal to 3.

$$A_1 = \begin{bmatrix} 1 & 2 \\ 2 & 1 \end{bmatrix} \qquad A_2 = \begin{bmatrix} 2 & 1 \\ 1 & 2 \end{bmatrix}$$

Figure 14-2

Theorem 14-1 *There are at most $n - 1$ Latin squares in a set of orthogonal Latin squares of order n.*

Proof Let $A_1, A_2, \ldots, A_r$ be a set of orthogonal Latin squares of order n. Notice that the orthogonality condition is not violated when the entries in the squares are renamed by permuting the numbers $1, 2, \ldots, n$. Let us rename the entries in such a way that the first row of each of the squares reads $1, 2, \ldots, n$; that is,

$$\begin{aligned} a_{11}^{(1)} &= a_{11}^{(2)} = \cdots = a_{11}^{(r)} = 1 \\ a_{12}^{(1)} &= a_{12}^{(2)} = \cdots = a_{12}^{(r)} = 2 \\ &\cdots\cdots\cdots\cdots\cdots \\ a_{1n}^{(1)} &= a_{1n}^{(2)} = \cdots = a_{1n}^{(r)} = n \end{aligned}$$

Let us consider now the entries in the second row and the first column of the Latin squares $a_{21}^{(1)}, a_{21}^{(2)}, \ldots, a_{21}^{(r)}$. None of these entries can be 1. Otherwise, the condition that every square is a Latin square is violated. Also, no two of these entries can be the same. Otherwise, the condition that the set of Latin squares is an orthogonal set is violated. It follows that there are at most $n - 1$ Latin squares in the set $A_1, A_2, \ldots, A_r$. ■

Theorem 14-1 gives an upper bound on the number of Latin squares in an orthogonal set. The following theorem asserts that such an upper bound can be achieved when n is larger than 2 and is a power of a prime number.

Theorem 14-2 *For $n \geq 3$ and $n = p^\alpha$, where p is a prime number and α is a positive integer, there exists a set of $n - 1$ orthogonal Latin squares of order n.*

Proof We prove the theorem by showing a construction procedure for a set of $n - 1$ orthogonal Latin squares. Let $b_1, b_2, b_3, \ldots, b_n$ denote the elements in the Galois field GF(p^α).† Let $+$ and $*$ denote the addition and multiplication operations, respectively, in the field. Also, let b_n be the additive identity, and let b_1 be the multiplicative identity in the field. We construct a set of $n - 1$ $n \times n$ squares $A_1, A_2, \ldots, A_{n-1}$ with entries

$$a_{ij}^{(e)} = b_e * b_i + b_j \qquad i,j = 1, 2, \ldots, n-1, n$$
$$e = 1, 2, \ldots, n-1$$

We notice first that each of $A_1, A_2, \ldots, A_{n-1}$ is a Latin square. Suppose that there are two entries in the ith row of A_e which are the same; that is, $a_{ij}^{(e)} = a_{ik}^{(e)}$ for some j and k. This means that

$$b_e * b_i + b_j = b_e * b_i + b_k$$

which gives

$$b_j = b_k$$

and

$$j = k$$

Suppose that there are two entries in the ith column of A_e which are the same; that is, $a_{ji}^{(e)} = a_{ki}^{(e)}$ for some j and k. This means that

$$b_e * b_j + b_i = b_e * b_k + b_i$$

which gives

$$b_e * b_j = b_e * b_k \tag{14-1}$$

Since b_e is not the additive identity ($e = 1, 2, \ldots, n-1$), there exists a multiplicative inverse of b_e in the field. Therefore, Eq.

† The definition of a field has been introduced in Chap. 7. Here, we make use of the fact that for any prime p and any positive integer α, there exists a (unique) field containing p^α elements. Such a field is usually called the Galois field with p^α elements and is denoted by GF(p^α). For a derivation of this result, see, for example, Birkhoff and MacLane [2].

(14-1) yields

$$b_j = b_k$$

and

$$j = k$$

We notice also that the set $A_1, A_2, \ldots, A_{n-1}$ is a set of orthogonal Latin squares. Suppose that in the Latin squares A_e and A_f, for some i, j, k, and l,

$$a_{ij}^{(e)} = a_{kl}^{(e)} \qquad \text{and} \qquad a_{ij}^{(f)} = a_{kl}^{(f)}$$

That is,

$$b_e * b_i + b_j = b_e * b_k + b_l \tag{14-2}$$

and

$$b_f * b_i + b_j = b_f * b_k + b_l \tag{14-3}$$

Subtracting Eq. (14-3) from Eq. (14-2),† we obtain

$$b_e * b_i - b_f * b_i = b_e * b_k - b_f * b_k$$

that is,

$$(b_e - b_f) * b_i = (b_e - b_f) * b_k$$

Since $b_e \neq b_f$ means that $b_e - b_f \neq b_n$,‡ there is a multiplicative inverse of $b_e - b_f$ in the field. It follows that

$$b_i = b_k$$

and

$$i = k$$

Equation (14-2) now becomes

$$b_e * b_i + b_j = b_e * b_i + b_l$$

It follows that

$$b_j = b_l$$

and

$$j = l \quad \blacksquare$$

† The subtraction of an element is defined to be the addition of its additive inverse.

‡ The reader is reminded that the additive inverse of every element in a field is unique, as proved in Sec. 5-2.

When n is not a power of a prime number, the question of the existence of a set of $n - 1$ orthogonal Latin squares has not been completely settled. However, the following theorems give a lower bound on the number of Latin squares in an orthogonal set.

Theorem 14-3 *If a set of r orthogonal Latin squares of order n and a set of r orthogonal Latin squares of order n' exist, then there exists a set of r Latin squares of order nn'.*

Proof Let $A_1, A_2, \ldots, A_r$ be a set of orthogonal Latin squares of order n, and let $B_1, B_2, \ldots, B_r$ be a set of orthogonal Latin squares of order n'. We construct a set of r squares $C_1, C_2, \ldots, C_r$ of order nn':

$$C_e = \begin{bmatrix} (a_{11}^{(e)},B_e) & (a_{12}^{(e)},B_e) & (a_{13}^{(e)},B_e) & \cdots & (a_{1n}^{(e)},B_e) \\ (a_{21}^{(e)},B_e) & (a_{22}^{(e)},B_e) & (a_{23}^{(e)},B_e) & \cdots & (a_{2n}^{(e)},B_e) \\ (a_{31}^{(e)},B_e) & (a_{32}^{(e)},B_e) & (a_{33}^{(e)},B_e) & \cdots & (a_{3n}^{(e)},B_e) \\ \cdots & \cdots & \cdots & \cdots & \cdots \\ (a_{n1}^{(e)},B_e) & (a_{n2}^{(e)},B_e) & (a_{n3}^{(e)},B_e) & \cdots & (a_{nn}^{(e)},B_e) \end{bmatrix}$$

for $e = 1, 2, \ldots, r$, where $(a_{ij}^{(e)},B_e)$ denotes an $n' \times n'$ matrix the entry in the kth row and the lth column of which is the ordered pair $(a_{ij}^{(e)},b_{kl}^{(e)})$ for $k = 1, 2, \ldots, n'$; $l = 1, 2, \ldots, n'$.

First, observe that the C's are Latin squares, because in every square any two ordered pairs in the same row or in the same column must differ either in the first component or in the second component. Next, we show that the set of Latin squares $C_1, C_2, \ldots, C_r$ is an orthogonal set. Suppose that in the Latin squares C_e and C_f, for some i, j, k, l, p, q, s, and t,

$$(a_{ij}^{(e)},b_{kl}^{(e)}) = (a_{pq}^{(e)},b_{st}^{(e)}) \tag{14-4}$$

and

$$(a_{ij}^{(f)},b_{kl}^{(f)}) = (a_{pq}^{(f)},b_{st}^{(f)}) \tag{14-5}$$

Equations (14-4) and (14-5) mean that

$$a_{ij}^{(e)} = a_{pq}^{(e)} \qquad b_{kl}^{(e)} = b_{st}^{(e)}$$

and

$$a_{ij}^{(f)} = a_{pq}^{(f)} \qquad b_{kl}^{(f)} = b_{st}^{(f)}$$

However, since A_e and A_f are orthogonal Latin squares,

$$i = p \qquad \text{and} \qquad j = q$$

Similarly, since B_e and B_f are orthogonal Latin squares,

$k = s$ and $l = t$ ■

Theorem 14-4 *Let $p_1^{\alpha_1} p_2^{\alpha_2} \cdots p_t^{\alpha_t}$ be the prime power decomposition of the positive integer n, where the p_i's are prime numbers and the α_i's are positive integers. Let r denote the smallest of the t quantities $(p_1^{\alpha_1} - 1)$, $(p_2^{\alpha_2} - 1)$, . . . , $(p_t^{\alpha_t} - 1)$. Then, there exists a set of r orthogonal Latin squares of order n.*

Proof According to Theorem 14-2, there exist a set of $p_1^{\alpha_1} - 1$ orthogonal Latin squares of order $p_1^{\alpha_1}$, a set of $p_2^{\alpha_2} - 1$ orthogonal Latin squares of order $p_2^{\alpha_2}$, . . . , and a set of $p_t^{\alpha_t} - 1$ orthogonal Latin squares of order $p_t^{\alpha_t}$. Let us arbitrarily select r orthogonal Latin squares from each of these sets. According to Theorem 14-3, they can be composed to yield a set of r orthogonal Latin squares of order n. ■

If in the prime power decomposition of the integer n, the power of the factor 2 is either equal to 0 or larger than 1, then according to Theorem 14-4, there exists a pair of orthogonal Latin squares of order n. In other words, Theorem 14-4 asserts the existence of a pair of orthogonal Latin squares of order n for $n \not\equiv 2 \pmod 4$. For the case $n \equiv 2 \pmod 4$, the result in Theorem 14-4 is inconclusive. As a matter of fact, the question of the existence of a pair of orthogonal Latin squares of order n for $n \equiv 2 \pmod 4$ has an interesting history. The question arises in the "problem of 36 officers" which was first proposed by Euler. There are six officers of six different ranks from each of six regiments. The problem asks for an arrangement of the 36 officers in a 6×6 square formation such that each row and each column in the formation will contain one and only one officer of each rank from each regiment. The problem is equivalent to that of finding a pair of 6×6 orthogonal Latin squares. One of the squares gives the ranks of the officers in the formation, and the other square gives the regiments which the officers in the formation are from. (In other words, an officer is identified by an ordered pair, the first component giving his rank and the second component giving his regiment.) Euler conjectured not only that there exists no pair of orthogonal Latin squares of order 6, but also that there exists no pair of orthogonal Latin squares of order n for $n \equiv 2 \pmod 4$. Around 1900, Tarry† verified that there is no pair of orthogonal Latin squares of order 6 by exhaustively examining all the 6×6 Latin squares in a systematic way. However, in 1960, Bose, Shrikhande, and Parker disproved

† See Tarry [15].

Euler's conjecture by showing that there exists a pair of orthogonal Latin squares of order n for $n \equiv 2 \pmod 4$ and $n > 6$. (Their proof is too lengthy to be included here.)†

In summary, we note that there exists a pair of orthogonal Latin squares of order n for any n except $n = 1$, 2, and 6.

14-4 BALANCED INCOMPLETE BLOCK DESIGNS

Let $X = \{x_1, x_2, \ldots, x_v\}$ be a set of v objects. A *balanced incomplete block design* of X is a collection of b k-subsets of X (the k-subsets denoted by $B_1, B_2, \ldots, B_b$ are also called the *blocks*) such that the following conditions are satisfied:

1. Each object appears in exactly r of the b blocks.
2. Every two objects appear simultaneously in exactly λ of the b blocks.
3. $k < v$.‡

It should be noted that the blocks are not necessarily distinct subsets.

As an example, Fig. 14-3 shows a balanced incomplete block design of the set $X = \{x_1, x_2, \ldots, x_9\}$ with $b = 12$, $v = 9$, $r = 4$, $k = 3$, and $\lambda = 1$.

B_1: x_1, x_2, x_3	B_2: x_4, x_5, x_6	B_3: x_7, x_8, x_9
B_4: x_1, x_4, x_7	B_5: x_2, x_5, x_8	B_6: x_3, x_6, x_9
B_7: x_1, x_5, x_9	B_8: x_2, x_6, x_7	B_9: x_3, x_4, x_8
B_{10}: x_1, x_6, x_8	B_{11}: x_2, x_4, x_9	B_{12}: x_3, x_5, x_7

Figure 14-3

As another example, the schedule in Table 14-5 is a balanced incomplete block design of the set $X = \{1, 2, \ldots, 6\}$ with the rows being the blocks. In this design, $b = 10$, $v = 6$, $r = 5$, $k = 3$, and $\lambda = 2$.

Since a balanced incomplete block design is characterized by the five parameters b, v, r, k, and λ, it is also called a *(b,v,r,k,λ)-configuration.* It is quite clear that not all five of the parameters are independent. In other words, it is not true that there exists a balanced incomplete block design for any arbitrary set of parameters b, v, r, k, and λ. However, there is no known sufficient condition on the existence of a certain (b,v,r,k,λ)-configuration. We shall show some relations among the

† See Hall [7].

‡ This condition excludes the degenerate case in which the design becomes a complete block design ($k = v$).

parameters b, v, r, k, λ that are necessary conditions for the existence of a corresponding (b,v,r,k,λ)-configuration.

Theorem 14-5 *In a balanced incomplete block design,* $bk = vr$, *and* $r(k - 1) = \lambda(v - 1)$.

Proof The term bk counts the total number of occurrences of the objects by counting the number of blocks, b, and the number of objects in each block, k. The term vr counts the total number of occurrences of the objects by counting the number of objects, v, and the number of occurrences of each of the objects, r. Therefore, $bk = vr$.

The term $r(k - 1)$ counts the total number of occurrences of 2-subsets that contain a particular object by counting the number of objects it pairs up with in a block, $k - 1$, and the number of blocks in which it appears, r. The term $\lambda(v - 1)$ does the same count by counting the number of distinct 2-subsets that contain the particular object, $v - 1$, and the number of occurrences of each distinct 2-subset, λ. ∎

Instead of a list of the k-subsets, a balanced incomplete block design can also be described by the *incidence matrix* Q, which is a $v \times b$ matrix with 0's and 1's as entries. The rows of the matrix correspond to the objects $x_1, x_2, \ldots, x_v$, and the columns of the matrix correspond to the blocks $B_1, B_2, \ldots, B_b$. The entry in the ith row and the jth column of Q is a 1 if the object x_i is in the block B_j and is a 0 otherwise. For example, the incidence matrix of the balanced incomplete block design in Fig. 14-3 is

	B_1	B_2	B_3	B_4	B_5	B_6	B_7	B_8	B_9	B_{10}	B_{11}	B_{12}
x_1	1	0	0	1	0	0	1	0	0	1	0	0
x_2	1	0	0	0	1	0	0	1	0	0	1	0
x_3	1	0	0	0	0	1	0	0	1	0	0	1
x_4	0	1	0	1	0	0	0	0	1	0	1	0
x_5	0	1	0	0	1	0	1	0	0	0	0	1
x_6	0	1	0	0	0	1	0	1	0	1	0	0
x_7	0	0	1	1	0	0	0	1	0	0	0	1
x_8	0	0	1	0	1	0	0	0	1	1	0	0
x_9	0	0	1	0	0	1	1	0	0	0	1	0

In many communication problems, when only the digits 0 and 1 can be transmitted through the communication channel, distinct messages are to be represented by sequences of 0's and 1's which are called *code words*. A *block code* consists of a set of code words of fixed length. The *distance* between two code words is the number of digits at which the two code words differ. For example, the distance between the code words 00000 and 00011 is 2, and the distance between the code words 01010 and 10101 is 5. The distance of a block code is defined as the minimal distance between all pairs of code words in the code. The distance of a block code is a significant characteristic of the code since it determines the number of erroneous digits that can be tolerated in the transmission of a code word. For example, a distance-3 block code can tolerate a single error in the transmission of a code word, because changing one of the digits in a code word will not cause any confusion with other code words. Similarly, a distance-5 block code can tolerate two errors in the transmission, and so on. It is interesting to observe that corresponding to a (b,v,r,k,λ)-configuration, there is a block code of distance $2(r - \lambda)$. If we look at the v rows of the incidence matrix of a (b,v,r,k,λ)-configuration as v binary words, we note that each word contains exactly r 1's. Since every two words have exactly λ 1's that are in the same positions, the distance between every two words is exactly $2(r - \lambda)$. The reader can check that corresponding to the (b,v,r,k,λ)-configuration in Fig. 14-3, there is a set of nine code words with the distance between every two code words being 6.

The next two theorems give further characterization of balanced incomplete block designs.

Theorem 14-6 *For a (b,v,r,k,λ)-configuration,*

$$QQ^T = (r - \lambda)I + \lambda J$$

where I is the $v \times v$ identity matrix, and J is the $v \times v$ matrix in which all the entries are 1's.

Proof Let t_{ij} denote the entry in the ith row and the jth column of QQ^T, which is a $v \times v$ matrix. Clearly, t_{ij} is the value of the inner product of the ith and the jth row of the incidence matrix Q.† For $i = j$, t_{ii} is the number of blocks that the ith object is in; that is, $t_{ii} = r$ for $i = 1, 2, \ldots, v$. For $i \neq j$, t_{ij} is the number of blocks in which both the ith and the jth object appear; that is, $t_{ij} = \lambda$ for $i = 1, 2, \ldots, v$; $j = 1, 2, \ldots, v$; and $i \neq j$. ∎

Theorem 14-7 *In a (b,v,r,k,λ)-configuration, $b \geq v$.*

Proof We notice first that in a (b,v,r,k,λ)-configuration, $r \neq \lambda$. According to the relation $r(k - 1) = \lambda(v - 1)$ proved in Theorem

† That is, $t_{ij} = \sum_{l=1}^{b} q_{il}q_{jl}$, where q_{il} is the entry in the ith row and the lth column of Q.

14-5 and the condition $k < v$ in the definition of a balanced incomplete block design, $r = [(v-1)/(k-1)]\lambda > \lambda$. Therefore, $r \neq \lambda$.

Suppose that $b < v$. Let us append $v - b$ columns of zeros to Q to form a $v \times v$ square matrix Q_1. Since $QQ^T = Q_1Q_1^T$ because the appended zeros do not change the values of the inner products of rows, we have $\det(QQ^T) = \det(Q_1Q_1^T) = \det(Q_1)\det(Q_1^T)$. Since Q_1 contains one or more columns of zeros, $\det(Q_1) = 0$. Therefore, we conclude that $\det(QQ^T) = 0$. According to Theorem 14-6,

$$\det(QQ^T) = \det \begin{bmatrix} r & \lambda & \lambda & \lambda & \cdots & \lambda & \lambda \\ \lambda & r & \lambda & \lambda & \cdots & \lambda & \lambda \\ \lambda & \lambda & r & \lambda & \cdots & \lambda & \lambda \\ \lambda & \lambda & \lambda & r & \cdots & \lambda & \lambda \\ \cdots & \cdots & \cdots & \cdots & \cdots & \cdots & \cdots \\ \lambda & \lambda & \lambda & \lambda & \cdots & r & \lambda \\ \lambda & \lambda & \lambda & \lambda & \cdots & \lambda & r \end{bmatrix}$$

Subtracting the first column from the second, third, . . . , vth column, we obtain

$$\det(QQ^T) = \det \begin{bmatrix} r & \lambda - r & \lambda - r & \lambda - r & \cdots & \lambda - r & \lambda - r \\ \lambda & r - \lambda & 0 & 0 & \cdots & 0 & 0 \\ \lambda & 0 & r - \lambda & 0 & \cdots & 0 & 0 \\ \lambda & 0 & 0 & r - \lambda & \cdots & 0 & 0 \\ \cdots & \cdots & \cdots & \cdots & \cdots & \cdots & \cdots \\ \lambda & 0 & 0 & 0 & \cdots & r - \lambda & 0 \\ \lambda & 0 & 0 & 0 & \cdots & 0 & r - \lambda \end{bmatrix}$$

Now, adding to the first row the second, third, . . . , vth row, we obtain

$$\det(QQ^T)$$

$$= \det \begin{bmatrix} r + (v-1)\lambda & 0 & 0 & 0 & \cdots & 0 & 0 \\ \lambda & r - \lambda & 0 & 0 & \cdots & 0 & 0 \\ \lambda & 0 & r - \lambda & 0 & \cdots & 0 & 0 \\ \lambda & 0 & 0 & r - \lambda & \cdots & 0 & 0 \\ \cdots & \cdots & \cdots & \cdots & \cdots & \cdots & \cdots \\ \lambda & 0 & 0 & 0 & \cdots & r - \lambda & 0 \\ \lambda & 0 & 0 & 0 & \cdots & 0 & r - \lambda \end{bmatrix}$$

$$= [r + (v-1)\lambda]\,(r - \lambda)^{v-1}$$

$$= rk(r - \lambda)^{v-1}$$

Since r, k, and $r - \lambda$ are all nonzero quantities, $\det(QQ^T)$ also assumes a nonzero value. This leads to a contradiction. ∎

A balanced incomplete block design is said to be a *symmetric balanced incomplete block design* if $b = v$ and $r = k$. A symmetric balanced incomplete block design is also called a (v,k,λ)-configuration since it is characterized by the three parameters v, k, and λ. For example, Fig. 14-4 shows a symmetric balanced incomplete block design of the set $X = \{x_1,x_2, \ldots ,x_7\}$ with $b = v = 7$, $r = k = 3$, and $\lambda = 1$.

B_1: x_1, x_2, x_4	B_2: x_2, x_3, x_5	B_3: x_3, x_4, x_6
B_4: x_4, x_5, x_7	B_5: x_5, x_6, x_1	B_6: x_6, x_7, x_2
B_7: x_7, x_1, x_3		

Figure 14-4

We now state, without the proof, a theorem characterizing (v,k,λ)-configurations.†

Theorem 14-8 *In a (v,k,λ)-configuration, if v is even, then $k - \lambda$ is the square of an integer; if v is odd, then the equation*

$$x^2 = (k - \lambda)y^2 + (-1)^{(v-1)/2}\lambda z^2$$

has a solution in integers x, y, and z, not all of which are zero.

14-5 CONSTRUCTION OF BLOCK DESIGNS

Several methods can be employed to construct block designs. However, each of these methods is useful only for a certain class of parameters. In this section, we discuss a method that can be categorized as a recursive method. We choose to present this method both because it is relatively simple and because the discussion also serves the purpose of illustrating the general approach of the other recursive methods.

First we show that nonsymmetric balanced incomplete block designs can be derived from a given symmetric balanced incomplete block design. To prove Theorems 14-9 and 14-10, which follow, we need to prove a lemma.

Lemma 14-1 *In a symmetric balanced incomplete block design, every two blocks have exactly λ objects in common.*

Proof For a symmetric balanced incomplete block design, the incidence matrix Q is a $v \times v$ square matrix. It was shown in the proof of Theorem 14-7 that $\det (QQ^T) = \det (Q) \det (Q^T) \neq 0$,

† See Hall [7], for example, for a proof.

which implies that $\det(Q) \neq 0$. Therefore, the matrix Q is nonsingular and its inverse Q^{-1} exists. Thus, according to Theorem 14-6,

$$\begin{aligned} Q^TQ &= (Q^{-1}Q)Q^TQ \\ &= Q^{-1}(QQ^T)Q \\ &= Q^{-1}[(k-\lambda)I + \lambda J]Q \\ &= (k-\lambda)Q^{-1}IQ + \lambda Q^{-1}JQ \\ &= (k-\lambda)I + \lambda Q^{-1}JQ \end{aligned} \tag{14-6}$$

To further simplify Eq. (14-6), we show that

$$Q^{-1}JQ = J \qquad \text{or} \qquad JQ = QJ$$

Because there are k 1's in each row of Q, we can write

$$QJ = kJ \tag{14-7}$$

or

$$\frac{1}{k}J = Q^{-1}J \tag{14-8}$$

Now,

$$\begin{aligned} QQ^TJ &= [(k-\lambda)I + \lambda J]J \\ &= [(k-\lambda) + \lambda v]J \end{aligned} \tag{14-9}$$

because $IJ = J$ and $JJ = vJ$. Equation (14-9) can be rewritten as

$$\begin{aligned} Q^TJ &= Q^{-1}[(k-\lambda) + \lambda v]J \\ &= [(k-\lambda) + \lambda v]Q^{-1}J \end{aligned}$$

which becomes

$$Q^TJ = \frac{1}{k}(k - \lambda + \lambda v)J \tag{14-10}$$

according to Eq. (14-8). Transposing both sides of Eq. (14-10), we obtain

$$JQ = \frac{1}{k}(k - \lambda + \lambda v)J \tag{14-11}$$

since $J^T = J$. It follows that

$$\begin{aligned} JQJ &= \frac{1}{k}(k - \lambda + \lambda v)JJ \\ &= \frac{1}{k}(k - \lambda + \lambda v)vJ \end{aligned} \tag{14-12}$$

Since $J(QJ) = J(kJ) = kJJ = kvJ$, Eq. (14-12) becomes

$$kvJ = \frac{1}{k}(k - \lambda + \lambda v)vJ$$

that is,

$$kv = \frac{1}{k}(k - \lambda + \lambda v)v$$

or

$$k^2 = k - \lambda + \lambda v$$

Therefore, Eq. (14-11) becomes

$$JQ = \frac{1}{k}(k^2J) = kJ \tag{14-13}$$

Combining Eqs. (14-7) and (14-13), we obtain

$$QJ = JQ$$

Now, Eq. (14-6) becomes

$$Q^TQ = (k - \lambda)I + \lambda J$$
$$= \begin{bmatrix} k & \lambda & \lambda & \cdots & \lambda \\ \lambda & k & \lambda & \cdots & \lambda \\ \cdot & \cdot & \cdot & \cdots & \cdot \\ \lambda & \lambda & \lambda & \cdots & k \end{bmatrix}$$

This means that the inner product of every two columns in Q is equal to λ, and the lemma follows immediately. ∎

Theorem 14-9 *If B_1 $B_2, \ldots, B_v$ are the blocks in a symmetric balanced incomplete block design of the set $X = \{x_1, x_2, \ldots, x_v\}$, then, for any i, $B_1 - B_i$, $B_2 - B_i, \ldots, B_{i-1} - B_i$, $B_{i+1} - B_i, \ldots, B_v - B_i$ form a balanced incomplete block design of the set $X - B_i$.*

Proof It is clear that the $v - 1$ blocks, $B_1 - B_i$, $B_2 - B_i, \ldots, B_v - B_i$, contain objects in the set $X - B_i$. Each of the objects in $X - B_i$ still appears in the blocks k times, and every pair of objects in $X - B_i$ still appears in λ of the $v - 1$ blocks. According to Lemma 14-1, each of the blocks, $B_1, B_2, \ldots, B_{i-1}, B_{i+1}, \ldots, B_v$, has λ objects that are in common with the objects in B_i. Thus, each of the blocks, $B_1 - B_i$, $B_2 - B_i, \ldots, B_{i-1} - B_i$, $B_{i+1} - B_i, \ldots, B_v - B_i$, contains $k - \lambda$ objects. In other words, the blocks $B_1 - B_i$, $B_2 - B_i, \ldots, B_{i-1} - B_i$, $B_{i+1} - B_i, \ldots, B_v - B_i$ form a $((v - 1), (v - k), k, (k - \lambda), \lambda)$-configuration. ∎

Theorem 14-10 *If $B_1, B_2, \ldots, B_v$ are the blocks in a symmetric balanced incomplete block design of the set $X = \{x_1,x_2, \ldots, x_v\}$, then, for any i, $B_1 \cap B_i, B_2 \cap B_i, \ldots, B_{i-1} \cap B_i, B_{i+1} \cap B_i, \ldots, B_v \cap B_i$ form a balanced incomplete block design of the set B_i.*

Proof The $v - 1$ blocks, $B_1 \cap B_i$, $B_2 \cap B_i, \ldots, B_{i-1} \cap B_i$, $B_{i+1} \cap B_i, \ldots, B_v \cap B_i$, contain objects in the subset B_i. Since there are only $v - 1$ blocks in the design, we see that each of the objects in B_i appears in $k - 1$ of the $v - 1$ blocks, and every pair of the objects in B_i appears in $\lambda - 1$ of the $v - 1$ blocks. According to Lemma 14-1, each of the blocks $B_1 \cap B_i$, $B_2 \cap B_i, \ldots, B_v \cap B_i$ contains λ objects. Therefore, the blocks $B_1 \cap B_i$, $B_2 \cap B_i, \ldots, B_v \cap B_i$ form a $((v - 1), k, (k - 1), \lambda, (\lambda - 1))$-configuration. ∎

According to Theorems 14-9 and 14-10, corresponding to each block in a (v,k,λ)-configuration, a $((v - 1), (v - k), k, (k - \lambda), \lambda)$-configuration, as well as a $((v - 1), k, (k - 1), \lambda, (\lambda - 1))$-configuration, can be derived. Consequently, we shall limit our discussion only to the construction of symmetric balanced incomplete block designs. (We by no means imply here that any balanced incomplete block design can be derived from a symmetric balanced incomplete block design.) As an example, suppose that we have constructed the symmetric balanced incomplete block design of the set $\{x_1,x_2, \ldots, x_{15}\}$ shown below ($v = b = 15$, $r = k = 7$, and $\lambda = 3$).

B_1: $x_1, x_2, x_3, x_4, x_5, x_6, x_7$

B_2: $x_1, x_2, x_3, x_8, x_9, x_{10}, x_{11}$

B_3: $x_1, x_2, x_3, x_{12}, x_{13}, x_{14}, x_{15}$

B_4: $x_1, x_4, x_5, x_8, x_9, x_{12}, x_{13}$

B_5: $x_1, x_4, x_5, x_{10}, x_{11}, x_{14}, x_{15}$

B_6: $x_1, x_6, x_7, x_8, x_9, x_{14}, x_{15}$

B_7: $x_1, x_6, x_7, x_{10}, x_{11}, x_{12}, x_{13}$

B_8: $x_2, x_4, x_6, x_8, x_{10}, x_{12}, x_{14}$

B_9: $x_2, x_4, x_7, x_8, x_{11}, x_{13}, x_{15}$

B_{10}: $x_2, x_5, x_6, x_9, x_{11}, x_{12}, x_{15}$

B_{11}: $x_2, x_5, x_7, x_9, x_{10}, x_{13}, x_{14}$

B_{12}: $x_3, x_4, x_6, x_9, x_{11}, x_{13}, x_{14}$

B_{13}: $x_3, x_4, x_7, x_9, x_{10}, x_{12}, x_{15}$

B_{14}: $x_3, x_5, x_6, x_8, x_{10}, x_{13}, x_{15}$

B_{15}: $x_3, x_5, x_7, x_8, x_{11}, x_{12}, x_{14}$

According to Theorem 14-9, we can derive the following balanced incomplete block design of the set $\{x_8, x_9, \ldots, x_{15}\}$ with $b = 14$, $v = 8$, $r = 7$, $k = 4$, and $\lambda = 3$:

$B_2 - B_1$: x_8, x_9, x_{10}, x_{11}

$B_3 - B_1$: x_{12}, x_{13}, x_{14}, x_{15}

$B_4 - B_1$: x_8, x_9, x_{12}, x_{13}

$B_5 - B_1$: x_{10}, x_{11}, x_{14}, x_{15}

$B_6 - B_1$: x_8, x_9, x_{14}, x_{15}

$B_7 - B_1$: x_{10}, x_{11}, x_{12}, x_{13}

$B_8 - B_1$: x_8, x_{10}, x_{12}, x_{14}

$B_9 - B_1$: x_8, x_{11}, x_{13}, x_{15}

$B_{10} - B_1$: x_9, x_{11}, x_{12}, x_{15}

$B_{11} - B_1$: x_9, x_{10}, x_{13}, x_{14}

$B_{12} - B_1$: x_9, x_{11}, x_{13}, x_{14}

$B_{13} - B_1$: x_9, x_{10}, x_{12}, x_{15}

$B_{14} - B_1$: x_8, x_{10}, x_{13}, x_{15}

$B_{15} - B_1$: x_8, x_{11}, x_{12}, x_{14}

Also, according to Theorem 14-10, we can derive the following balanced incomplete block design of the set $\{x_1, x_2, \ldots, x_7\}$ with $b = 14$, $v = 7$, $r = 6$, $k = 3$, and $\lambda = 2$:

$B_2 \cap B_1$: x_1, x_2, x_3

$B_3 \cap B_1$: x_1, x_2, x_3

$B_4 \cap B_1$: x_1, x_4, x_5

$B_5 \cap B_1$: x_1, x_4, x_5

$B_6 \cap B_1$: x_1, x_6, x_7

$B_7 \cap B_1$: x_1, x_6, x_7

$B_8 \cap B_1$: x_2, x_4, x_6

$B_9 \cap B_1$: x_2, x_4, x_7

$B_{10} \cap B_1$: x_2, x_5, x_6

$B_{11} \cap B_1$: x_2, x_5, x_7

$B_{12} \cap B_1$: x_3, x_4, x_6

$B_{13} \cap B_1$: x_3, x_4, x_7

$B_{14} \cap B_1$: x_3, x_5, x_6

$B_{15} \cap B_1$: x_3, x_5, x_7

A *Hadamard matrix of order* n is an $n \times n$ matrix H with $+1$'s and -1's as entries and is such that

$$HH^T = nI$$

In other words, the inner product of any two distinct rows is equal to 0, whereas the inner product of a row with itself is equal to n. Clearly, if a matrix is Hadamard, then permuting the rows and the columns of the matrix does not change the property. Also, multiplying a row or a column by -1 does not change the property either. A Hadamard matrix is said to be *normalized* if its first row and first column consist entirely of $+1$'s. By normalizing a Hadamard matrix, we shall mean multiplying those rows containing a -1 as the leftmost entry and those columns containing a -1 as the topmost entry by -1 so that the resultant matrix is normalized. For example, Fig. 14-5a shows a Hadamard matrix and Fig. 14-5b shows its normalized form.

$$\begin{bmatrix} 1 & -1 & 1 & 1 \\ 1 & -1 & -1 & -1 \\ -1 & -1 & -1 & 1 \\ 1 & 1 & -1 & 1 \end{bmatrix} \qquad \begin{bmatrix} 1 & 1 & 1 & 1 \\ 1 & 1 & -1 & -1 \\ 1 & -1 & 1 & -1 \\ 1 & -1 & -1 & 1 \end{bmatrix}$$

(a) (b)

Figure 14-5

We want to show that if an $n \times n$ matrix is Hadamard, then $n \equiv 0 \pmod 4$. In other words, a Hadamard matrix of order n can exist only if n is a multiple of 4. Let H denote a Hadamard matrix of order n. We first normalize the matrix. It is clear that except for the first row, each of the remaining rows must contain $n/2$ $+1$'s and $n/2$ -1's. We shall permute the columns in such a way that the first $n/2$ entries in the second row are $+1$'s and the remaining $n/2$ entries in the second row are -1's. Now, let us examine the third row. Suppose there are t $+1$'s [and $(n/2) - t$ -1's] in the first half of the row, t' $+1$'s [and $(n/2) - t'$ -1's] in the second half of the row. Then

$$t + t' = \frac{n}{2}$$

Because the inner product of the second and third rows must equal 0,

$$t + \left(\frac{n}{2} - t'\right) = \frac{n}{2}$$

That is,

$$t + t' = \frac{n}{2} \qquad \text{and} \qquad t - t' = 0$$

Therefore,

$$t = \frac{n}{4} \qquad \text{and} \qquad t' = \frac{n}{4}$$

Since t and t' must be integers, it follows that

$$n \equiv 0 \pmod 4$$

The following theorem relates a normalized Hadamard matrix to a symmetric balanced incomplete block design.

Theorem 14-11 *A normalized Hadamard matrix of order n $(n \geq 8)$ is equivalent to a (v,k,λ)-configuration with $v = n - 1$, $k = (n/2) - 1$,* and $\lambda = (n/4) - 1$.

Proof Let H be a normalized Hadamard matrix. Delete the first row and first column from H, replace the -1's by 0's, and denote the resultant matrix by Q. Clearly, each row in Q contains $(n/2) - 1$ 1's and $n/2$ 0's. Also, the inner product of every two rows is equal to $(n/4) - 1$. Thus, Q is the incident matrix of a (v,k,λ)-configuration with $v = n - 1$, $k = (n/2) - 1$, and

$$\lambda = \frac{n}{4} - 1 \quad \blacksquare$$

For example, the reader can check that from the normalized Hadamard matrix shown below, we can derive the symmetric balanced incomplete block design shown in Fig. 14-4.

$$\begin{bmatrix} 1 & 1 & 1 & 1 & 1 & 1 & 1 & 1 \\ 1 & 1 & 1 & -1 & 1 & -1 & -1 & -1 \\ 1 & -1 & 1 & 1 & -1 & 1 & -1 & -1 \\ 1 & -1 & -1 & 1 & 1 & -1 & 1 & -1 \\ 1 & -1 & -1 & -1 & 1 & 1 & -1 & 1 \\ 1 & 1 & -1 & -1 & -1 & 1 & 1 & -1 \\ 1 & -1 & 1 & -1 & -1 & -1 & 1 & 1 \\ 1 & 1 & -1 & 1 & -1 & -1 & -1 & 1 \end{bmatrix}$$

Let A_1 and A_2 be matrices of orders n and n', respectively. The *Kronecker product* of the two matrices denoted by $A_1 \times A_2$ is an nn' matrix defined by

$$A_1 \times A_2 = \begin{bmatrix} a_{11}^{(1)}A_2 & a_{12}^{(1)}A_2 & \cdots & a_{1n}^{(1)}A_2 \\ a_{21}^{(1)}A_2 & a_{22}^{(1)}A_2 & \cdots & a_{2n}^{(1)}A_2 \\ \cdots & \cdots & \cdots & \cdots \\ a_{n1}^{(1)}A_2 & a_{n2}^{(1)}A_2 & \cdots & a_{nn}^{(1)}A_2 \end{bmatrix}$$

where $a_{ij}^{(1)}A_2$ is an $n' \times n'$ matrix in which the entry in the kth row and the lth column is $a_{ij}^{(1)}a_{kl}^{(2)}$.

Theorem 14-12 *The Kronecker product of two Hadamard matrices is a Hadamard matrix.*

Proof Let H_1 and H_2 be two Hadamard matrices of orders n and n'. Clearly, the entries of $H_1 \times H_2$ are $+1$'s and -1's. Each row in $H_1 \times H_2$ contains $nn'/2$ $+1$'s and $nn'/2$ -1's because each row in $h_{ij}^{(1)}H_2$ contains $n'/2$ $+1$'s and $n'/2$ -1's. Therefore, the inner product of a row with itself is equal to nn'. The inner product of two distinct rows in $H_1 \times H_2$ is equal to the product of the inner product of the corresponding rows in H_1 and the inner product of the corresponding rows in H_2 and, therefore, is equal to zero. According to the definition, $H_1 \times H_2$ is indeed a Hadamard matrix. ■

According to Theorem 14-11, we can construct symmetric balanced incomplete block designs from Hadamard matrices, whereas the result in Theorem 14-12 enables us to construct Hadamard matrices of higher order from Hadamard matrices of lower order. Therefore, we can construct Hadamard matrices in a recursive manner and then obtain from them symmetric balanced incomplete block designs.

Notice that there are two limitations in such a construction procedure. First, a Hadamard matrix of order n exists only for $n \equiv 0 \pmod 4$. This immediately places a limitation on the values of v of the (v,k,λ)-configurations that can be constructed, which, in turn, places a limitation on the values of the parameters of the (b,v,r,k,λ)-configurations that can be constructed. Secondly, the existence of a Hadamard matrix of order n for every $n \equiv 0 \pmod 4$ has not been established because the result in Theorem 14-12 is not sufficient for the construction of a Hadamard matrix of order n for every $n \equiv 0 \pmod 4$. Hadamard matrices of order n for $n = 4, 8, 12, \ldots, 116$ have been constructed. Using these known matrices together with Theorem 14-12 one can construct many more (but not all) of the Hadamard matrices of order n, $n \equiv 0 \pmod 4$. For example, for $n = 120$, since n cannot be factored as the product of multiples of 4, the composition procedure based on Theorem 14-12 is not applicable.

14-6 SUMMARY AND REFERENCES

Because of its richness in mathematical content, block design is one of the more important topics in combinatorial mathematics. Not only is there an impressive accumulation of the efforts of many outstanding mathematicians, but also there are still many challenging open problems. As to the applications of these mathematical results, the reader has seen in this chapter that the subject of block design is closely related to the design of experiments, the construction of error-correcting codes, and so on. It was pointed out in Chap. 6 that our study of the theory of graphs aims at the understanding of the properties of a class of general structures which can be described abstractly by graphs. Block designs are simply another class of structures which we have found to be both useful and interesting.

Our presentation in this chapter is quite elementary. A more complete presentation of the subject material would require more background in algebra and number theory and will be beyond our scope of coverage. An excellent coverage of the subject can be found in Hall [7]. See also Chaps. 7, 8, and 9 of Ryser [14], and Chap. 13 of Beckenbach [1]. The falsity of Euler's conjecture was proved in several papers by Bose, Shrikhande, and Parker [3, 4, 5]. On the design of experiments and their statistical analysis, see the books by Quenouille [11], Cochran and Cox [6], and Mann [10].

1. Beckenbach, E. F. (ed.): "Applied Combinatorial Mathematics," John Wiley & Sons, Inc., New York, 1964.
2. Birkhoff, G., and S. MacLane: "A Survey of Modern Algebra," The Macmillan Company, New York, 1953.
3. Bose, R. C., and S. S. Shrikhande: On the Falsity of Euler's Conjecture About the Non-existence of Two Orthogonal Latin Squares of Order 4t + 2, *Proc. Nat. Acad. Sci. USA*, **45**:734–737 (1959).
4. Bose, R. C., and S. S. Shrikhande: On the Construction of Sets of Mutually Orthogonal Latin Squares and Falsity of a Conjecture of Euler, *Trans. Am. Math. Soc.*, **95**:191–209 (1960).
5. Bose, R. C., S. S. Shrikhande, and E. T. Parker: Further Results on the Construction of Mutually Orthogonal Latin Squares and the Falsity of Euler's Conjecture, *Can. J. Math.*, **12**:189–203 (1960).
6. Cochran, W. G., and G. M. Cox: "Experimental Designs," John Wiley & Sons, Inc., New York, 1950.
7. Hall, M., Jr.: "Combinatorial Theory," Blaisdell Co., Waltham, Mass., 1967.
8. Hamming, R. W.: "Numerical Methods for Scientists and Engineers," McGraw-Hill Book Company, New York, 1962.
9. Kendall, M. G., and B. Babington-Smith: Tables of Random Sampling Numbers, in "Tracts for Computers, No. 24," Cambridge University Press, London, 1939.
10. Mann, H. B.: "Analysis and Design of Experiments," Dover Publications, Inc., New York, 1949.
11. Quenouille, M. H.: "The Design and Analysis of Experiment," Hafner Publishing Company, Inc., New York, 1953.
12. Ralston, A., and H. S. Wilf (eds.): "Mathematical Methods for Digital Computers," vol. 2, John Wiley & Sons, Inc., New York, 1967.
13. RAND Corporation: "A Million Random Digits with 100,000 Normal Deviatives," The Free Press of Glencoe, New York, 1955.

14. Ryser, H. J.: "Combinatorial Mathematics," published by The Mathematical Association of America, distributed by John Wiley & Sons, Inc., New York, 1963.
15. Tarry, G.: Le Probleme de 36 Officieurs, *Compt. Rend. Assoc. Franc. Avance. Sci. Nat.*, **1**:122–123 (1900); **2**:170–203 (1901).

PROBLEMS

14-1. Let A be a $(t + 2) \times n^2$ array with the integers $1, 2, \ldots, n$ as entries ($n \geq 3$, and $t \geq 2$). Prove that if the columns of every $2 \times n^2$ subarray of A are the n^2 ordered pairs of the integers $1, 2, \ldots, n$, then there is a set of t orthogonal Latin squares of order n corresponding to A.

Hint: The columns of A can be permuted so that the first row will read $11 \cdots 1\ 22 \cdots 2 \cdots nn \cdots n$ and the second row will read $12 \cdots n12 \cdots n \cdots 12 \cdots n$. Each of the remaining rows can be made to correspond to a Latin square in the orthogonal set.

14-2. Show that there exists no (b,v,r,k,λ)-configuration with the following parameters:

(*a*) $b = 8, v = 6, r = 5, k = 3, \lambda = 2$

(*b*) $b = 22, v = 22, r = 7, k = 7, \lambda = 2$

14-3. (*a*) A balanced incomplete block design has the parameters $v = 15$, $k = 5$, and $\lambda = 2$. Determine b and r.

(*b*) A balanced incomplete block design has the parameters $v = 21$, $b = 28$, and $r = 8$. Determine k and λ.

14-4. Four of the seven blocks of a (7,3,1)-configuration are $\{1,2,4\}$, $\{2,3,5\}$, $\{3,4,6\}$, and $\{4,5,7\}$. Find the remaining blocks.

14-5. Prove that in a (v,k,λ)-configuration, if v is even, then $k - \lambda$ is the square of an integer. (This is the first part of Theorem 14-8.)

14-6. A *Steiner triple system of order v* is a (b,v,r,k,λ)-configuration with $k = 3$ and $\lambda = 1$.

(*a*) Show that $r = (v - 1)/2$ and $b = v(v - 1)/6$.

(*b*) Show that $v \equiv 1 \pmod 6$ or $v \equiv 3 \pmod 6$.

14-7. Let S_1 be a Steiner triple system of order v_1, and let S_2 be a Steiner triple system of order v_2. Let $\{a_1,a_2, \ldots ,a_{v_1}\}$ be the set of objects in S_1, and let $\{b_1,b_2, \ldots ,b_{v_2}\}$ be the set of objects in S_2. Let X be a set containing v_1v_2 objects c_{ij} ($i = 1, 2, \ldots, v_1$; $j = 1, 2, \ldots, v_2$). Let S be a set of 3-subsets of X, $\{c_{pq},c_{tu},c_{yz}\}$, each satisfying one of the following conditions:

1. $q = u = z$, and $\{a_p,a_t,a_y\}$ is in S_1.
2. $p = t = y$, and $\{b_q,b_u,b_z\}$ is in S_2.
3. $\{a_p,a_t,a_y\}$ is in S_1, and $\{b_q,b_u,b_z\}$ is in S_2.

Show that S is a Steiner triple system of order v_1v_2.

Index

Index